Meeting the Challenges of Climate Change to Tourism

Meeting the Challenges of Climate Change to Tourism: Case Studies of Best Practice

Edited by

Louis D'Amore and Patrick Kalifungwa

CAMBRIDGE SCHOLARS PUBLISHING

Meeting the Challenges of Climate Change to Tourism:
Case Studies of Best Practice,
Edited by Louis D'Amore and Patrick Kalifungwa

This book first published 2013

Cambridge Scholars Publishing

12 Back Chapman Street, Newcastle upon Tyne, NE6 2XX, UK

British Library Cataloguing in Publication Data
A catalogue record for this book is available from the British Library

ISBN (10): 1-4438-4893-X, ISBN (13): 978-1-4438-4893-0

TABLE OF CONTENTS

Part III: Destination Success Stories and Case Studies

Part VII: Supportive Academic Research

Part VIII: Declarations

LIST OF FIGURES

LIST OF TABLES

LIST OF ABBREVIATIONS

AMCEN	African Ministerial Conference on Environment
APTDC	Andhra Pradesh Tourism Development Corporation
CCCCC	Caribbean Community Climate Change Center
CCR	Crete's Culinary Sanctuaries
CHICOP	Chumbe Island Coral Park
CO_2	Carbon dioxide
COTS	Crown-of-Thorns Starfish
CTO	Caribbean Tourism Organization
DEFRA	Department for Environment, Food and Rural Affairs (UK)
DFID	Department for International Development (UK)
ENT	Estonian Nature Tours
ESMAP	World Bank Energy Sector Management Assistance Program
ETS	Emissions Trading Scheme
EUCC	Coastal and Marine Union
GDP	Gross Domestic Product
GHG	Greenhouse Gas
GSTC	Global Sustainable Tourism Council
GTA	Guyana Tourism Authority
GTZ	Deutsche Gesellschaft für Technische Zusammenarbeit
HCMI	Hotel Carbon Measurement Index
HES	Hotel Energy Solutions
IATA	International Air Transport Association
ICLEI	Local Governments for Sustainable Cities
ICZM	Integrated Coastal Zone Management
IHEI	International Hotel Environment Initiative
IHTEC	International Holistic Tourism Education Center
IIPT	International Institute for Peace through Tourism
IMF	International Monetary Fund
INCN	International Union for Conservation of Nature
IRENA	International Renewable Energy Association
IPCC	Intergovernmental Panel on Climate Change
LCA	Life Cycle Assessment

LDC	Least Developed Countries
LEED	Leadership in Energy, Environment and Design
MDG	Millennium Development Goal
MER NTZ	Misool Eco Resort No-Take Zone
MPA	Marine Protected Area
NGO	Non-Governmental Organization
NIMET	Nigerian Meteorological Agency
NORAD	Norwegian Agency for Development Cooperation
NAS	National Adaption Strategy
NHA	Non-Hotel Accommodation
NTZ	No-Take Zone
OPC	Ordinary Portland Cement
PATA	Pacific Asia Travel Association
PPC	Portland Pozzalana Cement
PPCR	Pilot Program on Climate Resilience
RETOSA	Regional Tourism Organization of Southern Africa
RT	Responsible Tourism
SADC	Southern African Development Community
SDPI	Sustainable Development Policy Institute
SIDA	Swedish International Development Cooperation Agency
SIDS	Small Island Developing State
SME	Small and Medium-Size Enterprise
TICOS	Tourism Industry Carbon Offset Service
UN	United Nations
UNCTAD	United Nations Conference on Trade and Development
UNDP	United Nations Development Programme
UNEP	United Nations Environment Programme
UNESCO	United Nations Educational, Scientific and Cultural Organization
UNFCCC	United Nations Framework Convention on Climate Change
UNREDD	United Nations Reduced Emission from Deforestation and Degradation
UNWTO	United Nations World Tourism Organization
US	United States
USAID	United States Agency for International Development
WDDPA	World Database of Protected Areas

WFP	World Food Program
WMA	Wildlife Management Areas
WMO	World Meteorological Organization
WSSD	World Summit on Sustainable Development
WTTC	World Travel and Tourism Council
WTWHA	Wet Tropics World Heritage Area
WWF	World Wide Fund for Nature

INTRODUCTION

LOUIS D'AMORE[1] AND PATRICK KALIFUNGWA[2]

The scientific evidence is clear – climate change is happening.

- 2010, 2011 and 2012 were three of the warmest years on record.
- Thirteen of the last sixteen years have been the warmest on record.
- On mountains and at the poles, glaciers are thinning and retreating – melting glaciers change the flow of rivers, adding to water stress for millions of people.
- Arctic sea-ice cover in 2012 was the lowest on record, 18% below the previous record set in 2007 and 49% below the 1979–2000 average.
- More than 25,000 square kilometers of ice shelf have been lost on the Antarctic Peninsula since 1947, with the most dramatic changes occurring since 1990. The Antarctic ice sheet contains 91% of Earth's glacier ice.
- Oceans are warming and the increased concentrations of carbon dioxide are causing an increase in ocean acidity, threatening coral reefs and small organisms at the base of the food chain.

These changes are affecting people, economies and ecosystems; and unless we act, we will see catastrophic consequences, including rising sea levels, droughts and famine, and the loss of up to a third of the world's plant and animal species. The longer we delay, the greater the risks, and the greater the risks of further impacts such as ocean acidification.

An ice-free Arctic will absorb more heat and accelerate warming, while more rapidly melting ice sheets will lead to higher sea-level rises. The world's most vulnerable people are likely to be affected earlier and more severely than others. Africa, for example, with approximately 14%

---

[1] IIPT Founder and President; Chancellor, Livingstone International University of Tourism Excellence and Business Management (LIUTEBM).

[2] Vice Chancellor, Livingstone International University of Tourism Excellence and Business Management (LIUTEBM).

of the world's population, contributes only 2.3% of global greenhouse gases (GHG) yet is particularly vulnerable to its effects.

Increase in Extreme Weather Events

Some of the extreme flooding that has occurred in the past few years has included: Pakistan, where monsoon rains resulted in one fifth of the country being flooded, 1,600 people killed, another 20 million displaced, and damage estimated at US$ 15 billion; China, where floods led to the evacuation of 15 million people, more than 3,000 persons killed, and damage estimated at US$ 50 billion; Australia, where massive flooding led to more than 20 deaths, affected more than 200,000 people, and had an estimated cost to the economy of $30 billion; the United States, where Hurricane Sandy resulted in 285 deaths and more than US$ 80 billion in damage. And 2012 was the third most active year for North Atlantic tropical cyclones.

Severe flooding was also experienced in Peru, Chile, Bolivia, Brazil, the Philippines, Bangladesh, Rwanda, Nigeria and the United Kingdom – for the latter, 2012 was the wettest year on record. And as we write this, the Maldives is being flooded from a sudden downpour of 252 millimeters (10 inches) of rain in less than two hours, leading to a surge in water levels in the capital city of Port Lewis, with twelve persons reported dead.

While some parts of the world have experienced flooding, others have experienced heat waves, severe drought and forest fires. 2010, 2011 and 2012 are the driest on record since 1950. A heat wave in Russia resulted in tens of thousands of acres of land destroyed by fire and claimed 15,000 lives – 7,000 deaths in Moscow alone. Carbon monoxide levels were two to three times higher than the level considered healthy. Grain output was slashed by 40%. The worst drought in sixty years affected East Africa, causing a severe food crisis across Somalia, Djibouti, Ethiopia and Kenya. In West Africa, Oxfam gave warnings of a "drought catastrophe" and the need for urgent action to prevent a humanitarian disaster in the Sahel Region affecting 13 million people.

2012 was the hottest year on record in the United States, with 62% of continental United States experiencing moderate to severe drought conditions, which in turn have resulted in dust storms and the loss of topsoil in several mid-western states. For farmers, these have been the worst conditions since record-keeping began in 1986. In addition, wildfires destroyed more acreage across the United States than in any other year since records began in the 1960s, and the Great Lakes were at their lowest level ever. Wildfires were also rampant in Chile and Brazil, while Argentina suffered a heat wave.

At the other end of the temperature spectrum, Europe experienced its worst cold spell in twenty-five years, causing 650 deaths, most of them in Russia, Ukraine and Poland where temperatures reached minus 49 degrees Fahrenheit. 2012 was the snowiest winter in Moscow in 100 years, paralyzing traffic for 3,200 km.

Some Consequences of Climate Change

> Clearly, we are endangering all species on earth.
> We are endangering the future of the human race.
> (Rajendra Pachauri, Chairman, Inter-Governmental Committee on Climate Change)

On average, weather-related disasters each year cause more than 100,000 deaths, more than US$ 100 billion in economic losses (more than the annual GDP of 133 countries), and leave 20 million persons displaced.

With climate change, some areas will experience less annual rainfall, while others will experience more. Seasonal rains may be unpredictable and arrive in sudden downpours causing severe flooding. Other threats include sea-level rise, which in turn leads to saline intrusion into coastal aquifers and damage to water infrastructure from coastal storms.

Water security is increasingly becoming a serious issue. Three billion people live in areas where water demand is greater than supply, and scientists are forecasting that 5.3 billion people (two-thirds of the world's population) will suffer water shortages by 2025. Water resources are particularly a problem for all of North Africa extending into the Middle East through to the Arabian Peninsula, Turkey, Iraq, Iran, Pakistan and India.

African nations make up thirty-six of the fifty nations most at risk for food security – nine are at extreme risk. It is estimated that US$ 7 billion a year is required to reduce climate impacts on agriculture in developing countries. To put this figure into perspective, it is less than one and a half days' global military expenditures.

Exposure to weather-related disasters such as heat waves, forest fires, cyclones, flooding, etc. means that virtually everyone is potentially at risk. Diseases transmitted by mosquitoes and ticks might increase in some areas and decrease in others. Forest fires increase the amount of pollutants, which in turn increases respiratory illnesses. Heavy rainfall and floods often lead to contamination of water supplies and unhealthy sanitation conditions. Drought and disasters reduce food supplies, leading to increased malnutrition and reduced capacity to fight infections. Malnutrition is the underlying cause of death of 30% of children under five.

Climate change is but one additional factor that is stressing our ecosystems. More than 83% of the world's population is currently living in countries where demands on nature exceed what the country's ecosystems can renew. We require the ecological capacity of one and a half planets to meet our current needs. More than 60% of the planet's ecosystems are being degraded or used unsustainably. A further increase in temperature of 1.5 to 2.5 degrees centigrade will result in 30% of all species being at high risk of extinction. We are currently losing one species every twenty minutes – 30,000 a year.

At the same time, world military expenditures in 2011 were US$ 1.7 trillion – more than the GDP of every nation in the world except nine. This is US$ 236 for every man, woman and child in the world. Less than one half of the world's military budget for one year would solve the climate change problem. Less than one half the world's current military expenditures would be sufficient to end the primary causes of poverty.

The Global Response to Climate Change

Achim Steiner, Executive Director of UNEP has stated: "We live in some of the most challenging times that perhaps any generation has faced – but also one of the most exciting moments where the possibilities of re-shaping and re-focusing towards a sustainable 21st century have never been more tangible."

Indeed, governments, industries, academic institutions and non-governmental organizations are responding to the challenge of climate change. There has been a substantial growth in knowledge and technology; increased sharing of information and successes; and the formation of new institutions and local, regional and global networks.

Noteworthy is the formation of the International Renewable Energy Association (IRENA), established in 2009, and now with 159 countries as members in addition to the European Union. IRENA assists countries in their transition to a sustainable energy future, and serves as the principal platform for international cooperation, a center of excellence and a repository of policy, technology, resource and financial knowledge on renewable energy.

Also of particular significance is the United Nations' recent declaration that 2014–2024 shall be the UN Decade of Sustainable Energy for All.

Goals of the decade include: universal access to modern energy by 2030; doubling the rate of energy efficiency improvement in 10 years; and doubling the share of the share of alternative energy by 2030.

In addition to numerous UN agencies and international organizations, there are a growing number of international networks that have formed in response to climate change and to promote green growth. These include, for example, Local Governments for Sustainable Cities ICLEI. Membership includes 12 mega-cities, 100 super-cities, 450 large cities and 45 small/medium-sized cities and towns in 84 countries.

As well as this, an increasing number of countries are developing national plans and strategies, including most recently China, which has developed a five-year plan for renewable energy. Amsterdam provides an interesting model for cities: it aims to be the world's first smart sustainable city. They have a holistic plan that includes waste management, green energy, bio-fuel, organic gardening and brewing, recycling and more.

In turn, these initiatives are supported by a growing knowledge base and open-access platforms to assist them in their efforts. These include: Guidebook to the Green Economy, Green Economy Index, Global Atlas of Renewable Energy Potential, IRENA Renewable Readiness Assessments, World Bank Energy Sector Management Assistance Program (ESMAP), World Wildlife Fund Global Foot Print Network, Climate Analysis Indicators Tool, We Adapt Platform, and Google's Earth Engine.

The Significance of Tourism

On December 13, 2012, the UN World Tourism Organization celebrated the arrival of the symbolic one-billionth international travel arrival in a single year. This was the culmination of more than six decades of steady growth from 25,000 international arrivals in 1950. Domestic tourism accounted for a further 5 billion tourism arrivals.

Travel and tourism is clearly a vital driver of the global economy. According to the World Travel and Tourism Council (WTTC), the direct contribution of travel and tourism to world GDP in 2012 was US$ 2.1 trillion. Taking into account its direct, indirect and induced impacts, WTTC estimates travel and tourism's total contribution to global GDP as US$ 6.6 trillion in 2012, contributing an estimated 260 million jobs, US$ 760 billion in investment, and US$ 1.2 trillion in exports. These numbers represented 9.3% of global GDP, 1 in 11 jobs, 5% of global investment and 5% of all exports.

WTTC is projecting a further 3.2% growth in 2013, faster than the 2.4% predicted for global economic growth. Longer-term prospects are even more positive, with annual growth forecast to be 4.4% per year over the ten years to 2022. The UN World Tourism Organization has forecasted that by 2030 there will be 1.8 billion tourism arrivals – nearly doubling the 1 billion arrivals in 2012. While it took more than sixty years to reach the 1 billion, the additional 800 million will be achieved in fewer than twenty years.

Of particular significance among these trends has been the growth of tourism to developing and emerging economies. In 1950, Western Europe and North America accounted for all but 3% of international arrivals. By 1990, the market share of tourism to developing and emerging countries had increased to 29%, and today it is nearly 50%. UNWTO has forecasted that travel to developing and emerging economies will grow at double the pace of advanced economies over the next twenty years, so that by 2030, developing and emerging economies will see more than 1 billion international arrivals.

Tourism is one of the few economic sectors through which the world's fifty least developed countries have managed to increase their participation in the world economy. It accounts for 45% of developing country exports and is a priority sector for their further integration into the global economy.

The Tourism Industry Response to Climate Change

This rapid growth in travel and tourism raises the issue of sustainability. Travel and tourism is highly dependent on non-renewable energy and is responsible for 5% of global greenhouse gas (GHG) emissions. As with the global response to climate change briefly outlined above, significant progress is being made towards sustainable tourism as well.

Over the past two decades, the industry has made significant steps towards minimizing environmental impacts, and in some cases enhancing the environment. One of the first of these initiatives was the development of Codes of Ethics and Guidelines for Sustainable Tourism developed by the International Institute for Peace through Tourism (IIPT) for the travel industry in Canada in 1993. This was followed by the PATA GreenLeaf Program, the introduction of Green Globe by the World Travel and Tourism Council (WTTC), and the International Hotel Environment Initiative (IHEI), now the International Tourism Partnership.

A number of initiatives have evolved since the early 1990s, and more have emerged in the last decade. Particularly significant has been the formation of the Global Sustainable Tourism Council (GSTC). Established in 2010, the GSTC serves as the international body for fostering increased knowledge and understanding of sustainable tourism practices, promoting the adoption of universal sustainable tourism principles and building demand for sustainable travel. At the core of these efforts are Global Sustainable Tourism Criteria, and Criteria for Destinations.

Specifically in the accommodation sector, the UNWTO has been helping hotels go green through their Hotel Energy Solutions (HES) project, and the WTTC, together with the International Tourism Partnership and a working group of industry members, has launched the Hotel Carbon Measurement Index (HCMI).

As well as these efforts, a growing number of online tools and resources specifically for tourism are available to help organizations develop plans and strategies related to climate change. These include: the GSTC criteria for destinations describing the minimum standards that a destination must reach in order to move toward social, cultural and environmental sustainability; Totem Tourism, which has a sustainable destination guide, sustainable tourism marketing guide, a briefing for Ministers of Tourism, and a 2012 report with hundreds of real-life examples of green successes, lists of green organizations, funding sources and articles on various subjects; and Climate Prepared, a tool kit designed to help a tourism business prepare for the effects of weather and climate variability.

The chapters that follow will provide further guidance and models of best practice in preparation for climate change. The chapters are based on the *Fifth IIPT African Conference: Meeting the Challenges of Climate Change to Tourism in Africa and the Developing World.* The conference was organized by the International Institute for Peace through Tourism, in partnership with the UNWTO.

We would like to express our appreciation to our hosts, the Zambia Ministry of Tourism, Environment and Natural Resources, the UN World Tourism Organization, our sponsors, including the UN Development Program and the Zambia Tourism Board, and particularly the speakers and participants at the conference.

PART I

GLOBAL PERSPECTIVES

CHAPTER ONE

MEETING THE CHALLENGES OF CLIMATE CHANGE TO TOURISM IN AFRICA AND THE DEVELOPING WORLD[1]

ZOLTÁN SOMOGYI[2]

Your Excellency Mr. Rupiah Banda, President of Zambia, Your Excellency Mrs. Catherine Namugala, Minister of Tourism, Environment and Natural Resources, distinguished guests, ladies and gentlemen,

On behalf of the World Tourism Organization (UNWTO) representing the tourism interests of 154 countries, I have great pleasure in welcoming you all to the 5th IIPT African Conference: "Meeting the Challenges of Climate Change to Tourism in Africa and the Developing World" in the beautiful and rapidly developing city of Lusaka.

Zambia boasts an enormous tourism potential, based on its unique natural resources, wildlife sanctuaries, its rich traditions and the well-known hospitality of its people. A large number of tourists from all over the world visit Zambia annually to witness the magnificent beauty of the Victoria Falls, one of the most celebrated tourist sites in the world. The country is the home to exotic wilderness that incarnates nearly the whole continent of Africa and fascinates travelers and tourists. That is why in order to guarantee the long-standing success of this destination, it is so important to keep untouched its natural wealth and to develop tourism on a sustainable and green basis.

First of all, let me praise the Government of Zambia for choosing such a timely and appropriate theme as the challenges of climate change. A conference on this subject is vivid proof of the fact that Zambia is developing tourism in line with the global trends, enhancing its sustainable development aimed at conserving the natural and cultural heritage. We at

---

[1] Welcome Address, 5th IIPT African Conference, Lusaka, Zambia, 15–20 May 2011.

[2] UNWTO (World Tourism Organization) Executive Director for Member Relations and Services.

the UNWTO are extremely pleased that our friends in Zambia, side by side with the world tourism community, attach particular attention to the issues of climate change and sustainable tourism development.

Before passing directly to the subject of the conference, let me begin by looking at what we know about the tourism industry of today:

- 25 million international arrivals in 1950
- 165 million in 1970
- 687 million in 2000
- 935 million by 2010
- 1.6 billion forecasted for 2020

Tourism's growth, even if it seems to be slowing down somewhat, is not approaching any asymptote; there is no evidence of saturation of demand, at least on a global level. International tourism receipts almost doubled in the same period – from US$482 billion in 2000 to almost US$900 billion a year.

At the same time, like those gases that physics tells us tend to fill all the space available to them when the conditions are right, tourism is spreading far and wide; in other words, it is becoming globalized. In 1950, the top fifteen destinations in the world accounted for 87% of foreign visitor arrivals, in 1970 for no more than 75% and in 2009 for only 56%. When there is too much pressure on a region, tourists travel to another one; it is what could be called "Mariotte's law of tourism." I am referring to the law formulated by the seventeenth-century French physicist and not to the American hotelier Marriott! Tourism growth today goes hand in hand with globalization.

Few places on this planet escape the curiosity of its inhabitants. Even the Antarctic is now visited. The complete exploration of Earth has not sated men's appetite for research and adventure or their desire to discover the customs of those who are like them – or unlike them.

Besides man's insatiable curiosity about the world in which he lives, three developments caused international tourism to explode: the increase in purchasing power, and in discretionary income in particular, of middle- and working-class households in the developed world; access to private motor cars and cheap air transport; and the expansion of free time, regulated and developed in many countries by social legislation in favor of employees. Benjamin Disraeli was to write in the nineteenth century: "Increased means and increased leisure are the two civilizers of man."

In the past fifty years, tourism and leisure have been a potent factor of economic development in the countries of Europe and North America with a long industrial tradition. The developing countries of Asia, the Middle East, Latin America and a few African ones have, in their turn, reaped the benefits; they have discovered that wherever tourism advances, poverty retreats.

An important contributor to the balance of payments (the contribution of tourism to economic activity worldwide is estimated at some 5%), tourism has proved to be a prodigious creator of added value, wealth and jobs. Especially in the countries of the South, it has opened up new prospects for entrepreneurship and individual fulfillment. Its function as a factor of both social and cultural integration has also been remarkable. At once the manifestation and instrument of increasingly widely shared cultural attitudes and consumer habits, tourism has fostered the growing unity of society and helped to integrate within it those who could otherwise have been excluded. Since it promotes encounters and dialogue between visitors and hosts, it has furthered their knowledge of each other and the mutual recognition of individuals and groups.

"Growth," "globalization" and "integration," then, are three words that we can without any hesitation associate with tourism. And three more can be added to the list – "multidisciplinary," "vulnerability" and "resilience."

Tourism is, by its very nature, multidisciplinary. We can think of it as a chain whose links are infrastructure, hospitality, freedom of movement across borders, and product quality – where the latter includes the quality of food and health, as well as the quality of the environment, staff training, transport, businesses' ability to respond, the efficiency of public services, the security of visitors, etc. The failure of any one of its links affects the tourism product as a whole.

The character of being multidisciplinary, along with the resulting fragility, constitutes the vulnerability of tourism – vulnerability to external shocks, to wars, to terrorism, to the disruption of transport infrastructure, to social movements or to environmental accidents.

As we all know, the global economic recession, the closure of air traffic in Europe due to the Icelandic volcanic eruption, such terrible disasters as the earthquake in Japan and the catastrophic nuclear accident in Fukushima, as well as the current situation in the Middle East, have caused a lot of harm to tourism flows and confirmed once again its vulnerability. International tourist arrivals declined by 4.2% in 2009 to 880 million. International tourism receipts reached US$852 billion (611 billion Euros) in 2009, corresponding to a decrease in real terms of 5.7% on 2008.

Though vulnerable, the paradox is that tourism is, at the same time, profoundly resilient. The past year clearly illustrated this dual nature. In 2010, world tourism recovered even more strongly than expected from the shock it suffered in 2008 and 2009 due to the global financial crisis and economic recession. The vast majority of destinations worldwide reported positive and often double-digit increases, sufficient to offset recent losses or bring them close to their target.

Worldwide, international tourism rebounded strongly, with international tourist arrivals up 6.7% throughout 2009 to 935 million. As a reflection of the economic conditions, recovery was particularly strong in emerging countries, where arrivals grew faster (+8%) than in advanced ones (+5%). Africa's 6% growth is an increase on the positive results of 2009. Supported by worldwide exposure created by the FIFA World Football Cup, hosted by South Africa, the region maintained momentum in 2010, achieving a total of 49 million arrivals.

This, then, is what tourism looks like today, summarized by a few key concepts:

- Growth
- Globalization
- Integration
- Multidisciplinary
- Vulnerability
- Resilience

To the above key concepts we can add the emergence of new destinations. The world has opened up. New destinations have emerged. Within the past ten years, we have witnessed the rise of emerging countries – and tourism has followed the same trend. In 2000, emerging economies held a 38% share of international tourist arrivals; by 2010 this had increased to 47%.

We also need to mention technology. Technology has transformed our sector. Tourists can now search for and book holidays on the Internet. When they arrive at their destination, they can check the weather, buy tickets for tourism attractions or download a city guide. E-visas, air and ground traffic handling, border systems, waste reduction and energy efficiency technologies: all these areas have developed rapidly throughout the past ten years and revolutionized the way we travel, creating a myriad of new opportunities.

Our tourism sector has faced major challenges in the past years, but we all know that the biggest challenges provide the biggest opportunities. Today world leaders are working together in ways that would have been unimaginable at any time in the past, to coordinate and collaborate with regards to their economies, their climate response and their development agenda.

We meet here against a backdrop of a recovering economy and significant geopolitical shifts.

Unbalanced economic recovery across the globe, faster in emerging economies but still subdued in many advanced ones, continues to pose risks for world economic stability and the desired sustained growth.

Global unemployment is at a record high for the third straight year since the start of the economic crisis. And at the current pace, it is estimated it will take up to five years to see employment back at pre-crisis levels.

Oil and food prices could rise even more sharply than currently predicted.

Interest rates in major advanced economies might start to rise again after having been maintained at unprecedented low levels for close to two years.

And, finally, we cannot forget our environmental challenge – this being the topic that has gathered us together today and that is most characteristic of the global society that we are living in. As the world works to leave the economic crisis behind, the energy/climate crisis continues to advance. Greenhouse gases have reached record levels and our natural capital is running dangerously low.

Despite all this, 2011 is set to be a year to consolidate the growth recovered in 2010. UNWTO forecasts international tourism to grow between 4 and 5%, this is above the long-term average of 4%.

Now – with recovery underway – it is time to look forward and ask what the next ten years hold for the sector. Should we expect the indefinite advance, clearly beneficial but unbridled, of world tourism? As things stand at present, this is undoubtedly the most likely scenario, given that the democratization of transport is so attractive and in view of people's strong inclinations towards leisure activities and their appetite for discovery, whetted by the enticements of modern means of communication.

But such a form of exponential growth would surely be unacceptable to many. International air transport, transformed by global alliances between airlines, is already seriously disrupted by crowded skies and airport facilities – a consequence not only of the vertiginous increase in traffic, but also of poorly controlled deregulation. The physical environment and host communities are already suffering from the anarchic development of sites or the unrestrained exploitation of destinations. While the pressure is mounting, there are obvious signs of rejection.

It is easy to imagine the consequences – for the most part intolerable – that a tripling of flows in the space of a generation, predicted by UNWTO, in spite of and beyond the current crisis, would bring in its wake if not properly managed:

- Unbearable pressure on natural environments, fragile areas, coastlines and mountains;
- Serious congestion in city centers and overcrowding at the most popular sites and monuments;
- Inadequate transport infrastructure, public facilities and services;
- Mounting demands, competing with those of local populations, on water and energy resources;
- Outmoded training systems; and
- Local cultures and traditions under threat.

What will happen in 2020 when international visitor arrivals exceed 1.6 billion and those of national tourists are greater still?

In the face of such strong and legitimate fears, the time has come for the international tourism community to make its voice heard through the international tourism stakeholders' gatherings like the 5th IIPT African Conference.

Tourism and travel is a vital contributor to the global economy and especially important for many developing countries. Tourism is an effective way of redistributing wealth and a catalyst for gender equality, cultural preservation and nature conservation. As a result, the sector is also contributing to the UN Millennium Development Goals.

Climate change is one of the most serious threats to society, the economy and the environment and has been an issue of international concern for decades.

Climate change is not some remote future event for tourism; its varied impacts are already becoming evident at destinations. As a result, the effects of a changing climate will have considerable impacts on tourism, travel businesses and destinations and they will need to adapt to climate change in order to minimize associated risks or capitalize upon new opportunities, in an economically, socially and environmentally sustainable manner.

Least developed countries (the 4th LDC Conference was held in Istanbul, 9–12 May 2011, also concentrating on tourism) and small-island developing states might be particularly affected. Coastal environments in general and especially in SIDS are vital for tourism-based regional and local economies; therefore, climate change affecting coastal zones puts their economies at risk.

In Africa, for instance, warming is very likely to be more significant than the global annual mean warming throughout the continent and in all seasons, with drier subtropical regions warming more than the moister tropics. Annual rainfall is likely to decrease in much of Mediterranean Africa and the northern Sahara, with a greater likelihood of decreasing rainfall the nearer one comes to the Mediterranean coast. Rainfall in southern Africa is likely to decrease in much of the winter rainfall region and western margins. There is likely to be an increase in annual mean rainfall in east Africa. It is still unclear how rainfall in the Sahel, the Guinean Coast and the southern Sahara will evolve.

There are a number of different impacts of climate change that will affect tourism destinations, the most severe being direct climate impacts (such as warmer summers, warmer winters, precipitation change (water supply), increased extreme events); indirect environmental change impacts (biodiversity loss (terrestrial and marine), sea level rise, disease); impact of mitigation policies on tourist mobility (affecting travel costs and destination choice); and indirect societal change impacts (global/regional economic impacts, increased security risks (social/governance disruption)).

UNWTO, as the leading UN agency on tourism, has taken up the challenge of devising how the tourist industry can adapt to and mitigate the impact of global warming and how the sector can incorporate the necessary changes, so that its growth is compatible with new patterns of energy consumption and new scenarios of low-carbon economies.

Since the Djerba Conference in 2003 we have witnessed or taken direct part in other milestone events regarding the realities and challenges of climate change. First of all, the Davos Conference in 2007, which UNWTO together with other UN agencies organized, and which concluded with a declaration that acknowledges the reality of climate change as well as its connection with tourism and the need for action from the tourist sector regarding mitigation of its greenhouse gas emissions, adaptation to climatic conditions, technological awareness and securing financial resources to help poor regions and countries. This was followed by another three meetings in 2007: the London Ministers' Summit on Tourism, the UNWTO Cartagena de Indias General Assembly and the UN

Climate Change Summit in Bali. Since then, the tourism sector has made significant advances in responding to the climate change imperatives.

Allow me also to present the steps taken to date in response to the Bali Roadmap and Action Plan, while addressing mitigation and adaptation strategies within the tourism and travel sector.

Progress has been made in promoting the urgent adoption of a range of sustainable tourism policies. UNWTO is leading by example and will pursue its efforts in reducing its own emissions by raising awareness among its staff members, encouraging more efficient travel and investing in energy-efficient material, among other initiatives. In addition, in the framework of the "greening the UN" initiative, the organization has calculated the GHG emissions generated by its own operations since 2008 and intends to compensate them in order to achieve climate neutrality.

UNWTO understands that reducing carbon dioxide and other GHG emissions is important for the long term; however, there is also a pressing need for the tourism and travel sector to adapt to the impacts of climate change in the short and medium term. In this regard, there is an increasing demand for accurate, detailed climate information to allow tourism businesses to minimize the risks and capitalize on any new opportunities. This data could limit or affect how nations, especially developing countries, adapt to climate change. In response to this need, the WMO, in collaboration with UNWTO, established an Expert Team on Climate and Tourism in order to improve the application of information in the tourism sector and, in 2007, launched a Climate and Tourism Information Exchange Service to enable tourism stakeholders' access to research and data. It was created as a repository for information and as a display for data, studies, policy papers, videos and other material related to tourism and climate change, which would facilitate tourism stakeholders working together to develop coordinated approaches to tackling mitigation and adaptation.

UNWTO has also developed and disseminated technical publications addressing climate change impacts and adaptation responses. The most important of these are "Climate Change and Tourism: Responding to Global Challenges" in support of the Davos Declaration, and "Climate Change Adaptation and Mitigation in the Tourism Sector: Frameworks, Tools and Practices" in coordination with the University of Oxford, the WMO and UNEP, released in 2008. This report provides a detailed and comprehensive synthesis of the state of knowledge about current and future predicted impacts of climate change on global tourism. UNWTO has also recently published a background paper "From Davos to Copenhagen and Beyond: Advancing Tourism's Response to Climate

Change," which provides an update of our work addressing climate change imperatives.

UNWTO has assisted countries through their National Tourism Administrations to access the GEF Special Climate Change Fund by approaching other multi- and bilateral funding sources and private-sector organizations in order to secure co-financing. For instance, a pilot project on tourism adaptation in the Maldives has been initiated through collaboration with UNEP and UNDP, and discussions initiated with SIDS in other regions.

In 2008, UNWTO launched the Hotel Energy Solutions Project, which is co-funded by the European Agency for Competitiveness and Innovation. UNWTO coordinates the project in partnership with UNEP, the International Hotel and Restaurant Association, the French Agency for Energy Management and the European Council for Renewable Energies. The international hotel sector is one of the largest drivers of global employment. However, in 2005, it was estimated that the accommodation sector accounted for 21% of the carbon dioxide emissions from tourism. The aim of this project is to provide energy-efficiency solutions and increased use of energy-efficient and renewable-energy technologies for the tourism accommodation sector across twenty-seven European Union countries. This is seen as a significant way to reduce the impact of the sector on climate change by mitigating emissions.

Furthermore, UNWTO is providing "in-kind" support and support aimed at raising funds for governments and regional initiatives for climate change adaptation and mitigation projects in the tourism sector for key regional initiatives involving governments of numerous countries. The CARIBSAVE Partnership, formed in 2008 between the Caribbean Community Climate Change Center (CCCCC) and the University of Oxford, and supported by a range of international partners including UNWTO, UNEP, UNDP, the World Bank, WWF and other international and regional partners, is addressing the impacts and challenges surrounding climate change, tourism, economic development and community livelihoods across the Caribbean Basin.

The global impacts of climate change call for an integrated and inter-sectoral response, which is especially important for UNWTO, given the transversal nature of tourism.

UNWTO has actively participated in major international events within the UN system's efforts to develop a coherent framework response to climate change. We have participated in the World Climate Conference-3 (2009), and the Fifteenth Conference of the Parties to the UN Framework Convention on Climate Change held at Copenhagen on 7–18 December 2009, in which UNWTO, together with the World Travel and Tourism Council (WTTC), hosted a side event on "Addressing the Challenges of Climate Change: Perspectives from the Tourism and the Travel Sector," which presented the steps taken so far in response to the Bali Roadmap and Action Plan, while addressing mitigation and adaptation strategies within the tourism and travel sector. The side event was a further step of the Davos Declaration process, and included interactive panels with tourism leaders from the public and the private sectors.

Most recently, UNWTO held a side event jointly with the Secretariat of Tourism of Mexico at the Sixteenth Conference of the Parties of the United Nations Framework Convention on Climate Change (COP 16), in Cancun, Mexico, from 29 November to 10 December 2010. The side event presented the initiatives undertaken so far to mitigate greenhouse gas emissions, promote adaptation in tourism businesses and destinations, invest in new technologies and support developing countries through financing, as well as the UNWTO-initiated Hotel Energy Solutions Project aimed at increasing energy efficiency and the use of renewable-energy technologies by SME hotels in the European Union.

The climate is changing and will continue to do so for the foreseeable future. Global warming is a scientific fact. This will result in social, economic and environmental impacts. The degree of these impacts will depend on to what extent nations, industry and individuals mitigate emissions and adapt to changes.

The tourism and travel sector must address the climate change issue and UNWTO has embraced the challenge by establishing collaborative partnerships, raising awareness, developing guidance and providing support. The moment has come for public and private sectors to join efforts in order to develop innovative low-carbon initiatives that will not only create employment but will also contribute to the economy.

According to the Davos Declaration on Climate Change and Tourism, the sector has to "rapidly respond to climate change, within the evolving UN framework and progressively reduce its greenhouse gas (GHG) emissions if it is to grow in a sustainable way." The concept of sustainable development does not refer only to economic growth, social progress and environmental preservation. It has an ethical and cultural dimension.

Today, we have gathered here together in order to underline once again the necessity of tourism development on a sustainable basis and, giving a glance at the future, do our best in order that our children, and the children of our children may enjoy traveling around our beautiful, fresh and green globe without any fear.

I wish success and fruitful results to the participants in the Fifth IIPT African Conference!

CHAPTER TWO

CLIMATE CHANGE AND TOURISM: ADVANCES IN KNOWLEDGE AND PRACTICE

SUSANNE BECKEN[1]

Abstract

There is plenty of evidence that tourism stakeholders, as well as tourists, are increasingly becoming aware of the challenges that climate change is posing for the sector. This chapter presents a number of case studies that illustrate tangible progress both in terms of reducing greenhouse gas emissions and adapting to changes in climatic or environmental conditions. The examples discussed in this chapter highlight initiatives at the policy level, and practical implementations. Learning at both of these levels is seen as the major next step in helping tourism to successfully address climate change. Key points extracted from the case studies include the need to avoid "unsustainable" tourism development, the importance of including carbon management as an integral part of business practice, and the benefits of taking a risk-management approach to tourism (at various levels).

Introduction

Tourism stakeholders, as well as tourists, have become increasingly aware of the challenges that climate change will pose for the sector. There is plenty of evidence that tourists care about, and are in principle willing to contribute to, climate-change initiatives, at least financially (Brouwer et al., 2008). Changes in tourist behavior toward more sustainable consumption patterns are less likely to happen at a large scale in the near future (Becken, 2007; Mair, 2011). However, despite some uncertainties around tourist demand and purchasing behavior, the greater debate on climate-change action in the tourism sector has now moved on from mere

[1] Land, Environment and People Research Center, Lincoln University, New Zealand.

discussions to exploring how to actually address climate change. The sector has focused on establishing frameworks and partnerships that facilitate policy making and implementation of measures that reduce tourism's greenhouse gas emissions (mitigation) as well as measures that help anticipate future climatic changes and increase tourism's resilience to these (adaptation). An example of this international effort is the "Climate Change – A Joint Approach to Addressing Climate Change" document prepared by the World Travel and Tourism Council (2010). This program details priorities for the tourism sector, as well as commitments and suggestions for cooperation with government agencies in relation to climate-change mitigation. The report also acknowledges that the "recent model of carbon-intensive tourism growth, consumption and production is now under review" (2010: 3). It is also suggested that tourism can and should take a leadership role in climate-change mitigation, especially in those countries where tourism dominates the national economy.

The last few years have seen a rapid increase in both academic and practically oriented enquiry into climate change and tourism. A number of books and reports have been published that draw together the collective knowledge in this field (e.g., Becken & Hay, 2007; Gössling & Upham, 2009; Jones & Phillips, 2011; UNWTO, UNEP & WMO, 2008). Detailed research recently undertaken includes, for example:

- Tourists' behavior in relation to changing climatic conditions (e.g., Hein et al., 2009; Rossello-Nadal et al., 2011);
- Impacts on the ski industry around the world, for example in North America (Scott et al., 2006), New Zealand (Hendrikx & Hreinsson, 2010), Australia (Hennessy et al., 2008), and Austria (Wolfsegger et al., 2008);
- Opportunities to reduce carbon footprints and increase energy efficiency (e.g., Chan et al., 2008; Dalton et al., 2008);
- Carbon offsetting in tourism (Gössling et al., 2007);
- New forms of tourism (e.g., slow tourism, Dickinson & Lumsdon, 2010); and
- Disaster-risk reduction and tourism (Becken et al., 2011; UNEP & CAST, 2008).

In addition to in-depth studies, there are also more and more toolkits and technical guides that are often produced by industry organizations and which typically have a very specific focus (e.g., improving environmental performance in the accommodation sector; see The International Tourism Partnership, 2010). These are very important for the day-to-day operational management of a business. Despite overall progress there still seems to be a gap between our increased understanding of the drivers and consequences of climate change on the one hand, and the technical advice on the other hand (Figure 2-1). This gap relates to specific climate-tourism policies (and institutional arrangements) that enable practical action at a larger scale. Thus, this chapter will provide an insight into progress made by policy makers and tourism businesses, drawing on case studies from around the world, including developing and developed countries. The importance of a sound conceptual understanding of climate change and tourism will be acknowledged, but the focus is on the roles of supportive policies and institutions in ensuring a strong "enabling environment" for practical responses, and on the practical responses themselves.

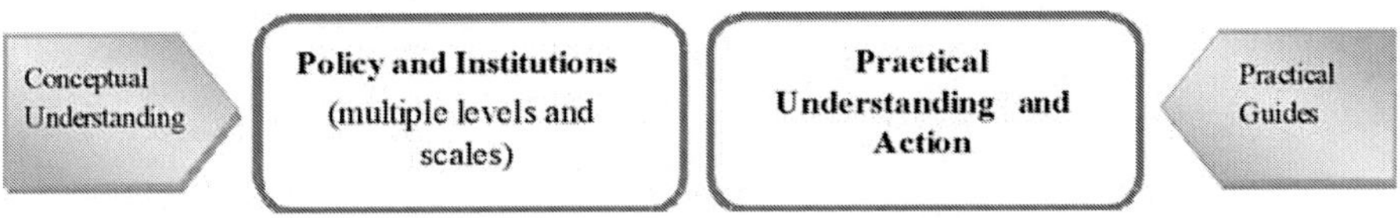

Figure 2-1. Filling the gap between our growing knowledge and understanding, and practical industry guides, by addressing policies, institutions and practical understanding (Becken & Hay, 2012).

The following discussion is structured into aspects and case studies related to mitigation policy and practice, followed by progress made in adaptation policy and practice.

Mitigation Policy

The explicit consideration of climate change as part of tourism policy dates back to the first United Nations World Tourism Organization conference on tourism and climate change in 2003 in Djerba. The Djerba Declaration was the first substantial policy document on climate change by the tourism sector. From then on, tourism policies were enhanced by adding climate-change mitigation, or specific policies were developed by tourism stakeholders at different levels to address climate change (Becken & Clapcott, 2011). Prominent examples include the inclusion of aviation into Europe's Emissions Trading Scheme (ETS) (see for example Anger & Koehler, 2010), and IATA's efforts and ambitions to reduce aviation's greenhouse gas emissions. IATA's climate change response hinges on four pillars: (1) technology, (2) operations, (3) infrastructure, and (4) economic measures (IATA, 2009). The difference between the European ETS and

IATA is their legal status. As for most industry initiatives, IATA's targets and goals are voluntary in nature and members are not obliged to comply. In contrast, the EU ETS comes with a clearly defined plan of emission reductions (caps), certificates and ways of auctioning. Imposing carbon-related costs or other taxes, such as the British Air Passenger Departure Duty, may act as economic disincentives and reduce tourist volumes. This might affect low-cost travel more than traditional carriers.

Figure 2-2. Passengers at Gatwick Airport, Great Britain. Photo credit: Susanne Becken.

Mitigation policies and actions by local or national-level organizations often relate to transport management and infrastructure, renewable portfolio standards, energy-efficiency programs, emission registries and cap-and-trade mechanisms. Local policy making, in particular, has gained increased recognition over the last few years, because local government agencies are often at the core of decisions relating to both climate-change mitigation and adaptation. Tourist destinations (e.g., through Regional Tourism Organizations) often, but not always, collaborate with local governments to develop policies and implement climate-change mitigation measures. Tourist destinations often wish to differentiate themselves by acting as leaders of climate-change policies, even if the country as a whole may lag behind. Initiatives such as the National Geographic Center of Sustainable Destinations (and Destination Stewardship surveys) or destination-based Sustainability Charters are evidence of this (e.g., in various regions in New Zealand).

A prominent example of local policy for climate-change mitigation is the Climate Change and Peak Oil Strategy developed by the Sunshine Coast Council in Australia. This strategy contains an Energy Transition Plan, a Carbon Neutral Plan and the Community Emissions Reduction Plan. The implications of higher energy costs were assessed in detail for five sectors: transport, construction, food supply, regional energy production and tourism. The tourism-specific risk assessment concludes that, while the Sunshine Coast's reliance on tourism is high, at this point in time its vulnerability to higher energy costs is relatively low. This is due to the very small proportion of international tourists (only 9% of arrivals) who arrive on long-haul flights. However, it is recognized that many Australian tourists arrive by low-cost carriers. As a result, assessment of airline viability is one of the recommendations made for further studies. Other useful examples of climate-change mitigation policies in Australia (both at a national and regional levels) are summarized in Zeppel and Beaumont (2011).

Private-sector policy initiatives often include voluntary initiatives or emissions targets. Motivations for business mitigation actions include: (i) influencing or pre-empting government action, (ii) creating financial benefits, and (iii) differentiating a company or its products (brands) from competitors. There are many good examples from the tourism industry that illustrate how environmental policies incorporate climate change, or how specific climate-change policies have been designed. Prominent case studies (amongst others) are:

- Accor, with its Environmental Charters;
- Fairmont hotels, with their Green Partnership program;
- Eurostar, with a 10-point plan to minimize climate impacts;
- Auckland International Airport, being a member of the global Carbon Disclosure Project;
- Thomas Cook group, setting environmental targets for fuel efficiency and carbon reduction;
- Aspen, Colorado (USA), reductions in greenhouse gas emissions and supply chain management; and
- Virgin Atlantic, with climate change being the priority issue in their environmental policy.

Mitigation Practice

Reducing greenhouse gas emissions from tourism can happen at many levels, starting with the individual tourist, and ranging to businesses, supply chains, destinations, countries, regions and the global tourism sector. Typical measures discussed in the literature and observed in practice include (i) visiting destinations that are closer, (ii) staying longer and making fewer trips, (iii) choosing carbon-efficient transport modes, (iv) driving efficiently, (v) staying in environmentally friendly accommodation and minimizing energy use (e.g., switching off lights and air conditioning when leaving rooms), (vi) avoiding energy-intensive recreational activities, and (vii) supporting low-carbon products on the ground (e.g., those using locally grown food).

There has been considerable debate on the option of carbon offsetting one's emissions, with some arguing that investing into offsetting makes a real contribution to reducing global emissions, while others accuse it of being a form of "greenwash" with high administrative costs, low transparency and limited accountability. Notwithstanding, a large number of tourism businesses employ carbon-offsetting programs, including many airlines (e.g., British Airways, Qantas, Pacific Blue), Intercity Coachlines in New Zealand, Thompson and First Choice, World Expeditions, Bluewater Adventures (Canada) and Christchurch International Airport. The Tourism Industry Carbon Offset Service (TICOS) in the United Kingdom is an industry-wide program to facilitate collective action by those involved in the tourism industry. TICOS collects financial contributions from customers to finance projects that both make carbon savings and have wider sustainable development benefits.

A key challenge for tourism relates to greenhouse gas emissions associated with transportation. Although there is a real and continuing problem of scale (especially due to growing volumes) with aviation emissions, many airlines have made substantial improvements over the last decade. A number of airlines are investing into bio-fuel; for example, the Dutch carrier KLM has announced the formation of a new company to develop sustainable bio-fuels called SkyEnergy. British Airways and the US-based Solena Group announced an agreement to build Europe's first plant to produce jet fuel from organic waste. Subject to regulatory approval, the plant is expected to go online in 2014, and produce about 72,700 liters (16 million gallons) of fuel annually, all of which will be sold to British Airways. Air New Zealand is involved in research on bio-fuel production from algae and plans to use one million barrels of sustainable fuel annually by 2013. Virgin Airlines, Air New Zealand, Continental and Japan Airlines have completed flight trials with bio-fuel. The use of bio-fuels is subject to debate as there is a risk of unsustainable production (e.g., competing with land for food production) and there also remain questions around realistic scales of production. Other improvements in aircraft technology and plans to increase operational efficiency through integrated air-traffic management are detailed in a roadmap developed by the International Air Travel Association (2009).

Figure 2-3. Deutsche Bahn was a sponsor of the Women's World Cup in 2011, also offering attractive "weltmeister bahncards" with a 25% discount on all train travel. Photo credit: Susanne Becken.

There are a large number of other examples of tourism business mitigation actions. The Dutch Railways, for example, found that the optimization of their brake energy recuperation settings saves them about 1% of total electricity use. Deutsche Bahn is involved in a number of "low-carbon" initiatives, including Fahrtziel Natur (train packages to 19 unique nature destinations) and attractive offers related to sports events (e.g., FIFA football events). Investments into electric vehicles have been made in Vietnam (in Hanoi), in Thailand (Green Island and Siaoliouciou), and hybrid cars are used in rental car fleets (e.g., Green Tomato, UK) and taxis (Green Taxis, NZ). More examples of industry initiatives can be found in Becken and Hay (forthcoming) and also in Gössling (2011).

Tourist destinations are also increasingly interested in carbon management and branding. For countries such as New Zealand, the brand of an environmentally clean and green ecotourism destination made it essential for decision makers to proactively engage in climate-change mitigation. A number of countries aspire to become the first carbon-neutral destinations in the world. Bonaire is one of them. Current activities in Bonaire include the auditing of GHG emissions and investments into renewable energy, energy efficiency and reforestation projects. In other cases, such as Växjö in Sweden, climate-change initiatives implemented by local communities for reasons other than tourism development have given a boost to visitor activity.

Adaptation Policy

The consequences of climate change are inevitable, and tourism, just like other sectors, needs to prepare for changes in climatic and environmental conditions to ensure its long-term viability. In some cases, this will entail the preparation of a policy and plan of action, and the establishment of new institutional arrangements. National and local governments in places where tourism makes a significant contribution to the economy and local livelihoods typically undertake such comprehensive responses. Large tourism corporations and industry organizations also develop policies in order to reduce the risks that climate change represents to their profitability. In many cases these initiatives will also seek to exploit any benefits climate change might bring to the tourism industry. Smaller tourism operators, and tourists themselves, should benefit from these higher-level policies, plans and institutional arrangements. In some cases, smaller businesses also prepare adaptation policies as part of their wider business plan.

Australia's "Tourism and Climate Change – A Framework for Action" (Department of Resources, Energy and Tourism, 2008) is a good example of a national policy that addresses the impacts of climate change on tourism. The rationale for the policy is that climate change will impose costs on the tourism industry due to tourism's dependence on natural assets and the built environment and their vulnerability to the physical impacts of climate change. The framework recognizes that these impacts will also flow through in the form of financial risks with rising insurance premiums, changes to business financing, and the need for businesses to incorporate climate change into their business models. The Framework aims to deliver five outcomes:

1. Improved understanding of the vulnerabilities of tourism to both the physical and economic impacts of climate change, in order to build the resilience and adaptive capacity of the industry and provide certainty for the purpose of future investment;

2. A tourism industry that is prepared for a carbon-constrained future and continues to make a substantial contribution to the Australian economy;

3. A repositioning of tourism marketing strategies to meet head-on the challenges and opportunities presented by climate change;

4. A fully informed tourism industry through consistent and effective industry outreach and communications; and

5. A nationally consistent, inclusive and cooperative approach to implementation.

There are also a number of examples of other regional or national adaptation strategies – for example, in the Caribbean, Germany (through the German Adaptation Strategy), Samoa (various climate policies and tourism development plans), and Israel (in relation to reducing water consumption from tourism). The Great Barrier Reef Tourism Climate Change Action Strategy (2009–2012) is a good example of a destination-based approach to climate change adaptation by tourism. It is also a good illustration of how resilience and risk management are integrated with carbon management and environmental sustainability.

Table 2-1. Examples of adaptation policies relevant to tourism in the Caribbean (Pulwarty et al., 2010).

Caribbean small-island developing states have articulated National Climate Change Adaptation Policies and Implementation Plans and many regional programs are now linking these plans to their resource and risk-management policies. The plans include:
1. Caribbean Environment Programme
2. International Coral Reef Action Network Project
3. Integrating Watershed and Coastal Area Management in Small Island Developing States
4. Caribbean Blue Flag Programme
5. Caribbean Global Water Partnership
6. Caribbean Regional Fisheries Mechanism
7. Caribbean Conservation Association

Another important area of climate-change adaptation relates to disaster risk reduction. There is some evidence that climate change will increase the frequency and intensity of certain hydro-meteorological events (at least in some regions) and tourism destinations will have to prepare for crises and disasters resulting from such events. A number of good examples exist where tourism and disaster management have been integrated. Following the destructions of Hurricane Ivan, for example, Grenada has invested into a Policy and Operational Framework for Mainstreaming Disaster Risk Reduction into the Post-Hurricane Ivan Reconstruction Process. Important elements include strengthening building codes and developing tools for integrating risk reduction into other planning. Phuket Province (in Thailand) has brought together tourism development planning with disaster prevention and management to create the Phuket Tourism Risk Management Strategy (UNEP & CAST, 2008).

Figure 2-4. Beach protection (sea wall) in light-colored rocks to reduce aesthetic disturbance (blending in with the white sand) at a beach resort in Antigua. Photo credit: R. Mahon.

Figure 2-5. Tourist restaurant in Cozumel (Mexico) destroyed by a hurricane. Photo credit: Susanne Becken.

Adaptation Practice

Table 2-2. Examples of adaptation strategies for various climatic impacts (Becken & Hay, 2007; Zeppel, 2011).

Impact	Adaptation	Implementing Actor/Agency
Reduced snow cover /amount	Artificial snow making, Snowfencing, Extend ski operations to higher altitudes, Re-design ski slopes, Strategic partnerships, Increase sale of season passes, Weather derivatives	Ski resorts
	Promote non-snow winter activities, Build attractions (ice rink, spa), Develop all-year tourism (i.e., summer activities), Insurance, Open higher-elevation ski runs, Subsidies for cableway operators to keep ski fields open	Destination, local government
Increased coral bleaching	Extend operations to other reefs, Assist reef propagation, Cancel dive tourism, Promote other reef tourism activities,	Operators
	Reduce impacts of ocean pollution, Ban fishing and coral collection on reefs, Close affected reef areas, Educate tourists	Government and non-government organizations
Hotter days/ heat waves	Install air conditioners and fans, Hotel pools and umbrellas, Plant more trees for shade, Provide drinking water, Cool buildings	Operators
	Develop artificial indoor beaches, Promote water-based or cool indoor activities, Expand shoulder seasons	Destinations
	Adjust building codes	Government
Increased cyclones/ hurricanes	Build to cyclone standards, Trim tree branches, Develop evacuation plans, Disaster insurance, Protect and maintain coastal native vegetation, Improve drainage (clear gutters)	Operators
	Provide cyclone warnings, Close damaged beach areas	Government/ destinations
Coastal erosion	Build coastal protection (sea/rock wall, groyne, dyke), Replenish beach sand (trucks, pumping), Revegetate/plant soil-binding vegetation in coastal areas, Ban development in at-risk zones, Establish building set-back limits well above mean sea-level	Government and destinations
Reduced water availability	Increased water storage, Recycle water, Desalination	Operators
	Encourage minimal water use by guests, Limit or set quotas on water use, Use trickle irrigation, Repair leaks, Use timers on taps, Plant drought-tolerant plants	Government
	Develop water savings policies (e.g., charging for usage), Provide incentives for water-efficient equipment	
Flooding	Build on higher ground, Disaster insurance, Pumping systems	Operators
	Install levees, dykes and drainage systems, Enhanced flood design and site standards, Close areas prone to flooding, Use alternative routes or areas	Local and Central Government

Adaptation to climate change and variability can take several forms: technical, managerial, behavioral and educational. Tourism adaptation measures often involve new product development, infrastructure management, environmental protection measures, marketing and risk reduction. These can occur at all levels, from individual businesses to destinations and regions (examples are shown in Table 2-2). A good example of a business approach to climate-change adaptation is the International Civil Aviation Organization briefing paper for airports. The purpose is to prepare airports for changing operating environments, for example due to sea level rises (issues for runways, taxiways, terminal buildings and access routes), temperature rises (e.g., decreased aircraft lift), precipitation changes (e.g., flooding of runways) and storms (e.g., closure of airports in severe events) (Gittens, 2010).

Conclusion

This chapter highlights that not only has our conceptual understanding of tourism and climate change increased rapidly, but so have the practical responses by those involved in the tourism sector (Becken & Hay, forthcoming). A wide range of examples have been introduced to highlight progress in both mitigation policy and practice. There are also many good examples where mitigation and adaptation are addressed in an integrated way (e.g., the business guide for climate change developed by the Australian Department of Resources, Energy and Tourism, 2009).

Three points stand out as key "lessons learned." One is that successfully addressing climate change begins with avoiding unsustainable practices. The more efficient use of resources (both energy and water, for example), the minimizing of impacts on the bio-physical environment (e.g., coral reefs or beach systems), the appropriate design of buildings (especially designed for local, climatic conditions) and the involvement of the local population are only a few examples of how climate-change responses become an integral part of sustainable development. Further, it is evident that reducing greenhouse gas emissions is no longer merely an ethical issue that defies business logics. The contrary is the case. Carbon management has now become part of good business practice that is essential to the economic bottom line. This is not only due to the cost savings associated with reducing energy demand and the favorable market feedback, but also because of a long-term strategic view of increasing energy costs and reducing dependency on fossil fuels (Becken, 2011).

Finally, it becomes clear that adapting to climate change is very closely linked to "future proofing," "risk management" and "disaster-risk reduction." All of these are highly relevant for tourism businesses, destinations and the tourism sector at a global level.

References

Anger, A., & J. Koehler (2010). Including aviation emissions in the EU ETS: Much ado about nothing? A review. *Transport Policy* 17:38–46.

Becken, S. (2007). Tourists' perception of international air travel's impact on the global climate and potential climate change policies. *Journal of Sustainable Tourism* 15, no. 4: 351–368.

—. (2011). A critical review of tourism and oil. *Annals of Tourism Research* 38, no. 2: 359–379.

Becken, S., and J. Hay (2007). *Tourism and Climate Change: Risks and Opportunities*. Clevedon: Channel View Publications.

—. (2012). *Tourism and Climate Change Mitigation and Adaptation: From Policy to Practice*. London: Earthscan.

Becken, S., & R. Clapcott (2010). Developing public policy for climate change in the tourism sector. *Journal of Policy Research in Tourism, Leisure and Events* 3, no. 1: 1–17.

Becken, S., J. Wilson & K. Hughey (2011). Planning for climate, weather and other natural disasters – Tourism in Northland. LEaP Research Paper 1. Available at www.lincoln.ac.nz/leap.

Brouwer, R., L. Brander & P. Van Beukering (2008). A convenient truth: Air travel passengers' willingness to pay to offset their CO_2 emissions. *Climatic Change* 90: 299–313.

Chan, W. W., L. M. Mak, Y. M. Chen, Y. H. Wang, H. R. Xie, G. Q. Hou & D. Li (2008). Energy saving and tourism sustainability: Solar control window film in hotel rooms. *Journal of Sustainable Tourism* 16, no. 5: 563–574.

Dalton, G. J., D. A. Lockington & T. E Baldock (2008). Feasibility analysis of stand-alone renewable energy supply options for a large hotel. *Renewable Energy* 33: 1475–1490.

Department of Resources, Energy and Tourism (2008). Tourism and Climate Change – A Framework for Action. Available at http://www.ret.gov.au/tourism/policy/Tourism%20and%20Climate%20Change/frameworkforaction/Pages/default.aspx.

—. (2009). Climate Change Guide. Mitigation and Adaptation Measures for Australian Tourism Operators. Canberra: Commonwealth of Australia. Available at http://www.ret.gov.au/tourism/Documents/Tourism%20and%20Climate%20Change/2141259A_TOURISM_CLIMATE_CHANGE_Workshop%20guide_v12.pdf.

Dickinson, J. & L. Lumsdon (2010). *Slow Travel and Tourism*. London: Earthscan.

Gittens, A. (2010). Airports and climate change adaptation. ICAO Council Briefing, Montreal, 16 April 2010.

Gössling, S., & P. Upham (2009). *Climate Change and Aviation*. London: Earthscan.

Gössling, S. (2010). *Carbon Management in Tourism: Mitigating the Impacts on Climate Change*. Oxford: Routledge.

Gössling, S., J. Broderick, P. Upham, J. P. Ceron, P. Peeters & W. Strasdas (2007). Voluntary carbon offsetting schemes for aviation: Efficiency, credibility and sustainable tourism. *Journal of Sustainable Tourism* 15, no. 3: 223–248.

Hein, L., M. Metzger & A. Moreno (2009). Potential impacts of climate change on tourism. A case study for Spain. *Current Opinion in Environmental Sustainability* 1: 170–178.

Hendrikx, J., & E. Ö. Hreinsson (2010). The potential impact of climate change on seasonal snow conditions in New Zealand. NIWA Client Report: CHC2010-153, November 2010.

Hennessy, K., P. H. Whetton, K. Walsh, I. N. Smith, J. M. Bathols, M. Hutchinson & J. Sharples (2008). Climate change effects on snow conditions in mainland Australia and adaptation at ski resorts through snowmaking. *Climate Research* 35: 255–270.

IATA (2009). The Technology Roadmap Report. Available at (02/01/2011) www.iata.org/SiteCollectionDocuments/Documents/Technology_Roadmap_May2009.pdf.

Jones, A. & M. Phillips (2011). *Disappearing Destinations: Climate Change and the Future Challenges for Coastal Tourism*. Wallingford: CABI.

Mair, J. (2011). Exploring air travellers' voluntary carbon offsetting. *Journal of Sustainable Tourism* 19, no. 2: 215–230.

Pulwarty, R. S., L. A. Nurse & U. Trotz (2010). Caribbean islands. Environment, November/December 2010. Available at http://www.environmentmagazine.org/Archives/Back%20Issues/November-December%202010/caribbean-islands-full.html.

Rossello-Nadal, J., A. Riera-Font & V. Cardenas (2011). The impact of weather variability on British outbound flows. *Climatic Change* 105: 281–292.

Scott, D., G. McBoyle, B. Mills & A. Minogue (2006). Climate change and the sustainability of ski-based tourism in eastern North America. *Journal of Sustainable Tourism* 14, no. 4: 376–398.

The International Tourism Partnership (2010). Environmental Management for Hotels. Available for purchase at http://www.tourismpartnership.org/Publications/EMH.html.

UNEP & CAST (2008). Disaster Risk Management for Coastal Tourism Destinations. Responding to Climate Change. A Practical Guide for Decision Makers. UNEP DTIE: Paris. Available at www.unep.fr/scp.

UNWTO, UNEP & WMO (2008). *Climate Change and Tourism: Responding to Global Challenges.* Madrid: United Nations World Tourism Organization; Paris: United Nations Environment Program; Geneva: World Meteorological Organization.

Wolfsegger, C., S. Gössling & D. Scott (2008). Climate change risk appraisal in the Austrian ski industry. *Tourism Review International* 12: 13–23.

World Travel & Tourism Council (2010). Climate Change. A Joint Approach to Addressing the Challenge. London. Available at www.wttc.org.

Zeppel, H. & N. Beaumont (2011). Green tourism futures: Climate change responses by Australian government tourism agencies. In *Tourism: Creating a Brilliant Blend*, proceedings of the 21st Annual Council for Australian University Tourism and Hospitality Education Conference (CAUTHE 2011), 8–11 Feb 2011, Adelaide, Australia. Available at http://eprints.usq.edu.au/18695/.

CHAPTER THREE

PERSPECTIVES ON GLOBAL CLIMATE CHANGE AND THE NIGERIAN RESPONSE

ABUBAKAR SADIQ[1]

This conference, with its theme of meeting the challenges of climate change to tourism in Africa and the developing world, couldn't have come at a better time, when viewed against the backdrop of the fragile ecosystems of the world that have come under threat by the forces of climate change occasioned by both natural and anthropogenic (human-induced) factors.

Climate change, a concept that describes the alteration or modification of the natural state of the climate, is a universal challenge to sustainable development. It is a long-term change in the distribution of weather conditions or in the distribution of weather events, and may be limited to a specific region or may occur across the whole earth. The United Nations Framework Convention on Climate Change defines it as a "change that is attributed directly or indirectly to human activity that alters the composition of the global atmosphere and that is in addition to natural climate variability observed within a comparable time period." In the latter sense, climate change is synonymous with global warming.

Some factors have been identified which predispose the climate to change. These factors are both natural and anthropogenic (human-induced). The natural elements that can shape the climate include variation in solar radiation, slight deviations in the earth's obit, mountain building and continental drift, and changes in greenhouse gas concentrations. These factors create an imbalance in the earth's crust as a result of compression and tension in the system, and elements such as the ocean and ice caps respond slowly in reaction to these climatic forces because of their large mass. In this respect, the climate system can take centuries or longer to fully respond to the imbalances.

---

[1] Ministry of Tourism, Culture and National Orientation, Federal Republic of Nigeria.

Anthropogenic factors can also change the environment. In some cases, the chain of causality of human influence on the climate is direct and unambiguous (for example, the effects of irrigation on local humidity), while in other instances it is less clear. Various hypotheses for human-induced climate change have been argued over for many years. Presently, the scientific consensus on climate change is that human activity is very likely the cause for the rapid increase in global average temperature.

Of most concern in these anthropogenic factors is the increase in carbon dioxide levels due to emissions from fossil-fuel combustion, followed by aerosols (particulate matter in the atmosphere) and cement manufacture. Other factors, including land use, ozone depletion, grazing and deforestation, are also of concern in the roles they play – both separately and in conjunction with other factors affecting the climate, microclimates, and measures of climate variables.

The frequency and intensity of extreme weather events are on the increase. In Nigeria, for example, the desert is encroaching from the northern part of the country, there are consequent effects on Lake Chad, and gullies are widening and floods are destroying lives and properties in the coastal areas. In other countries, coastal areas are under threat due to sea-level rise, hurricanes, earthquakes, tsunamis, increase in ultraviolet radiation, biodiversity loss and other meteorological hazards.

Tourism and the environment are symbiotic. Thus, sustainable growth in the tourism industry cannot take place in an environment with a polluted and fragmented ecology. Tourism is multi-dimensional in nature; hence, a holistic and multi-sectoral approach is required to tackle the menace posed by climate change. This informs the decision of the Commission for Sustainable Development, a UN agency responsible for the environment, to give adequate interest to eco-tourism development worldwide. It has become clear that all sectors have to address the climate-change challenge, without abandoning their responsibilities toward socio-economic issues. For tourism, the transformation toward the green economy is also a major task to be addressed. Hence, in 2007, the United Nations World Tourism Organization (UNWTO) was actively involved in helping the tourism sector to:

- *Assess the inter-relationship of climate change and tourism* through a science-based report on the economic, operational and market impacts as well as the challenges and opportunities.
- *Consider meaningful response patterns* based on multi-stakeholder evaluation and recommendations contained in the Davos Declaration framework. The Declaration (augmented by

ministerial decisions in London and Cartagena) sets out directions for change for the tourism sector and calls on stakeholders to establish a long-range low-carbon emission road map with immediate concrete action plans supporting the global response, and coherent with commitments to the Millennium Development Goals.

- *Participate in the global response framework* to include tourism input alongside other economic sectors in the recent UN Climate Summit in Bali, Indonesia, as well as within UN Secretary-General Ban Ki-moon's system-wide road map to support adaptation, mitigation, technology and finance.

The motive for international tourists to leave their country of origin depends, amongst other factors, on the condition of the climate. Changing climate and weather patterns at tourist destinations and tourist-generating countries can significantly affect the tourists in a number of ways:

(a) Tourists' comfort and their travel decisions.
(b) Tourism business, mostly in the host communities that could suffer setbacks.
(c) Reduction in tourist flow that in turn can affect employment and generate further poverty.
(d) Attractiveness of tourist destinations that could be negative.
(e) Adverse climate conditions would cause medical problems that affect the comfort and movement of tourists.

The prosperity of the tourism industry largely depends on favorable weather conditions. In instances where there have been vagaries of weather, or other events such as volcanoes, earthquakes, etc., tourism has paid dearly: safety is crucial to the tourist, who is in search of fun and relaxation but not at the expense of his or her life. Thus, responsible tourism through the establishment of the World Committee on Tourism Ethics was included in the program of the UNWTO. The reason for this is to encourage responsible tourism, safeguard the environment and protect the mutual interest of all stakeholders in the industry, including the local community.

It is of utmost importance that the environment be preserved for sustainable tourism to flourish. In this connection, governments and the private sector, as well as all stakeholders in the tourism sector, must work hand in hand to preserve the environment.

The frequency and intensity of extreme and unusual weather events, with the attendant destruction of lives and property, are increasing globally. Their impact on the environment can however be minimized if they are predicted in time and early-warning information on the impending disaster provided. This will no doubt assist stakeholders in the tourism, aviation, agricultural and environmental sectors to plan better and offer solutions to the envisaged problems.

In Nigeria, the government is taking seriously the challenges of climate change and the associated negative impact on the environment and eco-tourism. It is for this reason that the government, through the Ministry of Aviation, has continued to support the Nigerian Meteorological Agency (NIMET) in its concerted effort to improve services through capacity building and modernization of equipment and infrastructure.

To further strengthen the preservation of the environment and combat the problems of climate change, the following are recommended:

(i) The private sector should collaborate more strongly with the government on all aspects of environmental preservation.
(ii) There should be synergy among all agencies of government responsible for handling national disasters and they should be well equipped.
(iii) Capacity building to improve skills in tackling natural disasters, such as fire outbreaks, erosion menaces, etc., should be priority agenda items in the programs of all tiers of government.
(iv) Emissions of carbon dioxide and other gases into the atmosphere should be reduced to the barest minimum. Also, appropriate infrastructure facilities and legal frameworks should be put in place to promote a healthier atmosphere.

CHAPTER FOUR

THE URGENT NEED TO REDEFINE THE MEANING OF SUSTAINABILITY IN A CLIMATE-CHANGED WORLD

BRUCE PRIDEAUX[1]

Abstract

In the future, when climate change will force a radical rethink of the concept of ecological sustainability, tourism researchers will be forced to consider how sustainability of natural resources and the sustainability of the tourism industry are linked, with the latter having a high degree of dependency on the former. This paper seeks to draw both aspects of sustainability together and consider how the sustainability of protected areas and the tourism industry will be affected by climate change in the near to long-term future. In this paper, considerations about biological sustainability are limited to protected areas that are used as a key selling point by tourism destinations.

Introduction

From an ecological perspective, sustainability describes the ability of ecosystems to maintain ecological processes, productivity, functions and biodiversity into the future (Krebs, 2008). In practical terms, sustainability describes the ability to use resources in a manner that can be continued indefinitely in the future (Murphy and Price, 2005). The ability of the global system to grow sustainability into the future was first questioned by Danella and Dennis Meadows in their book *Limits to Growth* published in 1972. In a forceful reminder of the pivotal role of sustainability in future human welfare, Krebs (2008: 128), an eminent ecologist, observed, "like all populations, human populations are subject to the rule that critical resources limit population growth sooner or later. The carrying capacity of

---

[1] Professor at James Cook University, Cairns, Queensland, Australia.

Earth for humans may have already been exceeded, and the transition from a growing human population to a stable one is one of the most important problems of the century." If this is the reality, and there is every indication that human carrying capacity in a global sense is at or has exceeded global human sustainability, how should the global tourism industry respond? This question will become even more critical as global climate change begins to affect ecosystems on a global scale.

From a global perspective, sustainability occurs when the consumption of resources equals the rate that resources are replaced. The evidence points to a global community that is consuming resources at a rate that exceeds nature's ability to replace them (Krebs, 2008). Unfortunately, the tourism literature continues to ignore this global dimension in its debate about sustainability. Current debates about sustainability generally assume ecosystem stability and the ability of these systems to retain their biological integrity into the future. This view of nature ignores the fact that biological communities are not restricted by spatial measures, such as park borders, but respond to changes in climate (both natural and human induced) by adaptation, migration or extinction (Prideaux 2009).

Global climate patterns rarely remain constant over the long term and are, in geological terms, constantly changing through a series of short- and long-term cycles that are still not well understood by climatologists (IPCC, 2007). Currently, the Earth appears to be facing a human-induced or anthropogenic period of climate change that will lead to global warming. Because of continual changes in global climates, biological communities are constantly changing in response. Human-induced climate change is a new factor in this long-term cycle of climate change and will force plant and animal communities to respond at rates that are more rapid than in the past.

In the near future, when climate change will force a radical re-evaluation of the concept of ecological sustainability, tourism researchers will be forced to reconsider the meaning of sustainability in a tourism context given the high degree of dependency of some forms of tourism on natural resources, particularly protected areas. In the recent past, there was an assumption that specific ecosystems can be (are) protected by creating areas with borders that ensure long-term sustainability. This view is naive in that it ignores the reality that climate is rarely stable over the long term. For this reason, protected-area boundaries that are now able to protect the integrity of specific ecosystems may have to be altered over time as climate change affects ecosystem stability. This will be become increasingly difficult as the global population increases and the network of roads, human settlements and farming activity continues to expand in areas surrounding remaining natural areas, including protected areas.

This chapter examines aspects of ecological and tourism sustainability and considers how the sustainability of protected areas and the tourism industry will be affected by climate change in the near to long-term future. In this chapter, considerations about ecological sustainability are limited to protected areas that are used as a key selling point by tourism destinations. This chapter will examine how the Climate Change Impact Model (Prideaux et al. 2010) illustrated in Figure 4-2 may be used to examine changes in long-term sustainability of the Wet Tropics World Heritage Area (WTWHA) Australia as it begins to be affected by climate change.

The underlying premise of this chapter is that the current concept of sustainability, built on the belief that long-term stability is possible if proper management strategies are implemented, will not be valid in the future; and that rather than stability, we are entering a long-term period of instability that will be characterized by large-scale changes in local and global weather patterns, landscapes and sea levels. In periods of rapid change, such as the one that is now beginning to occur, destinations that rely on protected areas as one of their key selling factors have little option but to rapidly adapt. This is particularly true of nations where wildlife tourism is a key pillar of their appeal to tourists.

Climate Change, Weather and Tourism

The major mechanism through which climate change will impact the tourism industry generally – and protected areas specifically – is weather, described as the pattern of precipitation, temperature, humidity, cloud cover, wind speed and hours of sunlight. There are multiple interactions between weather and tourism. As Becken and Hay (2007: 7) note: "Climate is both a resource for the tourism experience and a risk." Climate is a key push factor for many visitors who seek favorable weather conditions. In the case of winter sports enthusiasts, snow is a key attraction, while for beach destinations, long hours of warm sunshine are desirable. However, weather is also a risk if the wind is too strong, there is too much rain or the temperature range is beyond desirable limits. In the future, climate change will cause changes to weather patterns that will in turn affect ecosystems and landscapes and the tourism industry they support.

In rainforest areas, for example, if rainfall declines, drying may occur and forest communities may be replaced by grasslands or the composition of the forests may change affecting the distribution of fauna. Other impacts caused by changing weather patterns include fire events that are capable of causing large-scale changes to plant and animal communities and the scenic values they provide.

Figure 4-1 highlights changes that are predicted to occur in the WTWHA rainforest. This region is World Heritage listed (1988) and extends along the northeast coastal zone of Queensland, encompassing an area of 894,000 hectares. The rainforest contains an almost complete record of the major stages in the evolution of plant life on earth with many rainforest species originating when Australia was still part of the Gondwana super continent. As temperatures increase, species richness and abundance will decline. In some cases, the decline will be a result of changes to local weather patterns that interrupt the growing cycles of plants leaving many animals with a shortage of food. In other cases, temperature-sensitive plants will die out and be replaced by other more temperature-tolerant species.

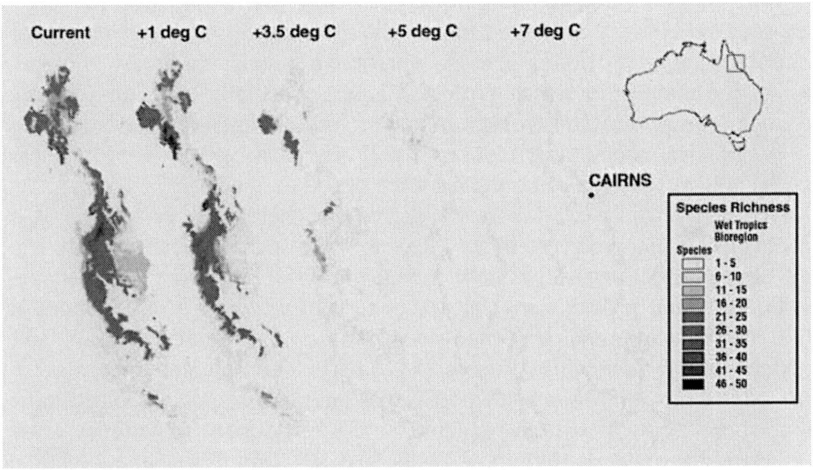

Figure 4-1. The decline in distribution of species richness of regionally endemic terrestrial vertebrates with increasing temperature. Source: Williams et al., 2003.

As Figure 4-1 illustrates, the weather element of climate change will cause substantial ecological changes and have a significant impact on the long-term sustainability of the present mix of flora and fauna that constitute the area's ecosystem. Over time, a new modified ecosystem will emerge that will include species that have migrated to the area to fill niches previously occupied by flora and fauna that will become extinct as temperature increases. It is not possible to protect the current ecosystem because there are no migration pathways for forest species to follow and no refuges for animals to escape to. Changing current protected boundaries will have little or no effect. In blunt terms, the rainforest system of the wet tropics as we currently know it is not biologically sustainable in the long term, while the extent of change that occurs will be almost entirely dependent on the effectiveness of global mitigation efforts.

The current concept of sustainability is therefore redundant and a future definition may have to include a willingness to allow the changes that are likely to occur to take place with as little interference as possible so that the new climate-modified ecosystem is given a chance to emerge and become established. Any new definition of ecological sustainability has to recognize that change is an ongoing process. Rather than resisting changes in biological systems caused by climate change through interference with natural processes, there is a strong case for allowing change to occur and during this process adapt the way that natural systems are currently used.

For example, if an ecosystem becomes more fragile, the level of use by the tourism industry will need to be adjusted, possibly by reducing visitor numbers and/or restricting access to particularly sensitive areas. This will affect the manner in which protected areas, especially those undergoing change, are used by the tourism industry. Any change in use, particularly if visitor numbers are reduced, will in turn affect the business sustainability of firms using the resource.

To understand how climate change will impact on the ecosystem, it is necessary to integrate scientific research with social-science research to develop a model that shows how changes in weather affect the landscape-scale protected-area systems that today constitute a significant element of the tourism product in many countries. The starting point for understanding how this process will affect the tourism sector is to investigate how the supply side (defined as landscapes, ambient temperature and scenic views) and the demand side (for tourism experiences) are affected. For example, Figure 4-2 illustrates a highly useful science-based tool that can be used as a starting point for future debates on the meaning of sustainability – both ecological and business. In effect, the model is able to give a prediction of how the forest will look from an ecological perspective at varying levels of temperature change.

One of the key factors in future debates about the concept of sustainability must be an understanding of the relationship between the level of global warming and the impact of that warming on the environment. It is not possible to predict how much global temperatures will rise in the remainder of this century because successful political intervention through mitigation may cap temperature increases at some point in the future, or alternatively the failure to achieve global consensus may lead to higher temperatures. Thus, the tourism industry will need to constantly re-evaluate how it uses ecosystems in general and from what philosophical perspective this will occur. In this manner, future sustainability will be a function of the rate of temperature increase; the impact that temperature increases will have on ecosystems measured by changes in abundance and distribution of species; and the sensitivity of the new ecosystems to human use.

The six-stage Climate Change Impact Model (Prideaux et al. 2010) illustrated in Figure 4-2 is one approach to assessing how the ecosystem and the landscapes it supports will change. The model is descriptive, designed to identify linkages that might otherwise go unnoticed, and has the capability of being used as an aid in planning. The assumption underlying the model is that changes to weather that are generated by climate change will modify ecosystems, patterns of human settlement, agricultural systems, economic patterns and tourism demand. Conceptually, the model relies on two sets of theory – the concept of consumer-push and destination-pull factors and a standard-economic-demand-and-supply model where a change in supply leads to a change in demand.

This chapter focuses only on the demand and supply for tourism use of protected areas. Various supply-side inputs are required to create a tourist experience that is subsequently marketed to and consumed by tourists. Inputs usually fall into these categories: the natural resource (landscapes, scenic views, flora and fauna and local weather patterns), management of the resource and infrastructure needed to support the tourism sector (e.g., transport networks, accommodation, walking tracks, ranger services and other services). Collectively, these inputs constitute the characteristic product of a protected area.

The consumption of the views and experiences available in the protected area by tourists may be measured as demand, which is a function of how successfully the destination has been marketed and the quality of the experiences available. If the quality of inputs changes as a consequence of ecological changes, the level of demand may change. If climate change has a negative impact, as in the case of the Wet WTWHA, the perceived quality of the protected area will change, possibly causing a shift in the level of demand. Also, the sustainability of the protected area measured by the number of visitors that may be sustained on an annual basis may also change, affecting the economic sustainability of the tourism sector that relies on the protected area. For example, areas of Africa that experience a severe decline in the abundance of the big-five animals will become less attractive to tourists and suffer a decline in business sustainability.

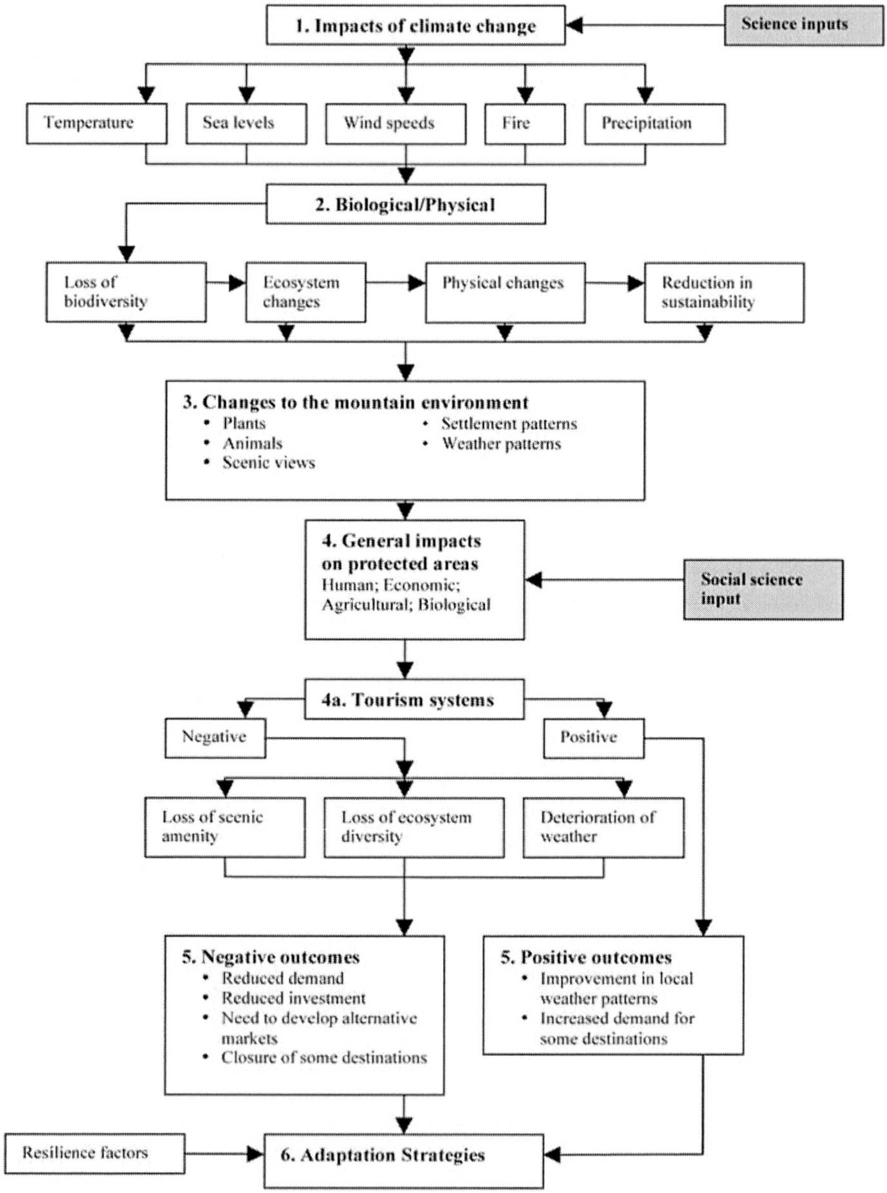

Figure 4-2. The six-step Climate Change Impact Model applied to protected areas.

In Figure 4-2, the first-order impacts of climate change are shown as changes to weather that include temperature (mean yearly average temperature variations based on day and night, and by season), precipitation (rain, snow, fog, clouds, etc.) and wind (including daily wind speeds and wind events, such as wind storms) as well as fire events and changes in sea level. Collectively these impacts cause ecological changes that may include reduced biodiversity, changes in flora and fauna and reduced sustainability and are shown as stage 2 "Biological/Physical" impacts.

The biological and physical impacts may cause visible changes to the natural ecosystem, which in turn may affect tourism demand. For example, a reduction in precipitation that leads to the retreat of rainforests, and advance of savannah forest will change the quality of the scenic views. Changes of this nature are illustrated by stage 3 "Changes to the environment." Local changes in weather may also affect the physical resources that define the nature and quality of the environment, which is often the major appeal of protected areas (Scott et al., 2007).

Impacts many include significant changes in plant communities, loss of iconic animal species, changes in settlement patterns and changes in agricultural systems. These changes are illustrated in stage 3 and may have a specific impact on aesthetic values and in some cases lead to a reduction in the perceived attractiveness of a specific region (Elsasser and Bürki, 2002) from a tourism perspective. Changes that are observed in stage 3 are shown as stage 4 "Impacts." In Figure 4-2, the impact on tourism is shown as stage 4a "Tourism impacts." From a tourism perspective, changes in weather may have either positive or negative impacts. In colder areas, for example, increased temperatures may generate increased demand (Scott et al. 2007; Richardson and Loomis 2004).

The impacts shown at stage 4a may be either positive or negative as illustrated in stage 5 "Possible tourism outcomes." Positive outcomes may generate increased tourism flow. Negative outcomes may lead to a reduction in demand, falling investment, the need to find alternative markets and even closure of some destinations. Strategies to adapt to climate change impacts are illustrated as stage 6 "Adaptation strategies."

The major feature of the model is its ability to illustrate linkages between the various elements of the physical and human systems operating in protected areas. As temperatures rise, impacts can be traced from initial biological and physical impacts through to tourism outcomes and adaptation strategies. Changes in the structure of the ecosystem may affect the nature of the tourism experience and the level of enjoyment experienced by visitors. While the model is primarily designed to show linkages and cause-and-effect loops, it also has a capacity to measure the scale of impact.

At each step of the model, it is possible to measure the nature and scale of impacts through a range of cause-and-effect relationships. Science-specific methods may be used in stages 1 and 3 to measure change in temperature, precipitation, wind and fire. These findings may then be used to predict possible and probable changes in stage 2 factors. Commencing from stage 4, social-science research methods, including surveys and the development of scenarios, can be employed to measure the size of potential changes in demand.

After potential outcomes on tourism demand have been identified in stage 5, "Adaptation strategies" (stage 6) can be developed to deal with the changes identified in stage 2. As part of this process, it is necessary to also consider resilience factors.

Conclusion

The argument presented in this chapter is that our current understanding of sustainability is built on the belief that ecosystems are stable over the long term. Even without human intervention, the global climate system is in a constant state of change. Current views on sustainability are therefore valid only in the short term, measured on a time scale that is more likely to be decades rather than centuries in length. As the pace of climate change increases, many ecosystems will undergo change that in most cases cannot be reversed by measures such as restricting tourism use.

In some ecosystems, this may mean a reduction of biodiversity and a significant change in species mix. In other ecosystems, such as some deserts, there is likely to be little noticeable change, at least to the landscape. The major difficulty for most protected-area managers and for destinations that rely on protected areas is the lack of scientific investigation into potential impacts. Without access to data of this nature, managers will find it difficult to make informed decisions on use and management. This will make it difficult to assess the desirable level of tourism use.

Many ecosystems will become vulnerable, and because they are in a state of change, it will be difficult to evaluate how human use will affect the ecological processes that are occurring. Human use may for example speed up undesirable impacts. For these reasons, it will be difficult to determine the extent to which tourists should be allowed to use areas of high biodiversity and fragility.

In the future, adaptation strategies will become critical, particularly if global efforts to reduce greenhouse emissions fail or are delayed as now appears to be the case. At the destination level, new adaptation strategies will be required to minimize adverse impacts. These should include a new understanding of sustainability. Before adaptation strategies can be developed, the potential impacts of climate change for each one-degree rise in temperature will need to be identified. This will require greater investment in scientific research, a more effective discussion between scientists, managers and tourism operators, a willingness of the supply side to respond proactively, and acceptance by consumers that they have an important role to play in both mitigation and adaptation. Some areas that are now viewed as being sustainable will be reclassified as fragile, requiring a reduction in tourism numbers that will in turn affect business sustainability.

References

Becken, S. & Hay, J. (2007). *Tourism and Climate Change: Risks and Opportunities*. Clevedon, England: Channel View Publications.

Elsasser, H. & Bürki, R. (2002). Climate change as a threat to tourism in the Alps. *Climatic Research* 20:253–257.

Intergovernmental Panel on Climate Change (2007). Summary for policy makers. In *Climate Change 2007: The Physical Science Basis*; *Contribution of Working Group 1 to the Fourth Assessment Report of the Intergovernmental Panel on Climate Change*, ed. S. Solomon, M. Qin, Z. Manning, M. Chen, K. Marquia, M. Averyt, M. Tignor & H. Miller. Cambridge: Cambridge University Press.

Krebs, C. (2008). *The Ecological World View*. Wallingford: CABi.

Meadows, D., & Meadows, D. (1972). *Limits to Growth*. Universal Book: New York.

Murphy, P. & Price, G. (2005). Tourism and sustainable development. In *Global Tourism*, ed. W. Theobald. Burlington, MA: Elsevier.

Prideaux, B. (2009). *Resort Destinations: Evolution, Management and Development*. Oxford: Butterworth Heinemann.

Prideaux, B., Coghlan, A. and McNamara, K. (2010) Assessing the impacts of climate change on mountain tourism destination using the climate change impact model. *Tourism Recreation Research* 35.

Richardson, R. & Loomis, J. (2004). Adaptive recreation planning and climate change: A contingent visitation approach. *Ecological Economics* 50:83–99.

Scott, D., Jones, B. & Konopek, J. (2007). Implications of climate and environmental change for nature-based tourism in the Canadian Rocky Mountains: A case study of Waterton Lakes National Park. *Tourism Management* 28: 570–579.

Williams, S., Bolitho, E. & Fox, S. (2003). Climate change in Australian tropical rainforests: An impending environmental catastrophe. *Proceedings of the Royal Society of London* (Series B) 270: 1887–1893.

CHAPTER FIVE

TOURISM AND RENEWABLE ENERGY: IN THE SIGN OF THE SUN

MAX HABERSTROH[1]

Energy and its Higher Purpose

We have all seen the headlines: houses in Alaska and Siberia collapse, as permafrost is melting; mountains lose their snow caps; glaciers in the Alps recede and disappear; polar ice breaks into pieces, leaving behind the familiar picture of a lonesome bear desperately rafting on the floe; hurricanes leave their deadly track in American towns and landscapes; dunes in the Gobi desert threaten fertile farmland; wells are drying out in the Sahara; droughts are occurring in parts of Africa, while floods hit interior and coastline cities – as witnessed currently in Thailand; and forests and houses fall prey to uncontrolled fire blazes. Increased volcano eruptions, sudden earthquakes and deadly tsunamis have become a constant challenge to modern civilization. We have become a threatened civilization, whose machinery is still largely kept in motion by oil and coal – air-polluting fossil energies that are running short.

As we are all aware, some experts say that there have always been climate changes throughout Earth's existence, and, as a matter of fact, civilizations have come and gone. The difference today is that, as scientists found out, these changes are to an important degree caused by human activities and, in return, affect a dramatically increased world population, disregarding whether they live in shacks of poignant poverty or operate the highly sensitive devices of a sophisticated lifestyle. Hence, the human impact has ceased to be regarded as negligible. Still, we must not forget that, being responsible for a great deal of public and private transport and running hotels and leisure facilities, travel and tourism certainly has its share of CO_2 and greenhouse gas emissions (GHG).

In an interview with the Washington Post, media mogul Ted Turner said: "I think the situation with global climate change is a life-or-death

[1] International Consultant on Sustainable Tourism Branding, Destination Management, Marketing.

issue for us that we have to get on right away … We'll be moving from subsidizing a dying smokestack industry to going to clean, renewable electrical power, locally produced. It will be terrific for our economy."

The matter of global climate change is not all about CO_2 and GHG emissions. When smog is in the air, people's health is at stake. Environmentalists point to diminishing rainforests, fossil and nuclear energy calamities and an almost pandemic spread of garbage. Hence, there is no conscious awareness of a "ticking clock."

The way our industrial society has developed until today is indisputably linked with fossil-energy production. How would the economy have developed had James Watt not invented the steam engine, using coal as its most appropriate energy source? What if, instead, he had shown up with a solar-powered steam engine, driven by a parabolic mirror, like the one presented at the Paris World Exhibition in 1878?

However, there was coal in abundance. Fossil energy and combustion, the "myth of the big" and centralization were characteristic of the Industrial Revolution.

Ultimately, energy means power. For Rudolf Diesel, who lived in the nineteenth and early twentieth century, this was a special challenge. His striving for independence from foreign oil was reflected in his book "Solidarism." At the Paris World Exhibition in 1900, Diesel's engine was driven by peanut oil – very impressive, but not very effective. However, Rudolf Diesel stuck to his mission. Today, there are diesel engines in everything from cars to construction machines to cruise ships.

To be truthful, the consequence of his invention is that Rudolf Diesel made an essential contribution to the proliferation of fossil energy, but the crucial thing that matters is that he wanted to serve humanity. In fact, Diesel's invention has influenced our lifestyle up until today.

A Trendsetting Vision to "Solar Revolution"

A trend-setting visionary in communication technology, Steve Jobs has created a new lifestyle. His vision was not to become a billionaire; his vision was for Apple to create a computer for everyone. He called it a "personal computer." Today, PC, the Internet, Facebook and mobile communication have become common ingredients of a modern lifestyle.

Another lifestyle trendsetter is indisputably travel and tourism. Tourism may lead to "prosperity and peace" – as the UNWTO says. Nonetheless, there is a more gloomy side of tourism. In his book *The Final Call*, Leo Hickman wonders who will really pay for the collateral damage caused by tourism.

Both in tourism and the energy business, it is easy to find dubious practices negatively affecting people. With millions more travelers predicted from emerging powerhouses like China and India, we have only a vague idea of tourism and its energy needs in times to come.

Every one anticipates that hotel heating, cooling and illumination are constantly available. But nobody likes being exposed to the sound of diesel-propelled power engines in holiday lodges. And hardly anyone would frown at being offered electric mobility from airport to hotel, or electric-shuttle services at trade fairs and exhibitions.

In many developing countries, access to both electrical power and electronic communication is insufficient or non-existent. Wood and charcoal are often the only energy sources in rural areas. People are out of reach of any existing power grid. Solar energy devices and mobile phones provide a solution.

Sometimes visions do come true, if the people in charge want it badly enough. The crucial point is leadership – "visionary" leadership. It's a matter of entrepreneurial capability and political will. In terms of energy, it's about harnessing the limitless power of the sun and realizing the great vision that "everyone can be an unlimited clean-energy producer."

Where are the protagonists who make this vision a cause for tourism stakeholders? Where are the tourism organizations that are ready to "energize" their concepts with a strong, new and cohesive approach – the "solar" approach? Where are those cross-corporate pioneers who set benchmarks of making renewable energy the pivot of sustainable tourism?

The sun is shown on so many national flags and coats of arms. The sun is an indispensable ingredient to wonderful holidays. The sun has always been the "face of tourism" on catalogues and advertisements. Providing the basic impulse for any kind of movement and life, the sun is a symbol of freedom, happiness, wealth, success, beauty and unification. The sun is the "face of energy," the symbol of an emerging Solar Age. Let's make the sun the interface of renewable energy and clean global travel and tourism!

It's time to say goodbye to the wait-and-see attitude – to say farewell to needless "combustion" and consumption. Let's set a new trend towards solidarity and authenticity in cross-corporate relationships! It's all up to pioneering decision makers. Renewable energy and sustainable travel and tourism can provide, as Tom Peters would say, the most "scintillating" venue of what is already under way – the "Solar Revolution."

World Tourism Foundation Founder Edward Beauchamp says: "The world's largest industry revolves around the smallest destination," that is, locally. The same holds true for renewable energy – due to its very nature of unlimited "power" and its capacity to produce "clean" energy locally, exactly where it is used.

Cross-promoting sustainable tourism and renewable energy would pave the way for win-win solutions so far unheard of. Let's just think about PATA's concept of "Total Tourism" (the benefits from covering the complimentary benefits of inbound and outbound tourism – introduced by the Pacific Asia Travel Association (PATA) years ago) being linked with renewable energy businesses. Could such a move not be instrumental in enhancing cross-industry investment and to mutually boosting exports in energy-related products and tourism?

As Climate Changes – So Does Tourism

Travel and tourism can become a promotional bandwagon, using renewable energy as key to sustainability of destinations and tourism operations. Let's make renewable energy the first principle of sustainable travel and tourism! This would provide tourism with the crucial strategic edge to effectively help counter global warming and ecological degradation.

There are more and more responsible stakeholders who understand that everywhere in the world millions of power stations could stand, different in size, technology and capacity, providing renewable energy to everything from private houses to cities. Renewable energy is predestined to become the first principle of sustainable tourism, since energy stands at the beginning of everything. As a matter of fact, renewable energy is not only "natural" – it is inherently "ethical."

Late renewable-energy pioneer Hermann Scheer's "energy autonomy" thesis points to what renewable energy is – decentralized, subsidiary, ecological and autonomous, including the innate higher purpose of nothing less than promoting peace. Travel and tourism positions itself as a paramount peace industry. Both industries share the same ideal.

We decide what should prevail: the authentic, the beautiful and the valuable, or mere mediocrity, hypocrisy and fallacy; the fossil-energy consumer or the renewable-energy user; the indifferent tourist or the responsible traveler; paradise lost or planet Earth regained. The options are more quantity or better quality – more caterpillars or beautiful butterflies.

CHAPTER SIX

MAINSTREAMING RESPONSIBLE TOURISM AT RIO+20 AND THE LONDON 2012 OLYMPICS

GORDON SILLENCE[1]

Abstract

A sustainable future requires fundamental mechanisms for human populations to create conditions for both personal and societal peace, one being dependent on the other. In the traditional spirit of the original Olympic Games, the United Nations has called for six weeks of world peace to accompany the Olympic competition. It is a positive and enlightened response to the current challenges of climate change, poverty, loss of biodiversity and socio-economic collapse. The opportunity now is that in 2012 there is a convergence of the Rio+20 Agenda 21 follow-up conference in June in Brazil and the London 2012 Olympic Games in the United Kingdom in August, just two months later. In preparing for these two large-scale events, tourism stakeholders can act as an internationally cohesive group whose common interest is to use responsible tourism as a means for sustainable development by creating true destination security whilst supporting the growth of the global green economy. A one-minute silence to mark the truce, a map of global safe passage and a collective effort to bring forty teenagers from areas of violence to the Games are being planned to make peace through tourism really happen. The current events in the Middle East confirm that we need to work harder for a global truce. The year 2012 will bring other issues of global security, whilst the Olympic Games themselves will make London a fortress city of fear in need of a positive affirmation of a world at peace and people at peace with each other and at peace within themselves. The London 2012 Olympic Peace Campaign sets a common vision for this. We now need to see well-

[1] Coordinator London 2012 Olympic Peace Campaign, Executive Director UN Type II DestiNet Partnership, Vice-President Ecotrans Network for Sustainable Tourism Development.

coordinated, well-resourced, timely top-down and bottom-up actions as the clock is counting down.

Introduction: Peace through Tourism and the Olympics

Figure 6-1. Olympic Rings of Light. The Olympics have been cited as a means to achieve the Millennium Goals.

For the ancient Greeks the Olympics were Sacred Games. When nations competed together in the Olympic tradition of "Ekecheiria," war was stopped for a specific period of time and safe passage guaranteed to athletes and spectators. The modern United Nations movement recalled this means of achieving peace on earth in the 1990s, calling for an Olympic Truce;[1] it is even stated as a means to end poverty in the Millennium Goals Declaration.[2] There have been minor political processes put in place to achieve it, although to this point they have proved ineffectual. The six weeks of world peace in 2012 that has been designated by the United Nations to accompany the Olympic (and Para-Olympics) competitions is a positive and practical response to the challenges of climate change, poverty, loss of biodiversity and socio-economic collapse. It is a monumental task, only capable of being executed by many professional people following this common vision and moral purpose. This article proposes a means to do justice to this process by calling on tourism stakeholders to work toward the truce in a spirit of enlightened self-interest; creating destination-security and safe passage is good for tourism business and contributes strongly to the process of overall sustainable development.

A Bottom-up and Top-down Move from a Culture of Violence to a Culture of Peace…

Like all global undertakings, no one individual can make this happen. It requires the personal commitment of a well-organized network of political, business, administrative and academic professionals together with a well-informed consumer audience, following a common vision. It should not remain as a rhetorical goodwill statement, but be turned into a real force for global harmony and prosperity. A concept of how to achieve this was developed for the 2004 Greek Olympics in the book *16 Days: The Role of the Olympic Truce in the Toolkit for Peace*.[3] Essentially, this concept demonstrates that all sections of society need to be involved in a process of moving us from a culture of violence to a culture of peace.

This move needs to happen on both a societal and personal level, from local to global levels, across all nations and regions, cultures and religions, genders and colors. It needs to be both a top-down and bottom-up multi-stakeholder process, drawing in all sectors of society across the globe. Sports, tourism, entertainment and health professionals can create big waves of public and political support. Celebrities, politicians and the business community need to be involved as well as citizens in local communities and through the workplace.

Alongside the governmental-level actions necessary for the truce to occur, a series of creative projects need to be established to balance a top-down approach so that everyone can participate in the six weeks of inner peace – a global peace effort. In other words, even if governments do not create the peace, individuals can create peace in themselves and their communities during this time, e.g., London Firemen for the Olympic Peace, Dancers for Olympic Truce, Businesses for Olympic Peace, Celebrities for the Truce, Christians for Truce Vigils, Buddhists for Olympic Peace, Olympic Athletes for Peace, Politicians for the Global Truce, etc.

Above all the media have a key role to play in taking up this message and making it a global issue so that politicians do their job and oversee the truce implementation process. Importantly, official Olympic sponsors should take on board the need for their support to see this happen.

The Top-down Process – Ending the Politics of Scarcity and Fear…

The role of the state, however, is paramount, as it is conflicting nation-states or power struggles within or against states that create most conflict and violence. At the UN World Summit on Sustainable Development (WSSD) in 2002, UNEP's Geo2 report showed world scenarios that put markets first through military-enforced "security."[4] What we need is a "Sustainability First" multi-sector, multi-stakeholder strategy based on the value system of sustainability – the just, fair, equitable and peaceful world of Agenda 21.

Those of us old enough to remember the concept of a "Peace Dividend" from reduced military budgets know the resources are there, and the world of scarcity is a creation of a powerful elite whose primary economic actions prolong inequality and foster insecurity in the preservation of their accumulated material wealth and the current socio-political power structure.

A state that develops citizens whose personal value systems show political responsibility, and operates a political system that takes care of individual citizens personally is within reach, but not achievable without more compassion in politics. Mahatma Gandhi's famous quote "Enough for everyone's need, but not enough for everyone's greed" was used at the opening ceremony of the WSSD, but was quickly buried in bureaucratic negotiations that enforced the status quo of the further concentration of capital at the expense of people and the planet. In 2012, we cannot afford to let politicians and corporations make the same mistake at Rio+20, and must hold them accountable to their Millennium declarations.

However clever the political rhetoric, the persistent poverty faced by the majority of global citizens, especially in Africa, and current events in the Middle East confirm that we need to work much harder for more permanent economic solutions than this phase of finance capitalism engenders. The year 2012 will bring other issues of global security, whilst the Olympic Games themselves will make London a fortress city of fear in need of a positive affirmation of a world at peace and people at peace with each other and at peace with themselves.

Making the policy work of the Agenda 21 practical – i.e., having a real and positive effect on markets and communities – is the task of governments at all levels. Rio+20 is another milestone on that road to sustainable development, but it needs corporate and political institutions to open their eyes and hearts to the impoverished reality of the world's majority population; address the issue of inequality by burying the myth of scarcity; and create culturally sensitive, knowledgeable and productive societies and citizens who can live in prosperity and peace. This is not utopian thinking, but rather the realization that with burning issues such as climate change, environmental degradation and social injustice outweighing materialist growth, we need a fundamental readjustment of the economic and political order. How is that possible?

The Tourism Sector Contribution – Mainstreaming Sustainability by Linking Large-scale Events…

Peace is a necessary condition for tourism; and tourism offers opportunities for economic, cultural and environmental recovery. Thus it makes sense for tourism stakeholders to cultivate the culture of peace. A world with an estimated forty major conflict zones is a world where there are not only forty more human-tragedy stories, but also forty fewer tourism destinations. Any Middle East country tourism stakeholder will tell you of the essential need for peace for tourism business to exist and prosper.

Figure 6-2. Destination security in Costa Rica. In stark contrast to Darfur, Costa Rica is a country with no army.

What is also in evidence, though, is the idea of using tourism as the economic rationale and driver for global peace and sustainability, with tourism stakeholders playing a key instrumental role in promoting a mindset of global awareness sensitive to the needs and plight as well as the differences and diversity of global cultures. To marvel at and appreciate the global environment and cultural diversity, as well as to provide business for local communities; to support and network rich and poor regions; to provide funds for nature and cultural conservation; to share knowledge and awareness of the best of what is local to a global audience – tourism can do all of this. In line with Rio+20's aim to deliver a global "Green Economy," we can see the strategic use of sustainable tourism development as a catalyst for socio-economic improvement.

Our opportunity now is that in 2012 there is a convergence of the Rio+20 Agenda 21 follow-up conference in June in Brazil and the London 2012 Olympic Games in the United Kingdom in August, just two months later. In preparing for these events, we can act as an international group of responsible stakeholders whose common interest is to implement the global truce, and, in this period of celebrated peace, showcase alongside each participating country's Olympic athletes the best nature and culture offers as outstanding examples of sustainable and responsible tourism development, and the best ways of accessing them through the travel and hospitality industries.

Mapping the Peace Process: London 2012 Olympic Peace Campaign

A mosaic of open travel and tourism destinations and routes that link to London from all over the world is being mapped in the London 2012 Olympic Peace Campaign. The Olympic athletes will leave these destinations and pass on these routes; their safe passage in preparation, traveling to, and returning from the Games should be every citizen's global right – to live and travel in peace anywhere on the globe at that time. In this way, a map of international destinations will highlight all the countries that work to achieve the right of peaceful passage for the six-week period stated in the UN Autumn Resolution, and respect a global truce. This map will be an international map of tourism that can immediately be used by the industry to provide income to those countries that are counting on tourism to develop their economies. Such a map of peaceful national tourism can be prepared for Rio+20 and then broadcast via the Olympic media machinery.

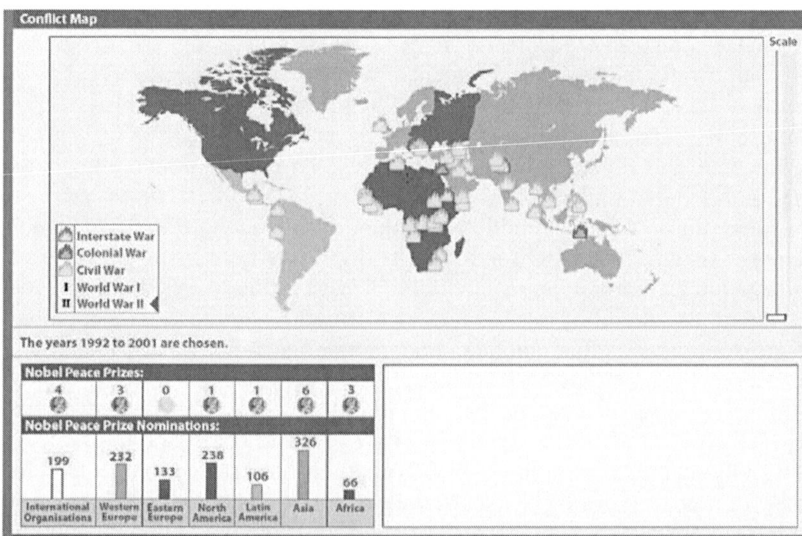

Figure 6-3. A map of UN peacekeeping operations at the turn of the millennium – more than 40 major conflict zones are listed.[5]

This all needs a degree of co-ordination that only now is possible in global culture, in the form of the Internet's ability to connect us all, as well as a set of individuals who recognize they can play a part in making it happen, professionally and voluntarily. If the aim is to truly realize six weeks of global peace during the Olympic competition period, then we need to start organizing now, and follow the London Olympic countdown clock, which started in March 2011. Yet (at 500 days to go) the formation of a "Coalition of the Willing" is not coalescing, as business as usual and our busy lives prevents the Olympian effort necessary to achieve this aim. Importantly, however, since the 5th IIPT African Conference, in Lusaka, Zambia, the International Institute for Peace through Tourism (IIPT) is supporting key elements of the campaign,[6] which will help other stakeholders come on board. We need to see personal commitment – like IIPT's – and well-coordinated, well-resourced timely actions by those in power as the clock is counting down.

Figure 6-4. The London 2012 Olympic Peace Campaign – aiming for six weeks of both inner peace and global peace.

The London 2012 Olympic Peace Campaign sets a common vision for this. The campaign's aims include:

1. *The effort from governments*: Every country attending the Olympics should support this autumn's UN Six-week Truce Resolution with an implementation strategy – with that process transparently mapped and made real.
2. *One-minute silence for global peace* at the Olympic opening ceremony to mark and honor the state of global peace on earth, and to remember those places and people who may have not been able to achieve the six-week truce.
3. *Youth for Olympic Peace Campaign*: One teenager from each conflict zone/area of violence will be sponsored to attend the

Olympics as youth peace ambassadors, and share the camaraderie and support of the world for those days.

4. *Responsible and Sustainable Tourism Business and Destination Support Travel Initiative*: Showcasing community-based responsible and sustainable tourism best practice offerings next to a country's Olympic icon. Responsible tourism can be targeted at the poorest nations to give a boost for local communities to benefit from global tourism opportunities for economic, cultural and environmental recovery.
5. *VISTAS Peace through Tourism Award*: A national award to highlight those countries that work most toward peace.
6. *The Olympic Rings of Light Inner Peace*: The promotion of inner peace together with global peace during the six weeks, so that we can have a mass movement from a culture of violence to a culture of peace alongside top-down political initiatives.
7. *Olympic peace parties*: Launching a global peace celebration during the London Olympics to mark this historic achievement.

The campaign is described in full on the DestiNet Portal for Sustainable Tourism Development.[7] It shows how the tourism sector can lead the call to implement the necessary top-down and bottom-up processes to end the worst climate change that has really happened in our lifetimes – that change to a climate of fear – so negative for tourists and host communities alike.

Conclusion

In recognition that it will be the culture and education programs of the London Organizing Committee that will promote the Truce in the United Kingdom, it would be useful to run campaigns in each country – to contact ministries of culture and education to get the commitment process going around the world. If you add sports, tourism and environment ministries, a multi-sector participative process can be set in motion, which is what is needed to spread the message to all sectors of society in time for the Games.

Each UN Member State, as an Olympic Competing Nation, can present signed declarations from either/all ministers of sports/tourism/culture/education/environment announcing their support for the UN agreed six weeks of peace during July, August and September 2012 and to run in-country campaigns to generate national awareness and participation. The UN resolution encourages such actions. States can then map their responsible and sustainable tourism activities and offers, and develop marketing campaigns to promote their contribution to the peace, to be showcased at Rio+20 in June 2012 and then mainstreamed during the London Olympics in 2012. United Kingdom and European travel agents and tour operators can promote community-based tourism, responsible tourism and sustainable tourism in the developing world, alongside Olympic athletes from those regions.

Figure 6-5. The Olympic Countdown Clock – time is running out on this historic opportunity but the clock can synchronize collective action.

The Olympic Countdown Clock gives us all a common timetable, but we need to commit and involve ourselves (and our organizations) now. Rio+20 in June 2012 will show the world's readiness to develop global peace as the basis for sustainable development. The aim is to have both top-down government and bottom-up citizen initiatives in place by then to achieve the peace. Tourism stakeholders have a central interest and role in this peace, creating a world of inter-cultural exchange and travel possibilities to support developing countries in their effort to improve their socio-economic situations. This is an Olympian effort, but even the warring nations of ancient Greece achieved it. Modern society can also reach a new height of civilization and make this happen if our politicians really want peace, but ordinary people and professionals must make the call for them to answer.

For further information or to support the campaign please contact Gordon Sillence, gordon.destinet@ecotrans.de, or go online to http://destinet.eu/topics/sustainable-consumption-and-production/peace-through-tourism or visit Facebook at http://www.facebook.com/pages/London-2012-Olympic-Peace-Campaign/211382448883873.

Notes

[1] See the UN Declaration on Sport and Peace; since 1993, the UN Assembly has unanimously adopted six Resolutions championing the cause of the Olympic Truce. See http://www.olympictruce.org/v2/index.html.
[2] The United Nations Millennium Declaration, adopted in New York in September 2000 by more than 150 Heads of State and Government, includes a paragraph on the promotion of peace and mutual understanding through the Olympic Truce.
[3] Rachel Briggs, Helen McCarthy and Alexis Zorbas (2004), *16 Days: The Role of the Olympic Truce in the Toolkit for Peace*, International Olympic Truce Center (Demos, ISBN 1841801259, 9781841801254).
[4] See http://destinet.eu/topics/sustainable-consumption-and-production/peace-through-tourism.
[5] See http://nobelprize.org/educational_games/peace/conflictmap/conflictmap.html. This moving map shows a history of conflict zones throughout the world.
[6] See http://www.iipt.org/newsletter/2011/august.html.
[7] See http://destinet.eu/topics/sustainable-consumption-and-production/peace-through-tourism.

PART II

NATIONAL AND REGIONAL PERSPECTIVES

CHAPTER SEVEN

CLIMATE CHANGE: CHALLENGES AND OPPORTUNITIES PRESENTED TO THE SOUTHERN AFRICA REGION

HONORABLE ENGINEER WALTER MZEMBI MP[1]

Climate change is here and now. It is not an approaching hazard that we can ring alarm bells about and try to make good our escape. That is why leading minds in this area have determined that we can only confront this challenge through two main ways, adaptation and mitigation, which we are all now familiar with.

I find it quite notable that the challenge of climate change compels all mankind to have a unity of purpose. This demand brings to the forefront the fact that all mankind are indeed one. It brings to center stage the need for harmony between man and nature, which has been the greatest preoccupation of human thought of all time and all cultures, including the Bantu or African thought of our region.

Indeed, almost all folklore – original, traditional and/or indigenous – has a narrative about how there was a huge flood, or huge fire, or some big bang followed by some major environmental change of one form or another. Fire and water feature very large in man's conception about his Creator (about God).

I find it a very sobering thought that at this stage of human history, with its great advancement in technology, mankind is being reminded that he can only survive within the context of one broad international agenda. Not only the absence of war is demanded for mankind's survival, but also the promulgation of a common agenda for survival.

This is an agenda that must be dominated by actions that lead to adaptation and mitigation. Because tourism contributes to the emission of greenhouse gases (GHG) through its air-travel and accommodation

---

[1] Minister of Tourism and Hospitality Industry, Zimbabwe.

activities, it has the responsibility to respond to the challenge of climate change.

For us in Africa, we must be pleased that the identified critical challenge for the tourism sector is to de-link the projected massive growth of the sector from increased energy use associated with GHG emissions, and ensure that growth contributes to the reduction of poverty amongst our people.

The identified major climate change impacts affecting the Africa region are: warmer summers, water scarcity, land biodiversity loss, marine biodiversity loss, increase in disease outbreaks and political destabilization.

It is the coupling of our response to the challenges of climate change to the programs of poverty reduction that led me to take the position (besides looking at the obvious and clear challenges of climate change, which are now well known and defined and quantified) that we must also, in this region, be preoccupied with what opportunities climate change presents to us.

I must highlight my satisfaction with the fact that the challenge of poverty, and the suffering it brings upon our people, is ranked together, and at par with, the challenges of climate change, and that in addressing one we must address the other. I must also, humbly, applaud the observation by our former UNWTO Secretary-General Francesco Frangialli that tourism is an important element in both climate change and poverty alleviation, and herein lies some massive opportunities for our region.

These are opportunities that we are best advised to exploit in a regionally integrated manner, and this I must emphasize to my fellow minister, Madame Namugala, Zambia's Minister of Tourism, and to other tourism ministers in the region. This regionally integrated approach augurs well, not only for the reigning and desired peace between our two countries and peoples, but also makes for sound strategy for the sustainable development of our region.

For our region, southern Africa, I believe, with people like Achim Steiner of the United Nations Environmental Program (UNEP), that far-sighted plans and action by our sector will send important signals to our government colleagues and private-sector partners that mitigation and adaptation to the climate change challenge makes economic and environmental sense. We must show the kind of leadership that can demonstrate that beside the challenges and problems associated with climate change, there are also abundant opportunities to alleviate our people's poverty through universally accepted adaptation strategies.

The leading opportunity offered to southern Africa's tourism by climate change is that the region currently enjoys what is regarded as very attractive climate and weather conditions. Zimbabwe has been judged, by leading international climate bodies, to be the country with the most attractive climate in the world. This must surely apply, to some measure, to countries that are adjacent to it. Climate conditions in Zambia, Botswana, Namibia and South Africa are not that different from those of Zimbabwe. We are therefore entering the climate change era from a relatively strong position in relation to other regions.

The current, comparatively attractive, climate conditions that we enjoy have not been fully exploited to attract international visitors to our region. One reason for this is that most of the countries in the region cannot, as individual destinations, justify the huge travel costs that must be incurred by long-haul travelers from the globe's major tourism source markets.

The obviously most constructive response to this is the putting together of regionally integrated tourist products and packages. Visitors coming to the magnificent Victoria Falls must be given the opportunity to visit the beautiful beaches of Mozambique's Indian Ocean coastline, for example.

Visitors visiting the ancient city of Great Zimbabwe, cynically referred to as the Zimbabwe Ruins in the past, must be afforded the opportunity to visit the historically related Mapungubwe of South Africa.

The second factor offered to our region as an opportunity by climate change is the variation of climate change impacts affecting the different tourism destinations of the world. As has been observed by others, what constitutes challenges and hazards in some places will actually amount to opportunities in other places.

Yet another factor regarding the nature of climate change in different parts of the world is that the developed countries of the world have already mounted a massive damaging assault on their natural environments, with disastrous consequences for global climate change, whilst the under-developed countries, particularly those of Africa, have continued in large measure with the original natural and complementary co-existence between mankind and nature, thereby preserving natural environments in their original, pristine conditions.

Again when we look at various climate change impacts that affect our region, we find that they are all amenable to mitigation and adaptation. The warmer summers will not get intolerable for a while, because we do not currently suffer conditions of extreme heat or the kind of heat waves that occasionally hit other regions. Water scarcity is another impact felt by our region; however, this is usually associated with periodic droughts, which tend to affect different parts of our region, and even our countries, with varying intensity. We therefore have the potential to both capture and store water during good rainfall seasons, and to move water from regions receiving surplus rain to those affected by drought.

One of the areas that the region should jointly research and invest in is the technology to store and transfer water, amongst others. We must be encouraged by the ability of countries in the Middle East to mount projects that draw and convert seawater to domestic and industrial use, and reclaim land from the sea.

As part of the several MOUs that we sign in the tourism sector, the region needs to establish expert, collaborative inter-country task forces to work on specific areas related to tourism and climate change.

It is very fortunate that most of our region is not characterized by extreme events, except for the occasional cyclone and wild fire, and we must exploit this fact to the fullest. The land bio-diversity loss that characterizes our region will, at its current rate, be quite amenable to mitigation.

The biggest negative impact on our region's tourism will be from the increase in the cost of air travel, as increasing mitigation policies will force prices up in a bid to lower carbon emissions of air travel; however, programs that increase the perceived value of the region's tourism product, particularly wildlife, can counter this.

Climate defines the length and quality of tourism seasons and plays a major role in destination choice and tourist spending. Our region's ideal year-round climate, our diverse natural resources, our freedom from extreme events and our still-thriving diverse, local, indigenous cultures, clearly give us a great advantage over other tourism hot spots.

It has been noted that climate is a principal resource for tourism, as it co-determines the suitability of locations for a wide range of tourist activities, is a principal driver of global seasonality in tourism demand, and has an important influence on operating costs, such as heating, cooling, snowmaking, irrigation, food and water supply and insurance costs. Our region is spared all this.

The length and quality of climate-dependent tourist seasons of other tourist destinations, give our destination, with its year-round good climate, a considerable advantage over them. This advantage is set to grow with the ongoing climate change.

Thus, whilst some studies indicate a likely shift of attractive climatic conditions for tourism toward higher latitudes and altitudes, I believe that such analysis does not take full account of Africa's competitiveness, under certain conditions, and the continent has the potential to be the globe's tourism "high ground" of the future.

This is largely attributable to the fact of the established vulnerability of current leading destinations, as a result of climate change. There are huge costs attached to snowmaking, for instance. Southern Africa cannot celebrate the projected negative increase of weather extremes in some destinations, but it is a fact that the resultant increase in cost of repairing increased infrastructure damage, additional infrastructure damage and additional emergency-preparedness requirements, will all add a shifting competitive advantage to regions like ours.

Anticipated increased mitigation policies and increased efforts at adaptation will have an impact on tourism mobility that creates several opportunities for the tourism economy of the southern Africa region. Such policies are bound to seek to reduce carbon emissions. This will lead to an increase in transport costs of air travel and it will foster environmental attitudes that will benefit the future of our tourism prospects in two major ways.

First, it will make the cost of long-haul travel so unattractive that the huge Africa region will be forced to pay greater attention to intra-continental domestic tourism. Such tourism can thrive without too much dependence on air travel. Intra-continental domestic tourism can rise on the back of transport modes like coach and rail that have substantially lower carbon emission rates.

Such a scenario not only makes economic sense for our region, but also presents us with unlimited opportunities for regional integration, with its many critical socio-political advantages for our region and its people; these include the peace and security and the singularity of purpose that the region needs in order to cohere and thrive.

However, as has been indicated elsewhere, there are some large knowledge gaps in this area and systematic regional-level assessments must be conducted to produce a more definitive statement on the net-economic and social impacts on the region's tourism economic sector.

Secondly, as the world seeks to grow its adaptive capacity, tourists themselves will have the greatest adaptive capacity with relative freedom to avoid destinations that currently impact, or in their construction have impacted, on climate change. Tourists can also simply avoid destinations with unfavorable climate conditions, or those that impact negatively on climate change in their mitigation efforts.

All this augurs well for southern Africa's under-developed tourism sector, which, in its development, has the opportunities to learn from the lessons of the developed world, and avoid trajectories that lead to negative climate change.

Africa has low levels of development of fixed immovable tourism assets, and this makes it possible for the continent to adapt to climate change as these can be developed with climate change in mind.

The continent's vulnerability is in the price of incorporating the huge costs of adaptation and mitigation into the development of new tourism infrastructure. This is where I believe Africa must impress, on the developed nations in particular, which have by far the largest carbon imprint on our planet, that the responsibility of incorporating climate change adaptation and mitigation measures into its future tourism infrastructure/development strategies must not lie with the individual countries of the region alone. The West should carry part, if not most of the burden.

In this regard, I must say some of us are rather dismayed with the West's half-hearted commitment to the Kyoto Protocol. Nor are we fully convinced about the carbon-emission-offsetting arrangement, whereby the developed industrialized countries literally "buy" their right to continue with their nefarious GHG-emitting activities for purposes of protecting their so-called rights to current levels of comfort.

I am for the "common but differentiated responsibilities" for reducing greenhouse emissions. This principle is indeed contained in the Kyoto Protocol, which places greater responsibility on industrialized countries such as the United States and those in the European Union, while making allowances for the fact that other countries are still developing.

Going forward, therefore, the African region must do all in its power to ensure that, as the world addresses the challenges of climate change, it does not in any way take away from its commitment to poverty alleviation. The poverty of our people and its associated suffering is a great indictment of our generation of Africa's leadership. We need to address it urgently.

Finally, I must, in conclusion, highlight the important role of the region's media in all these efforts. It has been observed that international perception of future impacts of climate change are likely to play the central role in the decision-making of tourists and tourism investors alike, as perceptions of climate conditions or environmental changes are just as important to consumer choices as the actual conditions.

Since perceptions of climate change in the southern Africa region will be heavily influenced by the nature of media coverage, the region's media operatives might therefore need special orientation for their coverage of the issues I have covered. It is important that they put an accent on the glowing opportunities that climate change offers the region, without, of course, overlooking the very real substantive challenges that we face. I believe that when all pertinent factors are taken into account, our region has the potential to become the “high ground” constituted by Noah’s ark during the biblical ages.

CHAPTER EIGHT

GROWING SUSTAINABLE TOURISM: THE SEYCHELLES APPROACH

ALAIN ST. ANGE[1]

Seychelles, by virtue of its small population of a mere 87,000 people, may be studied as a microcosm and a mirror for some of the issues of climate change.

Seychelles is an archipelago of some 115 sparkling islands, located between 4 and 10 degrees south of the equator in the western Indian Ocean. Only settled in 1756 by French settlers, it was first a French colony before being ceded to Britain in 1812, whose colony it remained until its independence in 1976. Today, the Republic of Seychelles has a multi-party political system with an executive president as head of state and government.

We have been fortunate to have, by history, the opportunity to make of our isolation and remoteness, a very real strength – to weld the many ethnic strains of our people into a vibrant but tranquil nation. We, in the Seychelles, pride ourselves on our claim that our islands are not only the land of perpetual summer, but also a place where harmony is a way of life. That stability is the foundation of our society and, of course, the pillar of one of the lifelines of our economy – our tourism industry.

I believe that this is a crucial starting point for any tourism industry; in order to be in a position to bring peace to the world through tourism, we first have to be able, and to be willing, to bring peace to ourselves, at home. At the same time, we must invest in the long-term prosperity and sustainability of our tourism industries and not forget the famous saying, "he who would conquer the world, must first conquer himself."

Like mother Africa, the Seychelles islands have many natural attributes. Nature has blessed Africa with so much sublime natural beauty, rich cultures, lush forests, spectacular savannahs, amazing flora and fauna, and a suite of activities that are the envy of the world and with all the space in the world to enjoy them.

---

[1] CEO, Seychelles Tourism Board.

Africa has untold potential to develop its respective tourism industries to the sky, capitalizing on the bounty of Mother Nature, but at the same time, ensuring that any exploitation is sustainable in nature and will guarantee that the integrity of the underlying attributes remain intact for future generations to enjoy and for future generations to benefit from. We must take pains to ensure that we do not commit the cardinal sin of succumbing to appetites for short-term profit and cut through the very branch upon which we are seated and upon which we depend for our livelihood – for our very existence!

In a world where there is hardly any country that is not competing in the tourism arena, tourism is about successfully raising one's profile, while ensuring that its house is in order and that its tourism products are, of course, worthy. In recent remarks, the secretary-general of the UN Conference on Trade and Development (UNCTAD) has emphasized the critical importance of quality in the development of a competitive tourism sector. My remarks come in the wake of the first public appearance of the UN Steering Committee on Tourism for Development, which reflects the increasing relevance of tourism in the development agenda and represents a clear commitment of the UN system in making tourism work for development.

How do we achieve this? How do we ensure that we stick out in the marketplace, head and shoulders above the competition? And also ensure that our industry remains healthy and vigorous in the long term?

A Case Study of Seychelles' Tourism

To answer some of these questions, I would like to present you with a short case study of Seychelles' tourism. Seychelles depends on its tourism industry. Tourism remains the pillar of the Seychelles economy. We need to succeed with our tourism industry.

In 2009, following the crash of financial sectors in America and Europe, and the resulting recessions, Seychelles experienced a serious dip in its visitor arrival figures and, therefore, in its revenue. It was decided that change was needed in the way the country was approaching its tourism industry, along with a serious streamlining and invigoration of institutions and also resources, both human and otherwise.

What then took place set the stage for the revitalization of Seychelles' tourism industry, and it paved the way for the successes that we have since enjoyed.

Firstly, government, in the new role of facilitator, entered into a new partnership with the private sector, which now assumed responsibility for its own industry. The composition of the board of the Seychelles Tourism Board was revisited to allocate a majority of board-member positions to private-sector players, and this as government guaranteed the continued funding of the board's activities and of its total budget.

With local stakeholders now in the driving seat, the resulting new synergy has allowed much-needed changes to be effected on the ground that, together, have enabled Seychelles to make greater and more efficient use of its resources across the board. Fresh, new policies have been adopted vis-à-vis open skies policy and airlift, which have resulted in a marked increase in the number of visitors arriving on our shores. New markets in the Middle East, Asia, Africa and the Americas are being explored together with new synergies for twin-hub vacations.

New initiatives designed to raise our profile internationally have included schemes such as appointing Seychelles Tourism Ambassadors – Seychellois nationals living abroad who are prepared to use their contacts and local knowledge of their country of residence to promote Seychelles as a tourism destination. We have Seychellois nationals everywhere in the world; and, as we are in Zambia, I have just appointed Seychellois Emily Luring yesterday as our latest Tourism Ambassador for Zambia.

We also have our Friends of Seychelles Press Program. This is for people who are members of the international press with loyalty to, and extensive knowledge of, Seychelles, and who are important conduits for disseminating news items about the islands.

To create even greater dynamism, President James Michel took the helm and the portfolio for tourism and asked the nation to refocus on a Seychelles' brand of tourism – that unique suite of attributes, natural beauty, culture, activities and events that make Seychelles what it is. He called on the entire country to stand behind its tourism industry and to get involved in making things happen.

The president reiterated what was becoming evident; although our islands have been blessed with perpetual summer, the most enticing waters and sublime beaches, these alone cannot continue to form the foundation of our tourism industry. Today's discerning traveler is also seeking "experiences" in terms of cultural encounters and activities, and any successful tourism industry must be in a position to provide these. We have to learn to expand and diversify our product base, while at the same time ensuring that whatever we do today in the name of enhancing our industry does not come at the expense of tomorrow.

Once we have identified the general direction that our tourism will take in terms of effectively showcasing all that we have to offer, providing quality products across the board, and after we have also targeted our markets, both traditional and emerging, we are still left with my earlier question: How can we differentiate ourselves from the competition? How can we raise our profile and create awareness about ourselves as a destination?

To achieve these things we must be prepared to abandon traditional viewpoints and entrenched positions and start thinking out of the box. An interesting place to start this process is perhaps by recognizing that the competition may be used to our (and their) advantage. In Seychelles, we are looking at developing tourism synergies with our neighbors on the African continent. Our recently launched "From the Big Five to the Best Five" campaign twins the attributes of East Africa and South Africa with those of Seychelles, creating an enticing combination of experiences for potential visitors to enjoy.

What can we do to get people interested in Africa? We, in Seychelles, believe that cultivating a raft of international events is one way of winning international participation and vital press coverage, and we have recently enjoyed both in ample measure with the launch of the first edition of the Seychelles "Carnaval International de Victoria" in March 2011.

Here, we saw some participation from our fellow African states in this highly successful event, which has since been reported around the world, while, sadly, others did not attend. I maintain that if our tourism is to flourish, Africa must do for Africa, standing alongside our African brothers and sisters and together doing all in our power to raise the profile of our continent. South Africa, Zimbabwe, and Zanzibar, Tanzania, were present at the Seychelles carnival of carnivals. The islands of Madagascar and La Reunion from the Indian Ocean islands group joined them. Thus it was an event for the community of nations in which twenty-one countries participated. As a continent, as people from Africa, we missed a great opportunity to see our cultural groups parade side by side with delegations from the best carnivals of the world.

The biggest press contingent ever that traveled to Seychelles covered this carnival of carnivals because it is unique. It remains the only carnival that brings together the best carnivals of the world and gets them to parade in one carnival procession alongside the cultural groups from the Community of Nations. The next edition of this unique event in Africa is March 2–4, 2012. In this edition, we are inviting participation from American Indians, Pacific islanders from Hawaii, Aborigines from Australia, the Pygmy people from Gabon, etc.

As we bring in new events for the Community of Nations to come together, we must sadly note that some great platforms have also come and gone (two that immediately spring to mind are the Cape to Rio Race and the Paris–Dakar Rally, which was abandoned for security reasons), but not without first leaving us with an important message, “tourism and insecurity cannot co-exist!”

A recent resounding success on the continent was South Africa’s hosting of the 2010 FIFA World Cup, which most definitely captured the world’s attention and imagination. Africa is crying out for a successor to that great event, and we should now do everything in our power to provide it. Let us put our heads together to come up with a worthy successor, for it is in such grand spectacles that bring us all together that the grandeur of Africa becomes resplendent for all to see and resonates around the planet.

We need to strive to maximize interest in Africa. In the domain of culture and music, for example, this would allow visitors to experience our unique culture. We all know that African music is very popular worldwide.

Have we as a continent really rallied to make an event of events of African music?

Popularizing culture increases yield from tourism and ensures that the tourism dollar is spread around.

I raise that as one example of an area we might all rally to further develop that will and could help change the perception of Africa.

It is through such events that we can win hearts, influence minds and genuinely effect change for the better. Have we not all held in our hearts the golden image of a world united by football? I am certain we all recognize that the lofty goals of oneness, harmony and, by extension, peace, can be realistically achieved in this and other ways. Our task is to find them, nurture them and bring them to reality.

Before we can hope to achieve that, there is another crucial component to the development of successful tourism that we shall ignore at our peril – sustainability. This is the internal-looking face of our tourism industry that asks us to examine how we manage this resource.

Long gone are the days when any of us can afford to ignore considerations of effective power, water and waste management, which underpin current and future levels of tourism development. We must ensure that our tourism industries benefit our local communities, taking every pain to mitigate the negative impacts of tourism development on our fragile environments and communities, the burden of increased foreign workforces, the threat of tourist numbers to our own cultures and ways of life, and to the very fabric of our societies.

Sustainability is about the continual checking of ourselves, by ourselves, as we hasten in the achievement of our goals. It is about the setting up of an efficient system of checks and balances to ensure that we do not get ahead of ourselves, but rather strive to ensure that all that we may do today, we shall still have the capability to do tomorrow.

Fellow Africans, dear participants, I would like to be able to state that these are the challenges of the future. But this is not the case. These are forces that demand an immediate and pressing response from us – here and now! And we will only surmount them by our putting, each of us, our own backyards in order, combating a lack of security whenever and wherever it threatens and by combining forces on the international stage to raise our individual and collective profiles for the betterment of our tourism industries, while regulating the development of those industries to ensure that they grow in a sustainable way.

Then, and only then, can we really hope to make a difference to our world through tourism.

CHAPTER NINE

MEETING THE CHALLENGES OF CLIMATE CHANGE TO TOURISM IN AFRICA: "THE CASE OF ZAMBIA"

VIOLA MORGAN[1]

The International Institute for Peace through Tourism has successfully concluded the 5th African Conference under the theme "Meeting the Challenges of Climate Change to Tourism in Africa and the Developing World."

The issues around climate change, central to the theme of the conference, are significant, and impact on all development and human endeavors. The government of Zambia is addressing this important phenomenon in its 6th National Development Plan (SNDP) and by developing a national climate-change-response strategy. There is a growing consensus that climate change is not just a meteorological or environmental concern relegated to science and experts, but a multifaceted issue cutting across several aspects of human life support systems, such as agriculture, water, forestry, industry, health and, of course, tourism. As such, climate change continues to be a topic that requires awareness and urgent attention, and its effective solution calls for a concerted, multidisciplinary and developmental approach.

Drawing from this premise, I will focus on three interrelated areas:

(i) The importance of tourism as a means toward reducing poverty and sustaining livelihoods
(ii) The link between climate change and tourism
(iii) The link with the HIV/AIDS response

Cutting across all these areas is the importance of peace.

[1] UNDP Country Officer, Zambia.

Tourism as a Means to Reduce Poverty and Sustain Livelihoods

Tourism is an important sector for the Zambian economy. From 1997 to 2008, international tourists averaged 5.4% of total exports. Furthermore, the tourism sector has been expanding. For instance, the number of international tourists in Zambia increased – quite impressively – by about 400%, from 163,000 arrivals in 1995 to 815,000 in 2010. Revenue from international tourists grew from $29 million (USD) in 1997 to $212 million in 2009.

What these numbers tell us is that tourism represents a very positive economic stimulus for the rest of the economy, through its direct linkages with the transport and hospitality industries, and indirect linkages with agriculture. Tourism is also a very labor-intensive industry that provides many jobs, in particular for women who depend on the sector for income to support their families, including sending children to school and paying for health services. It is estimated that globally between 60% and 70% of the sector's total labor supply are by women. In Zambia, there are numerous opportunities for employment generation in the handicraft industry, through museums, national parks and safaris. The potential of Zambia's tourism sector is immense. There are, as well, many investment opportunities through hotels, lodges or safari operations. Zambia has 19 national parks and 34 game areas, as well as 23 million hectares devoted to the conservation of an exceptional variety of animals.

However, tourism in Zambia has experienced some drawbacks. Despite its positive trend, the sector has been quite volatile, suffering from economic downturns. As a result, long-term jobs are not always guaranteed and uncertainties in households are ever-present, which hampers longer-term investments in critical areas such as in higher education.

To facilitate progress in the tourism sector there is a need to focus on supply-side constraints, such as access to electricity, better communications, infrastructure and a more competitive investment climate for the private sector to flourish. In many developing countries, the sector's main challenge relates to sourcing partnerships with foreign suppliers rather than encouraging local suppliers and investors. Zambia's domestic suppliers should be encouraged to flourish and if there are foreign investors, the benefits of the investment should also be retained or shared in the country through investments. An overall tourism entrepreneurial culture should be ingrained in advocacy campaigns and programs for poverty reduction and wealth creation.

Zambia faces competition, especially taking into account the country's main competitors in its niches – Victoria Falls and biodiversity. With regards to Victoria Falls, the main competition is Zimbabwe. On biodiversity, competitors for international tourism in sub-Saharan Africa in 2009 included Botswana (1.5 million tourist arrivals), Kenya (1.4 million), South Africa (9.9 million), Malawi (755,000) and Tanzania (714,000). Zambia's authorities, departments and statutory bodies should pursue ambitious, innovative marketing campaigns on the country's strategic niches in order to gain market share from those countries. Efforts in that regard are recognized and would require support at all levels.

Climate Change and Tourism

Let me now turn to climate change and its consequences on the tourism sector. Countries like Zambia, where tourism depends to a high degree on natural heritage and biodiversity, are highly vulnerable to the consequences of climate change. It is not difficult to imagine how increased climatic variability may exacerbate tourism volatility (for example, droughts that may affect tourists' visits to Victoria Falls). Changing climate and weather patterns at tourist-generating and destination countries can significantly affect tourists' travel decisions.

While the tourism sector is affected by climate change, it is also a significant contributor to climate change, mainly through greenhouse gas (GHG) emissions generated mostly through the transportation of tourists and their energy consumption at destination points. Some estimates have noted that the tourism sector contributes 5% to the global carbon dioxide emissions; therefore, it is imperative for the sector to respond to climate change within the evolving national responses, thus progressively reducing its carbon footprint so as to grow in an environmentally sustainable manner. Action in this regard could be focused on a number of initiatives, including the following:

- Mitigation of the sector's GHG emissions;
- Adaptation of tourism businesses and destinations to changing climate conditions;
- Application of existing and new technologies to improve energy efficiency;
- Mobilization of financial resources required for investment in climate-change adaptation and mitigation initiatives;
- Provision of technical and training support to tourism destinations and operators in developing countries (this includes awareness

raising for customers and staff members on climate change impacts and engaging them in response processes);
- Diversification into climate-focused products to reposition destinations and support systems; and
- Striving to conserve biodiversity, natural ecosystems and landscapes in ways that strengthen resilience to climate change and ensure a long-term sustainable use of the environmental resource base of tourism.

Perhaps a most noteworthy characteristic of the effects of climate change observed in many studies is the inequity factor. While greenhouse gas emissions emanate mostly from developed countries and the industrialized giants of the developing world, the consequences are disproportionately borne by the least-developed countries, where adaptive or resilience capacity is minimal or lacking.

For our part, UNDP, in partnership with UN agencies and cooperating partners, deriving from international commitments and plans, have supported the government to establish a Climate Change Facilitation Unit, tasked with formulating a national climate change strategy, now completed, and providing the most needed campaign for climate-change literacy. Along with other interlinked initiatives, our hope, with our partners, is to support the government in efforts to ensure that climate change is mainstreamed in all development endeavors; the flow of funding is sequenced; institutional, human resource and negotiation capacities are strengthened; priorities are laid out; and a broad understanding of the impact of climate change and adaptive and mitigating responses is secured.

Further, we are supporting and advocating for revisions of policies and acts, such as the Forestry Act, and legislation needed for enforcement and the timely implementation of all programs, including a national program on climate resilience and low-emission development. Programs such as the Pilot Program on Climate Resilience (PPCR) with support from the World Bank will include interventions on droughts and flooding; others, with support from the Global Environment Facility, will address areas such as crop diversification, conservation farming, water harvesting, alternative livelihoods and capacity building for early warning for communities. Still others, such as the UN Reduced Emission from Deforestation and Degradation (UNREDD), seek to reduce deforestation through increased productivity in agricultural systems and alternative energy sources. The task is enormous and will require commitment at all levels.

Climate Change, Tourism and HIV/AIDS

Another developmental link is between tourism and HIV/AIDS. Tourism is an industry characterized by high job mobility, which, as a factor, creates increased vulnerability to HIV/AIDS infection. Zambia has a prevalence rate of 14.3%. Since it tends to be significantly affected by HIV/AIDS, the tourism sector is a likely target for HIV/AIDS interventions in many countries. The tourist industry is at particular risk from the pandemic because of the mobility of the work force, the presence of sex tourists and the heavy reliance of many countries upon tourism revenues. Indeed, tourism is one of the largest and fastest-growing industries in many countries.

Some individuals have speculated that potential tourists' fear of AIDS could discourage them from visiting certain countries, while others have even proposed that tourism should be discouraged because the industry contributes to the spread of HIV. Empirical studies suggest some tourists often take more risks than they normally would at home. They tend to drink more, use drugs more and be generally more adventurous while on holiday. Such adventures often include taking sexual risks. When such tourists have sex with prostitutes, hotel staff and others in the local population, a bridge can be created for HIV to cross back and forth between the tourist's home country and the tourist destination.

For example, four years ago, the Center for Global Health and Development (CGHD) and the Farming Systems Association of Zambia looked at the impact of HIV/AIDS on Zambia's tourism industry, as represented by Sun International Zambia (the largest tourism company in Livingstone), small and medium-sized companies in the same city and the Zambia Wildlife Authority (ZAWA). In the case of Sun International, the study found for each employee who died of AIDS-related causes, the average cost to the company was 3.5 times that of the employee's average annual salary, but as early as 2005, many HIV-positive employees were already accessing effective treatment with antiretroviral therapy. For the smaller tourism companies, the loss of a skilled worker was estimated to cost just more than one full year's average annual salary, but there were relatively few AIDS-related deaths, and most managers ranked HIV well below other economic threats to their businesses.

Unlike these private companies, ZAWA experienced both high AIDS-related mortality and very high direct and productivity costs, losing roughly 5% of its total service delivery capacity as a result. It is important that HIV/AIDS education reaches all high-prevalence areas and tourism capitals of the country.

Action should focus on:

- Developing and implementing health, HIV and AIDS tourism sector and work place policies;
- Assessing at what stage employees become too ill to perform duties;
- Assessing the loss of skilled manpower;
- The adaptation of training programs to deal with illness progression;
- The need for additional training for inexperienced employees; and
- Death and impact on work performance.

Conclusion

Tourism should be harnessed as a vehicle for the expansion of local employment and poverty reduction – one that takes into account equity and equality considerations, including gender empowerment dimensions. For Zambia, endowed with natural resources and biodiversity backed by a rich culture, and the peace dividend, the potential is enormous. The promotion of the sector should be ambitious, while maintaining the equilibrium of ecosystems and reducing the sector's vulnerability to the adverse effects of climate change. These actions would contribute toward the country's attainment of the related Millennium Development Goals and the implementation of the 6th National Development Plan while maintaining peace within and peaceful co-existence with her neighbors.

CHAPTER TEN

CHALLENGE AND RESPONSE TO CLIMATE CHANGE BY ISLANDS OF THE INDIAN OCEAN

MAGA RAMASAMY[1]

This article focuses mainly on the island nations of the south Indian Ocean regrouped under the Indian Ocean Islands Commission and composed of five main island nations – Madagascar, Mauritius, Seychelles, Reunion Island and the Comoros Islands. These islands are working together on different issues of common interest and, among others, tourism and the effects of climate change.

At the Climate Change Science Compendium 2009, UN Secretary-General Ban Ki-moon stated:

> The science has become more irrevocable than ever: climate change is happening. The evidence is all around us. And unless we act, we will see catastrophic consequences, including rising sea levels, droughts and famine and the loss of up to a third of the world's plant and animal species.

Tourism for the five islands states is a major contributor to the economy of each country. It was recorded that nearly 1.8 million tourists visited the islands of the Indian Ocean last year of which 50% went to Mauritius as their main tourist destination.

The mystique associated with the islands of the Indian Ocean that makes them attractive to tourists is the blend of lifestyles, indigenous cultures, unique land formations, flora and fauna, and lagoon and coastal resources. To maintain this mystique, there is an important call to harmonize policies and actions in a responsible manner from social, environmental and economic perspectives – that is according to the concept of Corporate Social Responsibility (CSR) – a responsibility of all stakeholders, public, private and individual.

As a Small Island Developing State (SID), Mauritius, like the other islands of the Indian Ocean, is highly vulnerable to climate change and the resulting sea-level rise. As per records for more than five decades from the

---

[1] President, IIPT Indian Ocean Islands Chapter, and President, World Airline Clubs Association (WACA).

Mauritius Meteorological Services there has been on average a temperature increase of about 1.0°C since 1951. The rainfall deficit based on records of the last 100 years is 100mm. Mean sea-level rise has been 2.1cm during the last decade.

These changes have had negative impacts on agriculture, water, fisheries, erosion of beaches, degradation of the reef system and bleaching of corals, loss of wetlands and depletion of biodiversity, among others. According to the Fourth Assessment report of the Intergovernmental Panel on Climate Change (IPCC) in 2007, the detrimental impacts of Climate Change on Small Island Developing States (SIDS) are now established facts.

Based on current trends, the Mauritius Meteorological Services Department has also predicted further projected changes as follows:

- Decrease of 8% in annual rainfall
- Increase in heavy precipitation events with increased risk of flash flooding
- More frequent heat waves in summer
- Milder winters
- Increase in the number of intense, tropical cyclones
- Increase in duration of dry spells, and therefore higher demands for water
- Increase events of high-energy waves (tidal surges) impacting the shores of Mauritius

Considering that Mauritius has limited land and other resources, any further increase in global temperature will cause immense harm to the well being of the human population. Mauritius is most likely to become a water-stressed and water-scarce country. Agriculture and the marine ecosystem will be negatively impacted upon.

Despite being a low emitter of greenhouse gases (GHG), Mauritius has been facing the impacts of climate change and has been pursuing sustainable development for decades. The Government of Mauritius has for years realized that climate change is a global issue that poses an environmental, economic and social threat that impacts on every facet of government and public life – from finance and planning to water quality and quantity, agriculture, tourism, health, employment and transport. The challenge now is not just an adaptation to the impacts of climate change but also reduction in emissions of GHG. It is a comprehensive re-engagement with core objectives and principles of sustainable development.

At the international level, the Government of Mauritius ratified the United Nations Framework Convention on Climate Change (UNFCCC) in 1992 and the Kyoto Protocol (KP) in 2001, and has been an active participant in negotiation processes including work towards "legally binding agreements" in Mexico in 2010 (the second commitments period under the Kyoto Protocol and the Bali Action Plan on long-term Cooperative Action).

At the national level, the government announced its vision toward Building Green Mauritius through the "Maurice Ile Durable project" with the overall aim of the project to reduce its reliance on fossil fuels by shifting to local, renewable-energy sources.

In the project's current five-year program, "Government will participate actively in all efforts of the international community to address issues relating to climate change and sea-level rise." Moreover, it also indicates that in line with Maurice Ile Durable, "Government will support and promote the interests of Small Island Developing States." Furthermore, "Government will take steps to accelerate the process of regional integration and will also encourage cross-border initiatives in key sectors such as food security, climate change, infrastructure, ICT, renewable energy and marine resources development."

The Government of Mauritius is already participating in high-level negotiations as indicated earlier and is making its voice heard through negotiating blocs such as the Alliance of Small Island States (AOSIS), the Africa Group, and the Group of 77 towards reduction of greenhouse gas emissions by developed countries that are part of the UNFCCC.

With regard to the Copenhagen Accord of the 15th Conference of Parties to UNFCCC/KP in Copenhagen in December 2009, the Government of Mauritius has associated itself with the Accord with the following conditions:

- A legally binding agreement should be reached at the 16th Meeting of the Conference of Parties in Cancun.
- A second commitment shall have to be taken by the Parties under the Kyoto Protocol for a determined period beyond 2012.
- The financial assistance of 30 billion US dollars for the period 2010–2012 promised in the Copenhagen Accord should be in the form of grants and not loans.
- The quick-start funding to be made available forthwith and be monitored to ensure that it is not a repackaged fund that should otherwise be granted as traditional overseas development assistance.

- The Copenhagen Green Fund proposed in the Accord should be set up and made operational without any further delay.
- Priority for funding should be given to the most vulnerable countries such as Small Island Developing States (SIDS) and Least Developed Countries (LDCs).

The Government of Mauritius is also supporting regional initiatives and programs through the African Ministerial Conference on Environment (AMCEN) and the Commission on Sustainable Development (CSD) among others.

The present and future detrimental impacts of climate change on SIDS are now established facts, and the situation seems to be irreversible for a number of SIDS with topographies just above sea level. Some observed impacts of a warming climate include temperature, changing patterns of rainfall, sea-level rise, the continuing coral bleaching, alteration in stratification and circulation patterns of ocean currents and increased frequency of extreme weather events. The impacts of climate change are all around us.

Some concrete actions taken and implementation progress made by Mauritius for instance can be summarized as follows:

- Close monitoring of climate change and sea-level rise
- Reducing greenhouse gases (GHG) emissions
- Investing in renewable energy
- Sensitization and raising awareness

Research by Zinke et al. (2010), observed that soil erosion, land degradation and modern climate change pose a mounting threat to coastal nations in the western Indian Ocean. A better understanding of the entire ecosystem in the region is required to achieve improved sustainable management of the coastal environment. The vulnerability of the coral reefs in the region to climate change (increasing sea-surface temperatures) has become increasingly documented by the scientific community. To ensure the long-term viability of coral reefs, the atmospheric-CO_2 level must be reduced significantly.

An aviation perspective is also important, as the Indian Ocean islands are tourist destinations and most tourists to the region travel by air. According to a report by the International Air Transport Association (IATA), aviation is responsible for:

- 2% of global carbon-dioxide (CO_2) emissions;

- 12% CO_2 emissions from all transport sources, compared to 74% from road transport; and
- 3% of the total man-made contribution to climate change.

IATA also emphasizes that by 2020, airlines are aiming for at least an additional 25% improvement in fuel efficiency and CO_2 emissions through technology and operations enhancement.

Air Mauritius, being the national carrier of the Republic of Mauritius, is one of the leading airline companies in the region of the Indian Ocean islands. The airline company is conscious that deforestation is a huge contributor to global warming and is contributing in the initiative to save the earth for future generations. Air Mauritius, to be in line with the philosophy of the government, i.e., Maurice Ile Durable, MID, (Sustainable Mauritius) and that of IATA, has developed a green project called "One Take Off, One Tree." This program is in partnership with the Mauritian Wild Life Foundation (MWF). It helps the foundation to plant a tree for every Air Mauritius flight that takes off.

It's worth noting that Air Mauritius is fully committed to ensuring the Corporate Social Responsibility concept is well understood from the business point of view so as its operations and activities respect the triple-bottom-line approach to corporate responsibility. Air Seychelles and Air Austral, two other airline companies in the region, are also strong advocates for a Corporate Social Responsibility philosophy and approach.

Reference

Zinke, J., G. Grove & G. Brummer (2010). The Western Indian Ocean Marine Science Association Magazine (WIOMSA), no. 4 (June 2010).

CHAPTER ELEVEN

TRAINING TO CHANGE MINDSETS: THE PRACTICAL BOTTOM-UP APPROACH IN IRELAND 2005–2011

MARY MULVEY[1]

Introduction

The Greenbox Project in North West Ireland was a typical Cross-Border European-funded project (both Republic of Ireland and Northern Ireland). This region had suffered from the worst effects of war in Ireland, resulting in high unemployment/emigration up to the mid-1990s, as well as poor investment and low tourism numbers and a tired tourism infrastructure. The standard model of partnership was a simple process of getting agencies to sit on joint boards, and in most cases the projects didn't seek to change policies or do things any differently. The norm is that the pilot project winds up after the funding phase, and long-term sustainability or mainstreaming of projects is not planned.

Tourism in the Region and in Ireland

Owing to poor land and property values in the late 1980s and 1990s, this region, on the northwest coast of Ireland, became attractive for artists, craftspeople, filmmakers and organic farmers. This new influx of creative people also gave a unique support to many environmental projects, etc.

Presently, the Irish tourism industry faces many challenges. Throughout the 1990s Ireland outperformed most of the competition in attracting increasing numbers of overseas visitors, peaking in 2000 with almost 7 million overseas visitors. While the overall growth in aggregate visits to Ireland has been impressive, a number of significant trends are evident in the profile of the demand and the behavior pattern of overseas

[1] Ecotourism Ireland.

visitors to Ireland. One of the most obvious changes is the increasing concentration of visits to Dublin, while tourism to many of the traditional tourism areas has not grown, or in some cases, in particular the northwest, has declined.

The spatial or regional distribution of overseas tourism within Ireland continues to change and can be explained by a change in the composition of the visitors, principally amongst those coming from Britain: visitors coming for shorter trips with fewer visitors bringing their cars, and a decrease in the average number of regions visited on the trip. In addition, the increasing cost of visiting Ireland and competition from new lower-cost destinations in Central and Eastern Europe combined with availability of low air fares is placing huge competitive pressures on the Irish market.

Ireland has been marketed with a Green Image since the late 1960s but without any real understanding of sustainability – or later, of climate change. There were no policies in place. Greenbox challenged this and in some cases it took leadership by heads of organizations to gain some credibility.

The Economics of Ireland in the Mid-1990s

In the mid-90s, Ireland was in the full economic growth phase of the Celtic Tiger and the national government policies in most parts ignored tourism.[1] Environment planning was fast-tracked to enable further construction. High-tech American companies and property inflation were the norm of the day. Political leaders were skeptical of talk of climate change. Not until late 1996 were the first Green Party politicians elected and made Ministers of Departments of Environment, Agriculture, Renewable Energy and Resources.

The large construction industry in Ireland also resulted in the development of a number of hotels. Ireland now has 60,000 beds, almost 20,000 in excess of demand. Many were the trophy possession of large developers and builders.

The Greenbox

The Greenbox is a tourism development and marketing initiative, established in 2004 by the Western Development Tourism Program in conjunction with the Organic Centre, Rossinver Co. Leitrim. Ecotourism is tourism that is ecologically sustainable in terms of honoring the environment and impacting on it as little as possible. Responsible ecotourism includes programs/initiatives that minimize the negative aspects of conventional tourism on the environment and enhance the cultural integrity of local people.

The Greenbox strategic plan sought to deliver an Ecotourism destination – a defined region containing environmentally sustainable products, accommodation and attractions within a context of pristine natural resources. This was in response to the need to create a dynamic and unique large-scale tourism attractor for the northwest capable of combating stagnant tourism numbers and increasing the scale of the industry in the region. Examples of some ecotourism products include walking, cycling, canoeing, artistic and cultural activities, health and personal well being activities, etc. The Greenbox is Ireland's first ecotourism destination, covering the counties of Leitrim, West Cavan, Fermanagh, North Sligo, South Donegal and northwest Monaghan.

Greenbox Activities 2004–2009

The company engaged in a number of interrelated activities to achieve its objectives. In 2008, all core funding ceased and project was reduced from seven staff to one and remains in existence only on commercial activities.

This Capital Development Project provided capital funding and training for tourism product providers to develop and enhance their ecotourism products. The Greenbox initially secured funding of €1,367,925.00 under INTERREG IIIA Measure 1, Priority 4. The project commenced in April 2005 and ended in August 2008. Funding was provided on the basis that two flagship capital projects would be established (one north and one south) and of the establishment of 42 smaller projects, employing eco-friendly practices. The businesses that were focused on included accommodation, food, activities and community initiatives.

Project applications were invited from private individuals, limited companies, community groups, and co-operatives that provide a product or wish to provide a product that would meet the needs of both domestic and international eco-tourists. Projects were only awarded grant aid if they either met the ecotourism standard as set by the Greenbox or had the potential to become certified in the future. Projects also had to be located within the Greenbox geographic area.

Training and Capacity-building

Certificate in Ecotourism, the EU Flower and the Greenbox Eco-label: The majority of Capital Development Project promoters who received funding have gained an understanding of ecotourism, increased their tourism skills, and are now working with like-minded promoters implementing a broad range of environmentally sound standards. At least 20 of the Capital Development Project promoters have followed through on Greenbox Eco-label training. The Certificate has received FETAC approval and has also received international accreditation by Sustainable Travel International. Eleven of the Capital Development Program promoters have received the EU Flower standard (making a total of 24 EU Flowers in the region, the largest cluster in Europe). Ten received Capital Development Program funding.

A catalyst for regional tourism development: Greenbox has created a different and innovative product, based on the natural and cultural resources of the area and by its nature therefore is low impact and sustainable. The product is also one offering a competitive advantage to an underdeveloped area by promoting the very aspect which has historically prevented its development – isolation, harsh landscape (in terms of agriculture production), and poor infrastructure in terms of road network, rail and air access. Ironically it is because of the lack of investment in the project region in the past that it is now relatively unspoiled and can offer one of the best eco destinations.

Internationally recognized: The success of the project is evidenced by the fact that 24 of the 50 top eco-destinations in the Eco Escape Ireland (2008) book are located in the Greenbox, many of which received Capital Development Program support. The success of Greenbox has been recognized in a number of recent awards including:

- Responsible Tourism Award, December 2006, World Travel Market – highly commended
- People Category, May 2007, Northern Ireland Tourist Board – commended
- Tourism for Tomorrow, Destination Category, June 2007, World Travel Market – finalist (one of three)
- Green Apple Award, December 2007 – gold
- CEO Mary Mulvey Highly Commended, Shine Awards 2010.

Training to change mindsets: The success of the Greenbox model also included "hands-on" proactive building of relationships and linkages by the executive with key representatives in the public and private sector at international, national and regional levels. Facilitation of all the main players to work together was the essence of Greenbox right from the beginning. This work led in the main by the Greenbox CEO over the past six years has moved the initiative from one that once had little or no support, to one where confidence has been built and mainstream agencies are seeing as exciting, innovative and new.

Funding support from mainstream agencies still remains an issue, however. Greenbox was very aware of the need to help educate policy-makers and in 2006 and 2008 hosted two national conferences with a wide range of international speakers. The networking of all these policy-makers both from north and south of Ireland made a significant difference in bringing about policy changes. The publication "Ecotourism Handbook for Ireland" in 2008 provided an important educational tool at all levels of Irish tourism and is being revised in 2011 by the national Tourist Board Fáilte Ireland.

Quality Certification as a Tool of Regional Development

Using a high-quality Certification Eco-Label for Accommodation with an independent audit gave the project further credibility and publicity. The Greenbox Ecotourism Label was piloted for more than two years based on TIES guidelines and was also independently audited. The label is the foundation for Ecotourism Ireland that will be formally launched in 2012.

Conclusion

The challenge of seeking support for ecotourism and sustainability remains. Ireland is now being governed by the IMF, having lost its economic sovereignty in 2010. The un-sustainability of the hotel building boom has not helped Irish tourism but has brought out a new competitive pricing model of cheap hotels being run by the same banks that caused the financial crashes.

There has been a 15% increase in visitors to Ireland in 2011 and a renewed interest in tourism by the new government, which has increased its priority in an effort for national renewal. There still continues to be a fear of green tourism/ecotourism but people in rural areas are now considering possibilities for a more sustainable and environmentally responsible tourism.

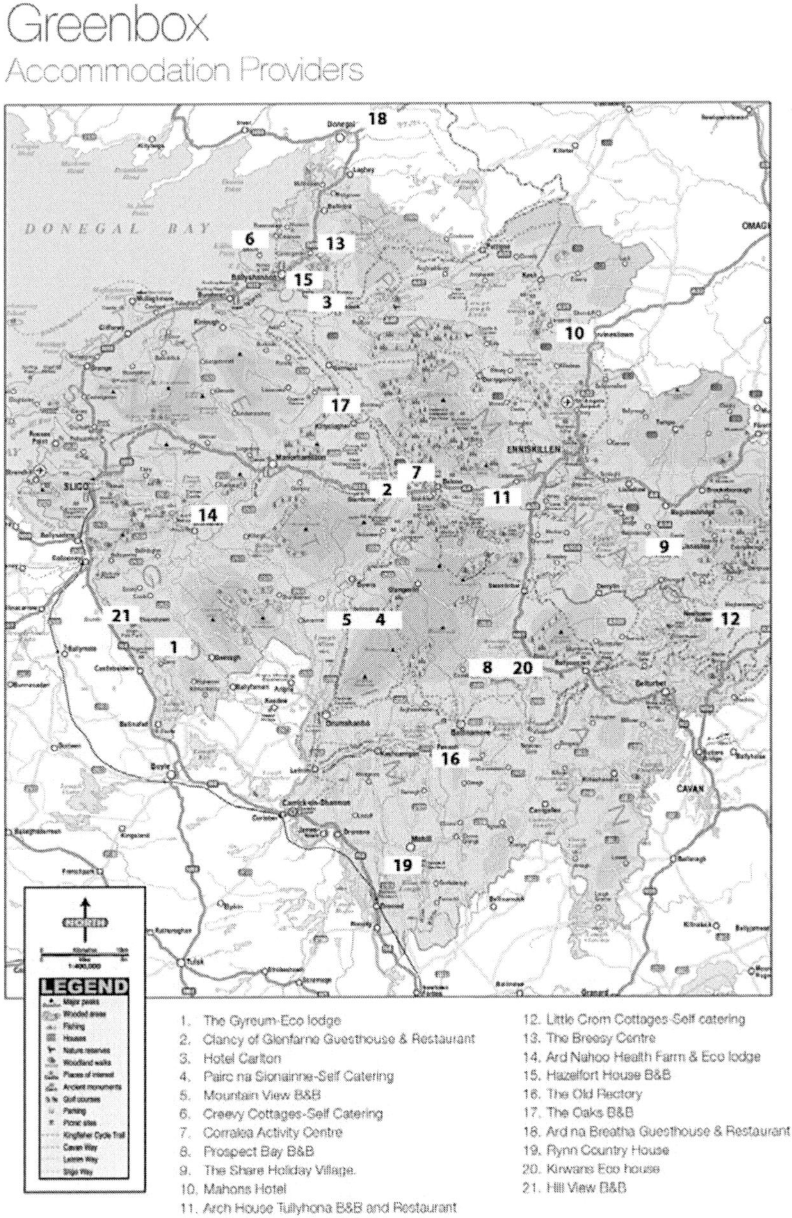

Figure 11-1. Map of north-west Ireland.

The lessons of Greenbox are that barriers to Green tourism exist and therefore flexibility in approaches is required. Training is essential including basic business skills for new entries in the field. The value of study trips, learning networks and sharing best practice has been invaluable. But more is required to respond to the challenges of climate change. The Mary Robinson Foundation model of "Climate Justice" is worthy of examination in this regard.[2]

Notes

[1] "Celtic Tiger" is a term used to describe the economy of Ireland during its period of rapid economic growth between 1995 and 2007. The expansion underwent a dramatic reversal from 2008, with GDP contracting by 14% and unemployment levels rising to 14% by 2010. Moody's proceeded to downgrade Ireland's government bond ratings to junk in mid-2011.

[2] "Climate Justice" is generally used as a term for viewing climate change as an ethical issue and considering how its causes and effects relate to concepts of justice, particularly social justice and environmental justice. For example examining issues such as equality, human rights and historical responsibility in relation to climate change. Many see recognition of the fact that those least responsible for climate change are currently experiencing (and will continue to experience) its greatest impacts as being central to climate justice. The term is also used with reference to legal systems, where justice is achieved through application and development of law in the area of climate change.

CHAPTER TWELVE

SUSTAINABILITY FIGHTS POVERTY

AGHA IQRAR HAROON[1]

The August 2010 floods in Pakistan resulted in approximately 20 million homeless persons and the collapse of the tourism industry. Swat Valley and Kaghan Valley faced total disaster due to the heavy rainfalls and flooding caused by deforestation in catchments. Experts are of the view that these floods were ultimately the result of global warming and climate change. According to the UNDP, poverty has increased in flood-affected areas because of the loss of livelihoods.

Pakistan constitutes one of the world's most extended high-altitude rainforest regions. Various forms of degradation and encroachment, mostly due to human intervention, seriously threaten this region. With its dramatic geological history, Pakistan spans a remarkable number of broad ecological regions.

Misguided economic policies have widened inequalities and forced mountainous people to exploit biodiversity at rates that are no longer sustainable. As a result, processes such as deforestation, overgrazing and soil erosion have become major threats to the remaining biodiversity of high lands in Pakistan. Poverty, lack of planning and appropriate policies have contributed to disproportionate pressures on resources resulting in degradation in uplands that ultimately cause more and more degradation at lowlands.

Some 80% of the 2010 floods impacted tourism destination areas, namely

- Upper and lower Swat and Kalam
- Kaghan Valley
- Gilgit and Baltistan

Kalam and the upper and lower Swat in northern Pakistan were popular tourist destinations, famous for their terrain, view and natural beauty. Kalam had some 400 hotels and restaurants before the floods; 101 hotels

---

[1] President and CEO, The Region Initiative (TRI).

were swept away in the raging waters. Most of the major hotels were either on the riverbed or overlooking it from the edges of River Swat.

"The More Degradation – The More Poverty"

The number of people affected by floods from 1990 to 2010 in Pakistan is reported to be approximately 44 million (according to Disaster Management Pakistan). Disasters caused by floods have been increasing, mainly because of the expansion of settlements and the growth of investment in flood plains.

Table 12-1. People affected by disasters, Pakistan.

Disaster	Date	Affected
Flood	2010	20,102,327
Flood	2005	7,000,450
Flood	1992	6,655,450
Flood	1992	6,184,418
Earthquake	2005	5,128,309
Drought	1999	2,200,000
Storm	2007	1,650,000
Flood	1996	1,300,000
Flood	2003	1,266,223
Flood	1995	1,255,000

Degradation in high mountain areas starts in high altitude watersheds where many factors, particularly deforestation, results in soil erosion. Degradation and soil erosion cause further degradation. Effects become causes and causes become effects in this "degradation cycle."

Vital economic resources, such as water, are usually managed with policies, institutions and practices that are disconnected from, or even in direct conflict with, those designed to protect forests, wetlands and other habitats from which the water comes. Moreover, watersheds face growing stress from rapid economic development, increasing human populations, and often wasteful use of natural resources.

Table 12-2. Economic damage.

Disaster	Date	Cost US$ Billion
Flood	2010	9.5
Earthquake	2005	5.2
Storm	2007	1.62
Flood	1992	1.0
Flood	2007	.33
Drought	1999	.25
Flood	2001	.25
Flood	2008	.11
Flood	1994	.09
Flood	2005	.03

The Cause-Effect Chain of Watershed Degradation

Removal of forests or other vegetation sharply reduces water retention and increases erosion, resulting in reduced water availability in dry seasons and more siltration downstream – resulting changes in river flows and sediment and pollutant loadings in dams.

An absence of trees results in loss of shelter for shrubs, which often die under the burning sun of May and June, just before monsoons hit the area and soils become more prone to water and wind erosion.

The Sustainable Development Policy Institute (SDPI), Islamabad, confirms in its June 2010 report that according to the World Wildlife Fund for Nature "while forests currently cover only 2.5% of the country's land, Pakistan has the highest annual deforestation rate in Asia. The WWF report warns that if the current deforestation rate of 2.1%, and the trend of land conversion from forest to other uses is not checked, the country will not be able to meet its international commitments under the Millennium Development Goals to increase its forest cover from 2.5% to 6% by 2015."

The June report says that more than 61,000 hectares (approximately 151,500 acres) of forest have been converted to non-forest use in the country since its inception.

Kaghan Valley Case Regarding Watershed Degradation

Taking the example of Kaghan Valley, we can see that massive forest cutting around Malika Perbat (Queen of Mountains), Lake Saiful Maluk, Saren Valley, Shogran, Sharan and Lalazar Meadow caused massive landslides in Kaghan after 1985. Moreover, illegal mining by local property owners for precious stones in the Jhalghat area also caused havoc, as they were using dynamite for mining, resulting in instant erosion and landslides in adjacent areas. These are areas from where water flows in small springs and creates countless small water channels flowing into the Kunhar River that joins with the Jhelum River and form the source of waters for the Mangla Dam – the second largest dam in Pakistan.

It may be mentioned that Kaghan Valley is a prime example of mountain feudal lord ownership, as more than 70% of the valley belongs to only two families, who are related. Therefore, the utilization of land is totally at their disposal with minimal government regulation.

Kaghan Valley has faced three massive and devastating floods within two decades, causing the deaths of more than 3,000 people and a number of villages to be flushed away due ultimately to heavy deforestation at watersheds.

Kaghan's economy depends upon fruit growing and tourism and both sectors have faced major negative impacts due to these floods and countless mudslides.

Wealth and Degradation Linkage in Kaghan Valley

Kaghan was a "paradise" until the early 1980s, when big landholders did not have money for infrastructure and were dependent on moderate tourism activities, fruit growing and farming. However, after the Afghan War in 1979, this area became a route for drug and weapon smuggling and local landlords made considerable revenue from these activities.

This provided the landholders with considerable money and power, resulting in massive infrastructure development and uncontrolled forest cutting. There were no regulations in place to prevent excessive logging and small landholders were powerless to voice any objections.

With earnings from timber sales, these landholders developed motels and hotels throughout the Kaghan Valley. Non-governmental organizations (NGOs) failed to play any role to stop this deforestation and unplanned growth of motels, transportation and housing, as they believed that they (as NGOs) could not work without the support of "important locals and local leadership," and in major cases the latter were the very ones responsible for the degradation.

This situation suggests that the following are the major socio-economic factors for degradations within the Kaghan Valley:

- Political systems provide shelter to big and influential landholders in mountain areas in return for their political support.
- Lack of appropriate planning by departments responsible for watershed management.
- Lack of coordination among major actors of planning and execution of watershed work.
- NGOs that are not bold enough to counter "important locals" when they are playing havoc with the environment and mountain cultures.
- Failure by "poverty-elimination programs" run by international bodies in mountain areas, that did not provide energy resources and economic development opportunities to poor people living in mountain areas.

There is however a positive model in another valley of Pakistan where local communities fought the "timber mafia" and did not allow environmental deterioration to occur. It is now an attractive destination for both domestic and international tourists.

The Example of Raikot Sarai

Fairy Meadows is one of the few remaining places in the northern region with awesome natural beauty and unspoiled forests and credit for this goes to the local communities. There was pressure on Raikot and Fairy Meadows from timber mafia but local communities joined hands and fought against this mafia, saving their areas from massive deforestation. Recent heavy rains did not affect these areas, in contrast to those where deforestation had been massive.

We can say sustainability combats poverty.

Recommendations

- Small credit schemes may be introduced for farmers of mountain areas.
- Energy resources like electricity and kerosene oil may be subsidized for people living in remote valleys to minimize use of expensive wood for fuel purposes.
- International Donor Agencies (IDA) should implement project-monitoring systems rather than depending on government departmental reports or NGO submissions.
- Third-party investigations may be done through professional consultants and chartered accountants to ascertain the facts.
- Governments and professionals, including engineers, town planners, professors and consultants, may establish Sustainable Tourism Planning (STP) departments and organizations.
- Laws may be amended to provide for greater community-level participation in activities supporting sustainable tourism development.
- The laws relating to communal ownership and access to sensitive ecosystems like water towers and watersheds should be reviewed and revised so as to protect and encourage natural resource management systems.

CHAPTER THIRTEEN

MEETING THE CHALLENGES OF CLIMATE CHANGE TO TOURISM: NIGERIA'S EXPERIENCE

MUNZALI A. DANTATA[1]

Abstract

Nigeria is a country with about a million square kilometers of landmass, blessed with untapped natural, human and tourism resources. Human activities affecting the land are leading to insecurity and restiveness, which in turn is impacting tourist flow. Findings show that land degradation, erosion, desertification, oil spillage, gas flaring, deforestation, low harvest, joblessness, socio-economic and political unrest are all key issues that impact on tourism. The paper thus concludes that unless human activities begin to incorporate environmentally responsible policies and actions, Nigeria will not prosper according to its potential. Early government and Non-Governmental Organization (NGO) intervention with the United Nations resolutions on climate change and tourism best practices are required to achieve sustainable tourism development in the country.

Background

Nigeria covers a total of 923,768km^2 along the West African Gulf of Guinea and is an important center of biodiversity with tropical rainforest, coastal plains, mangrove and savannah zones. Its population of 150 million makes it the most populous country of Africa.

Nigeria's economy is primarily based on petroleum with relatively high growth rates in both commerce and industry. A population explosion has led to a high rate of urbanization and material consumption that, together with mining and industrial activity, has led to a despoiled environment.

---

[1] Director General, National Institute for Hospitality and Tourism, Nigeria.

According to the World Travel and Tourism Council (WTTC), travel and tourism revenues were \$10.67 billion in 2007 increasing to \$12.35 billion in 2008. Travel and tourism clearly has excellent prospects for growth in Nigeria and should be encouraged to supplement the petroleum industry as a source of revenue and foreign exchange.

Problem

Nigeria's environmental issues are centered in the Niger Delta region, which has been experiencing oil spills, desert encroachment in the north, soil degradation and deforestation, all of which are having an effect on climate change and global warming leading to a poor agricultural harvest, poverty, diseases and low per-capita income.

The aim of this paper is to suggest strategies and policies for enhancing the preservation of Nigeria's abundant natural resources and tourism assets as a basis for sustainable tourism development at both the domestic and international level, thereby contributing to Nigeria's Vision 2020.

The methodology employed for this paper included the reviews of documents from conventions, legislative acts and resolutions passed through government and non-governmental organizations on climate, environment and tourism.

Reviews

Climate Change: Nigeria emits about 0.1 tons of carbon per person per year, as compared to United States' and European Union's 7 tons and 3 tons respectively. Yet carbon emissions are a threat to Nigeria's citizens and the less industrialized neighboring nations of West Africa. Nigeria requires a strategy to achieve high economic growth and poverty reduction together with a low-carbon economy in meeting the targets of the country's Vision 2020. Sustainable tourism development, together with agriculture and protection of wildlife can contribute to such a strategy.

Desertification: Some 55 million people in Nigeria have been affected by increasing desertification and desert encroachment in the northern and south-eastern parts of the country due to climate change visible through intense heat, loss of arable land and soil degradation and gully erosion. The desert is advancing at a rate of 600 meters per annum (Ishaya, 2010).

Gas Flaring: Refineries and oil exploration in the Niger Delta region account for 50% of Nigeria's greenhouse gases, a result of which is the loss of soil nutrients due to intense temperature rise and lack of vegetation cover, making agro-economic activities difficult for smaller farmers who lack access to improved crop varieties and facilities.

Coastal Areas: Heat emitted from rivers and the Atlantic Ocean is also adversely affecting the coast of Nigeria, leading to the loss of aquatic life.

Poor Disbursement of Ecological Funds: The ecological funds disbursed to states to prevent and deal with environmental disasters have often been misappropriated. Newly elected President Jonathan has created the Ministry of Environment to direct and manage these funds to achieve their intended purpose.

Nigeria's Tourism Sector

Nigeria is quickly growing into an industrialized nation with increased demand for industrial and domestic products. This is resulting in significant growth in international business arrivals. As petroleum is the main economic sector for Nigeria, tourism has received little attention.

The UNWTO Tourism Master Plan for Nigeria (2006) highlights the fact that the tourism sector lacks reliable statistics and market information; essential databases to implement the Tourism Satellite Account (TSA) are not available; there is a lack of information on frontier arrivals; and existing incentives and investment procedures are not attracting investment in the tourism sector.

Reports on the marketing approaches indicate that current marketing activities are under-funded, inadequate and ineffective; the tourist products are not organized or packaged for the marketplace; tourist attractions are unknown to the international travel trade; and Nigeria's negative image abroad is not being addressed.

In light of the above, it can be concluded that tourism in Nigeria is still in its infant stage.

Nigeria's Tourism Product and the Challenges of Climate Change

Issues that must be addressed by Nigeria in meeting the challenges of climate change and to promote sustainable tourism development include the following:

- Lack of adequate political will to legislate and regulate human activities related to climate change, the environment and the tourism industry at large.

- Lack of adequate funding to address desertification and de-forestation in affected areas.
- Lack of rain causing drought and death of wildlife that in turn is resulting in the encroachment of herdsmen into game reserves and national parks.
- Flooding that is eroding road networks, and negatively affecting wildlife and human socio-economic activities as experienced in Kano, Jigawa, Taraba, Nasarawa and Sokoto States.
- Mining activities and the emission of radioactive gases at the Jos Plateau, Jagindi Kaduna, Nasarawa Eggon/Keana Nassarawa, Nkalagu Enugu and Zamfara States.
- The dumping of toxic waste in Nigeria's sea ports and industrial cities including Lagos, Port Harcourt, Kano, Onitsha, Kaduna and other capital towns. These have resulted in mysterious ailments proving difficult to cure.
- Business travel and movements by air and land emit high levels of carbons through aircrafts and automobiles.
- Tourist flows to Nigeria for ecotourism, safari and related activities is negligible compared to business travel.
- Nigeria's industrial cities such as Lagos, Kano, Ibadan, Aba, Onitsha, Port Harcourt and the cosmopolitan cities of Sokoto, Maiduguri, Lafia that utilize automobiles to the fullest constitute much of the air/smoke pollution.
- Insecurity through the activities of restive youth such as the Niger Delta Militants in the South-South and the Boko Haram Sect in the northern parts of Nigeria threatens the inflow of tourists.

Prospects for Sustainable Tourism Development in Nigeria

Nigeria is blessed with rich and abundant natural resources that are mostly untapped. These can be seen from the rich coastal mangrove and rainforest to the savannah regions.

There are also opportunities that abound in government efforts in the promotion of tourism at both domestic and international levels. Some of both the public and private sectors' initiatives are as follows:

Public Sector

- The environmental and tourism policies of Nigeria give priority to ensuring environmental sustainability of the tourism resources.
- The establishment of the Climate Change Unit under the Federal Ministry of Environment at the federal level that oversees and regulates all industrial and rural development.
- Federal Ministry of Tourism and Culture that promotes the utilization of environmental resources for ecotourism.
- Federal Ministry of Agriculture and Forestry implementing re-forestation and agricultural management programs, as well as park, garden and forestry reserve activities.
- Ecology funds being made available to States in billions of Naira on a yearly basis to combat erosion, desert encroachment, gullies, petroleum oil spills, etc.

Private Sector

- The World Wildlife Conservation Federation with their activities in the forest reserves and national parks.
- National Conservation Societies are coordinating environmental activities in all states of the federation.

Strategic Approaches

The works of Balmford, Bereford, Nadoo, Walpole, and Manica (2009), with further reference to Buckley (2009) and Volume II of the UNWTO's Nigeria Tourism Development Master Plan (2006), addressed the issues identified above, and determined the strategies and actions government must take if tourism is to become a profitable and sustainable economic activity. Sustainable tourism development must also be linked to the government's overall policy of development through institutional capacity building.

The development of a vibrant and sustainable tourism industry will rest on a range of actions and implementation processes. Nigeria's tourism master plan provides key recommendations from which the action and implementation program can evolve and includes:

Type of Tourism: That Nigeria adopts a clear identifiable positioning as a cultural and regional conference destination in the tourism marketplace.

Tourism Clusters: The development of tourism should be concentrated in five clusters: Tropical Rainforest, Conference Capital, Atlantic Gateway, Scenic Nature and Sahara Gateway.

Product Development: Market-led quality products should be developed for international, regional and domestic tourists .

Flagship Tourism Projects: Within each tourism cluster, flagship projects should be developed to act as tourism icons and as a catalyst for all further development of tourism within the cluster. The Flagship Projects proposed are:

- Tinapa Business, Leisure and Conference Resort (Tropical Rainforest)
- Cross River Tropical Rainforest (Tropical Rainforest)
- Development of Conference, Meetings and Event Management Capability (Conference Capital)
- Farin Ruwa Eco Tourism Project (Conference Capital)
- Olokola Cultural Resort (Atlantic Gateway)
- Museum of Traditional Nigerian Architecture (Scenic Nature and Ancient City of Kano (Sahara Gateway)

The Environment: Each state government should appoint a Waste Management Officer to establish and implement a waste management system involving the general public, local governments and private waste management enterprises.

The development of tourism products should be preceded by an environment impact study. Planning and protection of cultural assets and heritage sites require the co-ordination of the three tiers of government and particularly local governments.

Conclusion

Nigeria's economy is primarily based on petroleum with relatively high growth in both commerce and industry. The population explosion has led to a high rate of urbanization and consumption that, together with mining and industrial activity, has led to a soiled environment and the need for improved waste management.

The solution is an inter-disciplinary approach between Nigerian experts in environment, natural resource planning and tourism management to develop policies and a master plan that preserves biodiversity and fosters sustainable tourism development contributing to Nigeria's Vision 2020.

Recommendations

For the tourism industry in Nigeria to prosper and meet the challenges of climate change, it is recommended that:

1. The Nigeria Tourism Development Master Plan, rolled out since 2006, but yet to be implemented, should be reviewed, updated and implemented to make tourism a major economic sector. This will help arrest climate change issues and create employment to supplement the petroleum sector.
2. Current conditions for the funding of tourism microenterprise and SME should be reviewed and simplified – to be amended as they are too complicated and difficult for SMEs.
3. The government and NGOs should develop a coordinated plan that meets the requirements of the UN Climate Change Convention Resolutions.
4. Political will is required to develop legislation in support of the greening process and regulation of protected areas.
5. The enhancement of Nigeria's wildlife, agricultural and ecotourism resources should be consistent with sustainable tourism development efforts.
6. Industrial cities such as Lagos, Kano, Ibadan, Port Harcourt, Onitsha, Aba, Kaduna and related towns should be audited and monitored against industrial chemicals, toxic waste discharge and waste management practices.
7. Petroleum spillages and toxic-waste dumping in especially the Ogoni area of the Niger Delta regions of Nigeria should be stopped and the land made productive for ecotourism.
8. Economic activities such as forestry, mining, bush burning, oil drilling and industrial activities should be regulated to mitigate gaseous emissions.

References

Balmford, A., Bereford, J., Nadoo, R. Walpole, M. and Manica, A. (2009). A global perspective on trends in nature based tourism. *PLOS Biology* 7, no. 6.

Becken, S. and Hay, J. E. (2007). *Tourism and Climate Change: Risk and Opportunities*. Cleveland: Channel View.

Buckley, R. (2009). Evaluating the net effect of ecotourism on the environment. *Journal of Sustainable Tourism* 17:643–672s.

Federal Republic of Nigeria (FRN). (2005). National Tourism Policy for Nigeria. Abuja: Govt. Press.

Gossling, S. and Hall, C. M. (2006). *Tourism and Global Environmental Change*. London: Routledge.

Gosling, S., Peeters, C. M. and Scott, D. (2010). The future of tourism: Can tourism growth and climate policy be reconciled? A climate change mitigation perspective. *The Recreation Research* 35, no. 2: 119–130.

Idowu (2010). Climate change: What Nigeria must do. *Daily Trust*. Abuja: Media Trust.

Ishaya, D. (2010). Climate change: 55 million affected by desertification. Abuja: Media Trust.

United Nations World Tourism Organization (UNWTO) (2006). Nigeria Tourism Development Master Plan. Madrid: UNWTO Pub.

World Travel and Tourism Council (WTTC) (2009). Trends in Travel and Tourism Activities. Available from www.wttc.org.

PART III

DESTINATION SUCCESS STORIES AND CASE STUDIES

CHAPTER FOURTEEN

SURAMA VILLAGE ECO-LODGE, GUYANA: MERGING TRADITIONAL AND MODERN APPROACHES TO PRESERVE BIODIVERSITY AND CREATE SUSTAINABLE LIVELIHOODS

JUDY KARWACKI[1]

Introduction

Surama Village Eco-lodge (www.suramaecolodge.com) is located at the edge of the vast North Rupununi wetlands, one of the world's greatest well-conserved, high-biodiversity hotspots. Here, where the Rupununi River joins the mighty Essequibo River and flows out to the Atlantic sea, the primary rainforest is still largely unspoiled.

Figure 14-1. Jaguar. Photo credit: FotoNatura.

[1] Small Planet Consulting, www.smallplanet.travel.

A recently released book *Rupununi: Rediscovering a Lost World* recounts that in the nineteenth and early twentieth centuries the Rupununi was as well known as the Galapagos, Serengeti and Ngorongoro are today. The Rupununi was visited and written about by well-known early explorers Sir Walter Raleigh and Charles Waterton, and later by naturalists Sir David Attenborough and Gerald Durrell and travel writer Evelyn Waugh.

The rainforest, rivers and seasonally flooded savannah of the Rupununi are home to an extraordinary mix of fauna and flora. Some of the world's largest species still thrive here – jaguar, giant anteater, giant river otter, giant river turtle, black caiman, tapir, green anaconda and arapaima. The Rupununi is a bird watcher's paradise offering a diversity of parrots, cotingas, toucans and antbirds, and specialties like Guianan cock-of-the-rock, harpy eagle, white-winged potoo, rufous-winged ground-cuckoo and crimson fruitcrow.

Surama Village Eco-lodge provides an outstanding example of a community-based tourism enterprise that merges traditional knowledge and customs with modern learning and techniques to successfully conserve and enhance the biodiversity of the North Rupununi and to create sustainable livelihoods. This chapter highlights Surama's leadership and best practices in biodiversity protection and sustainable community tourism development.

About Surama

Surama is home to 300 Makushi indigenous people, called Amerindians in Guyana. The village is set in five square miles of savannah, surrounded by the rainforest upon which Surama's residents depend for survival, and ringed by the foothills of the Pakariama Mountain range.

Surama started the eco-tourism enterprise in 1994 with a vision to "develop, own and manage a community-based eco-tourism business by constructively using the natural resources and our traditional culture in a socially appropriate manner." The community-based enterprise borrowed one million GYD (approximately $5,000 USD) to build the Eco-lodge. The loan was repaid early and the community's tourism economy has thrived since then. Surama now calls tourism the "Green Gold Rush," and fully embraces sustainable tourism concepts and core values to create sustainable livelihoods and preserve Guyana's ecosystems and indigenous traditions.

The success of the eco-tourism enterprise has led to international recognition for Surama's leadership in sustainable community-owned tourism development:

- In 2011, Surama was named by National Geographic Traveler as amongst the best lodges in South America and as the most authentic and unique hotel in Guyana.
- Also in 2011, Surama Village Eco-lodge was chosen as the joint winner of the 2011 Caribbean Excellence in Sustainable Tourism Award. The award sponsors, the Caribbean Tourism Organization (CTO) and TravelMole, selected Surama because the Eco-lodge harnesses the economic power of the visitor industry in a responsible and sustainable way to create wealth and to benefit the community and the environment surrounding it.

Figure 14-2. Surama Benabs. Photo credit: Phil Palmer.

- In 2010, Sydney Allicock, the visionary Executive Director of the Eco-lodge, was chosen as the Public and Civic Contributions Laureate for the prestigious Anthony N. Sabga Caribbean Awards for Excellence (ANSCAFE). He was selected for his leadership as a pioneer of community tourism in Guyana, implementing a sustainable tourism project that has been copied by other Amerindian communities and considered a model for indigenous tourism development.

- In 2009, Surama took top honors as the Responsible Tourism Showcase Honoree for the U.S.-based Educational Travel Community, the highly respected 25-year-old membership organization for zoos, museums, alumni associations and other institutions dedicated to promoting lifelong learning through travel.
- Surama has received two Guyana tourism awards – the President's Award for great community effort on community-based tourism, and the Tourism Pioneer Award from the Guyana Tourism Authority (GTA) for an outstanding contribution toward community-based tourism development.

Sustainable Land-use Planning and Management

Surama shares a boundary with the Iwokrama International Center for Rainforest Development and Conservation (ICC), a relationship that plays an important role in Surama's approach to conservation and sustainable livelihoods development.

Iwokrama is a one-million acre rainforest preserve that the Government of Guyana "gifted" to the world after the Rio Summit in 1992 as a living laboratory for scientific research, eco-tourism and sustainable tropical forest management. Dubbed the "Greenheart of Guyana," Iwokrama is home to a vast array of flora and fauna.

From the beginning, Surama was active in the development of Iwokrama. Community leader Fred Allicock led the site selection process and was the manager of Iwokrama for many years. His son, Sydney Allicock, has sat on the Board of Iwokrama for many years to represent the interests of the Makushi communities. Surama has played key roles in surveying and monitoring the one-million acre forest preserve, building the research station, working with international scientists and researchers and training the members of surrounding indigenous villages as rangers.

The leaders of Surama were impressed with the Iwokrama model, which led to Surama embracing sustainable tourism development and adopting a land-use plan that divides the community's lands into wildlife areas and sustainable resource-use areas. Twenty square kilometers of community land have been set aside as protected, and a formal environmental inventory of the area was conducted in 2010 with the support of the International Union for Conservation of Nature (IUCN).

The Eco-lodge is a catalyst for uniting the community to protect their natural resources and use them to generate sustainable livelihoods. The Eco-lodge conducts all of its programs in a sustainable manner, emphasizing environmental protection and minimal impact. Direct efforts are taken to foster conservation awareness and natural-resource management practices at the household level are made through environmental awareness outreach programs. Surama households make a conscious effort to reduce their exploitation of the forests for food, fuel and building materials.

Sustainable Community-based Tourism

The Surama Village Eco-lodge is built using local materials and traditional building techniques. The Eco-lodge has four traditional benab guest cabins, a lodge with four guest rooms and a large dining and gathering benab. Surama also operates remote satellite jungle camps along the Burro Burro River. The Eco-lodge accommodates small groups only, reducing the impact of visitors in the wilderness areas and on every day community life.

Surama's authentic guides lead visitors on fascinating and interactive village experiences ranging from mountain treks, medicinal walks, rainforest hikes, bird-watching treks, community tours and dugout canoe excursions. The programs offered include a jungle survival course, delivered in partnership with Bushmaster Tours, where visitors learn how to survive in the jungle, from building fire and shelter to avoiding dangerous animals and finding and cooking food. The Eco-lodge also has an Amerindian spa featuring traditional herbal steam baths and massages using botanicals hand-made from ingredients of the rainforest.

Surama's residents – especially the children – are highly involved in the eco-tourism enterprise. They lead visitors on nature and culture hikes, welcome them into their classrooms and give performances showcasing the Makuhi culture.

Tourism has become the single largest driver of economic activity in the area. More than seventy people (one quarter of the community) are employed either directly as hospitality staff, tour guides, cooks, artisans and drivers, or indirectly as farmers, hunters, fishermen and construction and maintenance workers. More than 60% of the community's income is now generated through sustainable tourism, and 75% of Surama's households derive income from tourism.

Figure 14-3. Surama Children. Photo credit: Kevin Loughlin.

Training and Skills for Sustainable Development

Surama has long prioritized training and development to support conservation and the development of sustainable livelihoods. Since 1998, together with other neighboring Makushi communities, Surama has had a Junior Wildlife Club for children six to eighteen years. The objectives are to develop awareness and understanding of the tropical rainforest environment and skills in natural-resource management and conservation. Throughout the years, through partnerships with international partners – such Audubon, Rainforest Alliance and Iwokrama – club members have received training in bird and wildlife watching and field research and monitoring techniques, such as using mist nets to safely catch birds, using sound recording equipment, setting automatic camera traps, rain-fall monitoring and setting research transects.

On-going junior wildlife club activities include assisting with flora and fauna inventories and collections, wildlife-knowledge competitions and wildlife-related artwork. Cultural activities include learning traditional skills, skits and cultural presentations. There is an annual Junior Wildlife Festival where club members compete in traditional skills such as basket weaving, cotton spinning and archery. Many of the club members have grown up to become environmental, political and community leaders.

Community members also take ranger, tour-guide and birding-guide training programs through Iwokrama and gain experience working as rangers, guides and scientific-research assistants there. Members involved in the Eco-lodge take numerous hospitality and tourism training programs offered by Iwokrama, the Guyana Tourism Authority (GTA), the Guyana Tourism and Hospitality Association and other organizations.

To support the development of sustainable tourism in Guyana, Surama often helps with training of trainees from Iwokrama programs, Amerindian villages and non-Amerindian tourism enterprises.

Natural and Cultural Heritage Protection

Surama supports the protection of Guyana's notable biodiversity in many ways. As mentioned above, almost all the young people of Surama are active in the conservation effort and belong to the Junior Wildlife Club. In addition, many community members have participated in a range of training and international scientific research programs that work to save biological diversity.

Wildlife trapping has been banned and enforced since 1994. Harvesting of endangered tree species such as the bullet wood, letter wood, greenheart and waramaden are banned, while reforestation and monitoring programs have been introduced. Through the Makushi Research Unit (MRU), a biodiversity project was conducted involving the comprehensive study of traditional uses of local plants and animals and the ways in which these plants and animals interact.

When it comes to cultural heritage, Surama's philosophy is that eco-tourism exists to support the conservation of the Makushi culture. To begin with, Eco-lodge staff members work on a rotating basis so that they can continue to live the traditional community lifestyle.

Authentic traditional skills and knowledge are showcased in all of the wildlife, nature and cultural tourism experiences offered for visitors, as outlined earlier. A women's cassava-making project supplies both an interesting tour and educates guests about agricultural traditions. A Women's Activity Center was created as a place where women come together for traditional craft making, sewing and embroidery, creating items to sell to visitors. A culture group was formed to help motivate and preserve the culture of the Makushi people especially among the young generation. The group comprises more than twenty people who perform dances, songs, poems and skits that delight tourists and enlighten them about the Makushi way of life. A Makushi Language Program is taught in school, which grew out of the Makushi Ethnobotany project.

Figure 14-4. Surama Cultural Group. Photo credit: John Canning.

Financial Sustainability

Apart from direct employment from eco-tourism, the community benefits from economic activity derived from products and services that support Surama's tourism micro-industry. Profits from the lodge have been used for refurnishing the lodge (using furniture built by Surama youth woodworking initiative), establishing base camps, buying vehicles to transport guests and supplies, installing a satellite internet service and installing solar power.

Visitor fees are combined with employee contributions to subsidize a local investment fund that contributes to community health and education infrastructure and programs, school supplies, maintenance of community facilities, solar power, construction of a windmill, university scholarships for Surama teens and the purchase of a truck for importing goods and exporting cash crops and other goods to the market.

The fund subsidizes a Community Credit Scheme, which provides micro-loans to local entrepreneurs. This fund addresses poverty reduction and is central to the community's general philosophy of self-sufficiency.

Eco-tourism income in Surama reduces risks associated with traditional subsistence agriculture. Further, the Eco-lodge's success in creating employment opportunities has significantly reduced incentives for men to leave the village for long periods to mine or cut timber elsewhere.

Leadership and Partnerships

In order to support environmental, cultural and social sustainability, Surama has taken on many leadership roles and undertaken a range of partnerships. For example, Surama community leaders have played a leading role in the establishment of the Bina Hill Institute, established in 2001 to develop training, research and other resources in North Rupununi. Its programs focus on natural-resource management, traditional knowledge systems and building capacity for both occupational and economic development. BHI also is the coordinating center for the marketing and sales of the local tourism role model for Amerindian communities in Guyana, inspiring and mentoring at least six other communities to start sustainable tourism enterprises.

Surama has played a leading role in the establishment of the Makushi Research Unit, a Rupununi region women's group that has collected a wealth of information on Amerindian lifestyles and ethno-botanical knowledge and published a number of books and educational resources based on local knowledge.

Surama is part of a regional consortium (the Community and Tourism Services, or CATS) that operates the Iwokrama Canopy Walkway and Atta Rainforest Lodge. A partnership between Surama and Eerepami Regenwaldstiftung, a German organization, supports remote-area development, biodiversity conservation and the exchange of science, culture and education between Germany and Guyana. Surama also participates in international cultural exchanges, including two exchanges with indigenous groups from Canada.

Conclusion

Surama Village Eco-lodge is demonstrating that investing in conservation, indigenous tourism, research, education and local development provides a variety of positive payoffs. The case study illustrates the importance of strategic alliances and collaborative agreements between conservation organizations, indigenous communities and the tourism sector. The benefits include community development, biodiversity protection and enhancement, enriched and higher value tourism experiences, sustainable livelihoods and long-term sustainable development.

CHAPTER FIFTEEN

TOURISM DEPARTMENT, GOVERNMENT OF ANDHRA PRADESH: SUSTAINABLE ECO-TOURISM – A CASE STUDY FROM INDIA

JAYESH RANJAN[1]

About Andhra Pradesh

Andhra Pradesh is a southern state of India that has been bountifully endowed by nature – in the form of hills and valleys of the Eastern Ghats. The state has a population of about 80 million, of which 7% population is of tribes who live in forest tracts. The state has a long coastline of close to 1,000 kilometers. It has immense natural resources in the form of five major rivers and dense tracts of forests. The economy is primarily agrarian, with more than 65% of people living in villages. Nevertheless, the cities are also vibrant, and the capital city of Hyderabad is an international hub for IT, biotech, manufacturing and knowledge-based activities. The state has rich cultural and religious traditions, and also a strong Buddhist heritage. Because of the presence of some of the holiest shrines, the state receives the largest number of domestic tourists in the region.

[1] Secretary, Tourism Department, Government of Andhra Pradesh.

Eco-tourism Initiatives in Andhra Pradesh

Andhra Pradesh Tourism Development Corporation took the initiative of identifying potential eco-tourism spots in the state. It has successfully implemented a number of popular eco-tourism projects by involving the local community, at Belum caves, revamping of Borra caves, and an exotic jungle destination (Jungle Bells at Tyda in Visakhapatnam District). The above initiatives have yielded excellent results, as the above projects support rural livelihoods and provide opportunity to the tourist for interacting with the local communities in sharing the local knowledge about primitive culture and traditions and bio-diversity in the area. The projects also have thrown open opportunities to the local tribal youth, ethnic minorities and women who participate in operating the eco-tourism projects on a community basis and sharing the benefits.

Community-based Eco-tourism

Basically all eco-tourism projects are rural-based and provide opportunity to the tourist to know about the environmental assets of the rural communities, local traditions, art forms and ethnic cuisine. Conscious efforts have been made to preserve the natural beauty and environment while creating infrastructure and facilities for tourists that fit in with the natural surroundings.

Andhra Pradesh is encouraging community-based eco-tourism in convergence with various stakeholders from the government departments and the community. The Forest Department under their Community Forest Management Program supports the initiative by providing opportunities for jungle walks, forest retreats, wildlife tourism, bird watching and trekking.

New Eco-tourism Initiatives

Andhra Pradesh Tourism Development Corporation (APTDC), in coordination with the AP Forest Department, has identified eight new eco-tourism destinations, namely at Maredumilli (East Godavari District), Nelapattu (Nellore District), Mamandur, Talakona, Nanniyal (Chittoor District), Balapalli (Kadapa District), Ettipotala (Guntur District) and Kambala Konda (Visakhapatnam District). The operation model followed in these projects is as follows:

- Eco-tourism projects generally would involve nature treks, wildlife tourism and jungle stays.
- Local Forest Protection Committees will provide accommodation, guide services, security and food.
- The AP Forest Department will provide infrastructure under Community Forest Management (CFM) and also help in management of eco-camps through local forest officials.
- Transport to and from the destination will be provided by APTDC/tourists.

All eight new initiatives have been made operational within a period of the last five years, and among them, the Maredumilli Eco-tourism Project has shown the most encouraging results.

About the Community-based Eco-tourism Center at Maredumilli

The Maredumilli Forests of East Godavari District, which form one of the less disturbed forest tracts of Eastern Ghats, are endowed with a rich biodiversity of flora and fauna and are home to important wildlife like the tiger, bison, panther, varieties of deer, birds and butterflies. This forest forms catchments for River "Godavari" and River "Shabari." It is also home to a number of primitive tribes like Konda Reddy, Konda Koyas, Valmikies, Gonds, Nayaka Podus, etc.

The Maredumilli Community Conservation and Eco-tourism Area is situated on Maredumilli-Bhadrachalam Road, nearly four kilometers away from Maredumilli Village, in the midst of undisturbed Eastern Ghats, which are home to rich flora and fauna. The area has many streams flowing over the undulating rocks in the deep woods and any visitor can feel the thrilling experience by living in nature and visiting the primitive tribal habitations of Konda Reddys and knowing their way of life.

Management Structure

The eco-tourism project is managed by the local, indigenous tribal community (Konda Reddy) of Valamuru Somireddypalem and Addaraveedhi Forest Protection Committees, with the active support of the Forest Department. The project was successfully completed in a short span of time due to the commitment of the highly motivated community coupled with the timely financial assistance from the Andhra Pradesh Community Forest Management Project. This also provided a unique opportunity for the native ethnic community to spread the message of conservation to the masses and provide opportunity to enjoy the nature in its pristine glory.

Community-based Eco-tourism Center infrastructure was created by the Forest Department under the Community Forest Management Project. The operations at this center are totally managed by local community members comprising of tribal community of Konda Reddys. Andhra Pradesh Tourism Development Corporation provides marketing support and training of the tribes in hospitality-related activities like food and beverage service, housekeeping, guide services, etc.

Revenue Sharing

The Forest Department under the Community Forest Management Program has arrived at a revenue-sharing mechanism with the local tribal community. Guide service, food and beverage, housekeeping and security are undertaken by the community members. The above activities are packaged in such a way that the total revenue derived is apportioned between the maintenance of infrastructure and services rendered by the community. The amount for the services rendered is shared equally between the thirty tribal families of Valamuru Somireddypalem and Addaraveedhi Villages.

The members from the thirty families divide the various activities by turns and manage the center, thereby providing a wholesome experience of living with nature to the tourist. The highlight of this rural tourism center, located in the heart of Eastern Ghat forests, is the incident-free operations within the last four and a half years in the extremist-infested forest tracts of Andhra Pradesh. The convergence of the activities by different departments and total involvement of community has paved the way for successful operation of the rural eco-tourism center.

Enhancement of Sustainable Livelihood

The focus of this tourism project is on sustainable livelihoods and addressing issues of poverty eradication through community mobilization around income and employment-generation activities related to tourism. It combines the cultural, craft and ecological dimensions of rural life and holds tremendous potential for income generation and enhancement of the basic quality of life in the rural areas.

This innovative concept of tourism is targeted at two distinct segments: firstly, to create interest among the host communities who are an integral part of the society so that they learn to take pride in their own identity and in their environment, culture and ecology; and secondly, to attract visitors who are interested in and fascinated by the nature, history, culture, artistic skills and lives of tribal societies, and who wish to enjoy and appreciate nature in all its forms without abusing, manipulating or destroying it, thereby ensuring its preservation for future generations.

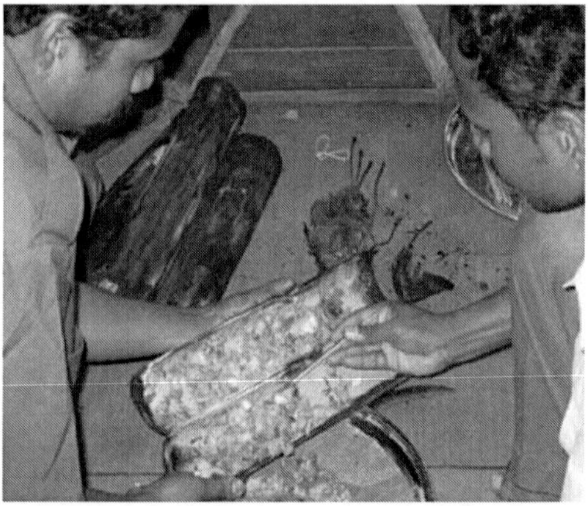

The main objectives of this innovative tourism project are: to distribute the benefits of tourism; reduce poverty; provide for sustainable development; encourage gender equity; and minimize adverse ecological impacts. It also aims at development of the local economy, revival and regeneration of tribal arts and crafts, preservation of culture and bringing awareness of how tourism benefits the local communities.

Local communities are involved in infrastructure development by using local skills and vernacular architecture; governance and management; setting up and strengthening traditional economic activities and tourism-related enterprises, including host stays, hospitality services, and guiding; and familiarizing tourists with indigenous flora, fauna and medicinal herbal plants. The transformative tourism project also aims at a complete and authentic tourist experience for the visitor managed by the local community from local cuisine to traditional activities (such as archery). Tourists return with a better understanding of the tribal society living in forest areas as well as the culture and bio-diversity of Eastern Ghats.

Benefits

The local tribal communities derive many economic and social benefits from this innovative project both directly and indirectly. These benefits can be listed as:

Economic Benefits

- Revenue to the 30 local families involved in the tourism project that is additional to their routine earnings.
- Direct employment for 30 families, including women, in one way or the other through the project.
- Sale of tribal handicrafts made of bamboo.
- Income from sales of forest products such as Honey, Gooseberry, Jackfruits and Toddy.
- Tips from visitors while performing tribal folk dances.

Forest Department Benefits

- Effective protection of forest and wildlife with involvement of tribal communities through Forest Protection Committees.
- Weaning away the tribal community from practicing shifting cultivation that is unsustainable and harmful to the environment.

Social Benefits

The local communities also get social benefits out of interaction with the visitors. As tourism activity increases in the area, the local people exhibit their indigenous traditional skills and disseminate these traditional knowledge systems to the visitors.

The local tribes also get trained in interpreting the flora, especially the medicinal plants, as well as the habits and habitat of the wild fauna, traditional therapies and healing systems.

Social Change

The most significant development in the area is a total return to peace. This area was till recently a stronghold of left-wing guerilla extremists operating in Eastern Ghats forest tracts. People used to be scared even to drive through the forest, let alone stay there. Now, with the tribal community taking full responsibility of running the ecotourism center, there is an unbelievable change in the safety and security perspective of the tourists. The tourists have the opportunity to stay in the jungle that was once considered not safe. Jungle Star Eco-camp is now a sought-after destination for birdwatchers, trekkers, butterfly lovers, wildlife enthusiasts and people interested in jungle stays.

Marketing Initiatives

The Forest Department directly markets this rural tourism initiative through their state and district hierarchy. APTDC has developed extensive marketing initiatives through its Information and Reservation Centers. The destination is also marketed through the APTDC website and several other supporting materials, such as brochures, handouts, etc.

Jungle Star Eco-camp

The highlight of the eco-tourism project is the Jungle Star campsite, which is located adjoining the Pamuleru River with the stream flowing on three sides overlooking the Vali-Sugriva hill that is believed to be the battleground during the epic Ramayana period.

The camp is managed by the local Konda Reddy tribal community, which has been trained in housekeeping, food and beverage production and servicing, guiding and security. The tribal community takes up the whole responsibility of the safety and security of the tourists of the camp. The community also organizes tribal folk dances in the evening and traditional archery for the tourists.

The tourists will get the experience of staying in the deep jungles with basic facilities without disturbing nature. The camp will be managed in line with the dictum: "take only memories and leave only footprints."

Other Places around the Center

1. Medicinal Plants Conservation Area: One gets the opportunity to explore nature and learn about the bio-diversity of Eastern Ghats with special emphasis on the medicinal plant wealth of this area. Local tribes have been trained by the Forest Department in the identification of the medicinal plants and their use. Guided treks are organized for the tourists in an area of 260 hectares of undisturbed semi-evergreen forest. About 203 plant species, including medicinal and rare plants, have been identified in this area. Tourists can have an insight into coffee cultivation through a visit to the luxuriant coffee estates managed by the Forest Department.

2. Jalatarangini: The area is rich in springs and rivulets with cascading waterfalls giving an enchanting experience to tourists. One such waterfall, called "Jalatarangini," which is very close to the campsite, is often the favorite spot of tourists.

3. Madanikunj: The Forest Department has also developed a wildlife-viewing place called Madanikunj with tropical plants and bamboo grooves abutting the Pamuleru River. The place is frequented by wild animals such as tigers, panthers, bison, peacocks, red jungle fowl and other birdlife, in addition to varieties of butterflies.

CHAPTER SIXTEEN

CHUMBE ISLAND CORAL PARK AN ECO-TOURISM DESTINATION RESPONDING TO CHALLENGES OF CLIMATE CHANGE

LINA MTWANA NORDLUND,[1]
ELEANOR CARTER[2]
AND SIBYLLE RIEDMILLER[3]

Abstract

Tourism can play a decisive role in addressing climate change, as it is one of the world's largest economic sectors. Chumbe Island Coral Park in Zanzibar, Tanzania, is a leading example of a small eco-tourism enterprise with not-for-profit objectives that is managing and conserving the unique marine and terrestrial ecosystems of Chumbe Island, and both directly and indirectly responding to the challenges of climate change.

This paper describes the establishment of a full "no-take" marine sanctuary and a forest reserve on Chumbe that effectively reduces the anthropogenic threats to the area and promotes reef resistance, forest resistance and resilience in the face of climate change pressures. The mission of the company – to conserve the area for conservation and education purposes – has been considered at every step of the development process, from project conceptualization to lodge design, to staff recruitment and training, to tailored education programs and the promotion of wider entrepreneurial endeavors.

This paper highlights the options for investment and management decisions, designs and technologies that exist for tourism developers to

[1] Chumbe Island Coral Park, Tanzania. ARONIA Coastal Zone Research Team, Åbo Academy University & Novia University of Applied Sciences, Finland.
[2] Former Project Manager and on-going Advisor to CHICOP; Marine Program Director, RARE Conservation, Indonesia/Timor Leste.
[3] Chumbe Island Coral Park, Tanzania.

ensure that their investments help minimize (or at least avoid exacerbating) threats to the coastal environment and, consequently, threats to the communities dependent upon these coastal resources. Indeed, for the sake of sustainability and thus long-term profitability, tourism investments in coastal areas can no longer afford to ignore the particular challenges of climate change.

Introduction

As one of the world's largest economic sectors and an important part of modern society, tourism can play a great role in addressing climate change (Cabrini et al., 2008). Chumbe Island Coral Park in Zanzibar, Tanzania is a leading example of a small eco-tourism business with not-for-profit objectives that manages and conserves the unique marine and terrestrial ecosystems of Chumbe Island, and with this, both directly and indirectly responds to the challenges of climate change.

Chumbe Island Coral Park, Ltd. (CHICOP) was established in 1992 and opened for eco-tourism in 1998. From 1991–1994, CHICOP successfully negotiated with the government of Zanzibar that the western coral reef and forest on the island be gazetted as a park that would be managed and funded by the company.

The company purpose is "to manage, for conservation and educational purposes, the Chumbe Island Reef Sanctuary and the Forest Reserve. This is supported by eco-tourism activities which are directly related to the non-consumptive use of the natural resources."

The company objectives are non-commercial, while operations follow commercial principles. Since 1995 CHICOP has been recognized as a privately managed marine protected area (MPA) in the World Database of Protected Areas (WDPA) and was, in 2000, also distinguished as a Member of the UNEP Global500 Forum.

The park includes the Chumbe Reef Sanctuary and the Chumbe Forest Reserve, a Visitors' Center and an Eco-Lodge on Chumbe Island. Management is based on consecutive Management Plans 1995–2005 and 2006–2016.

When the project started in the early 1990s, Zanzibar had no marine protected areas and no policy, legislation or institutions to manage them. Chumbe not only became the first MPA in Zanzibar and Tanzania, but also developed the first full "no-take zone" (NTZ) for marine conservation. This marine zone is permanently closed to all fishing activities and provides a vital refuge for biodiversity conservation, and a key site for fish and marine species to breed.

Permitted uses of the park include recreation (swimming, snorkeling and underwater photography), education and research. Extractive and destructive activities, such as fishing, anchorage, and the collection of specimens (even for research) are not allowed. Research is coordinated with the Institute of Marine Sciences, University of Dar es Salaam. A detailed monitoring program closely records all activities in the park and any infringements of park regulations have been recorded and controlled since 1992. Rangers are permanently stationed on the island to patrol the no-take zone.

Figure 16-1. The eco-bungalows on Chumbe Island. Photo credit: Oskar Henriksson.

Tourism operations on the island started in 1998, and overnight capacity does not exceed approximately 5,000 visitors/year. Seven eco-bungalows are available for guests, and no further construction of overnight facilities is planned (Figure 16-1). Day visitation to the park is also closely regulated to ensure visitor numbers remain within the carrying capacity of the island. The Visitors' Center hosts a classroom for local schoolchildren, exhibits about the islands' marine and terrestrial ecology and a restaurant.

The Park rangers offer daily guided snorkeling tours in the Reef Sanctuary, and walks along inter-tidal, mangrove pools and forest trails.

Chumbe Island combines sustainable tourism with sustainable conservation area management. While most protected areas around the world are dependent on, and often short of financial support from governments or donor agencies, the revenue generated from small-scale but high-value eco-tourism on Chumbe Island provides for the park management and the conservation and education. Since 2000, CHICOP has been able to fully fund these efforts through the revenue from eco-tourism, and along the way has won many prestigious awards. About a third of operational costs are spent on conservation management staff and programs. Although a not-for profit venture, Chumbe is taxed like any other tourism enterprise, with no exemption or tax incentive for the conservation and education work done (Riedmiller, 2012).

Chumbe Island provides an exemplar model for tourism developers, investors and managers to ensure enterprises minimize climate change impacts and are sustainable in the face of climate change pressures.

Responding to the Challenges of Climate Change: Protecting the Coral Reef

Threats to the Coral Reef and Implications for Eco-tourism

Climate change is recognized as one of the greatest threats to coral reefs. The most serious and immediate phenomenon is coral bleaching (Obura & Grimsdith, 2009), which occurs when important symbiotic relationships in the coral community break down due to warmer water temperatures. Once stressed in this way coral communities turn white and can die within weeks.

Ensuring the integrity and survival of the coral reef environment is critical to the conservation of the Chumbe MPA, and is essential to maintain the very environment for which tourists are traveling, and paying money, to see. The sustainability of the program depends on the sustainability of the coral reef.

Reducing Anthropogenic Threats through Creating and Actively Managing a "No-take" Marine-protected Area

There are two key properties that determine the ability of coral communities to persist and survive in the face of rising temperatures: their sensitivity and recovery potential (Obura & Grimsdith, 2009). A reef's sensitivity to stress can depend on many factors, i.e., the coral communities' dynamics, physical/chemical factors acting on the reef habitat and local conditions (such as crown of thorns starfish outbreaks, that can weaken the reef) (Obura & Grimsdith, 2009). A reef's resistance to such stress is its ability to remain relatively unchanged in the face of such a major disturbance, such as bleaching (Grimsditch et al., 2006)

Sensitivity to stresses, such as rising temperatures, can be greatly exacerbated when the reef is concurrently exposed to other stresses. Anthropogenic activities on coral reefs are a major cause of stress (Keller et al., 2009) and destructive practices (such as seine netting, dynamite fishing, etc.) considerably weaken a reef's resistance to climate change stresses (such as bleaching). Over-extraction of fish and other species (as occurs on most reefs worldwide) also exacerbates stress, as the removal of top predator fish (generally the most important species commercially) alters the reef-community dynamics, changing ecological relationships, and potentially leads to algal invasion on reefs and Crown-of-Thorns Starfish outbreaks. This in turn degrades the reef's ability to resist climate change pressures.

Therefore, a key management response to ensuring that a reef area maintains its integrity is the reduction of additional stress factors. Likewise a reef's recovery potential, or resilience to bounce back and recover after any stressful event (such as bleaching), is promoted greatly when anthropogenic threats are removed (Salm & West, 2003).

The establishment of a full no-take marine sanctuary at Chumbe effectively reduces the anthropogenic threats to the area and promotes both reef resistance and resilience in the face of climate change pressures. Maintaining strong ecological integrity in the area, and carrying out consistent monitoring of the reef sanctuary has been essential to managing climate change pressures (Figure 16-2). For example, this proved pivotal in the 1998 bleaching event that occurred globally, in which the Chumbe sanctuary suffered losses of only approximately 20% acroporus species, and showed considerable recovery within two weeks of the event, compared to other reefs around Zanzibar that suffered far greater proportional losses of reef species, with limited recovery.

By maintaining the reef's resilience in this way, key benefits are ensured, not only for the sustainability of the enterprise but also for local fishers, including food security and biodiversity conservation.

Chumbe Helps Restock Locally Depleted Fisheries and Promote Recovery of Degraded Coral Reef Ecosystems

Chumbe Island is located upstream of the most important fishing grounds opposite Zanzibar town. The sanctuary provides a protected breeding ground for fish, corals and other species, which "spill over" to re-colonize nearby over-fished and degraded areas. As the predominant ocean current in the Zanzibar channel is northerly, larvae of corals and other marine organisms as well as juvenile and adult fish tend to migrate to the northern heavily fished reefs.

Studies into fish-size-frequency-distribution patterns show that fish are achieving significantly greater size inside the reserve (Tyler, 2005). The presence of these larger, older fish promotes greater exponential breeding and increases in biomass, leading to the spillover effect (Hughes, 2007). Qualitative studies have shown that 83% of fishers report perceived greater catches proximal to the MPA due to this spillover effect (Tyler, 2005).

Figure 16-2. The very healthy and resilient reef at Chumbe Island. Photo credit: Oskar Henriksson.

Chumbe contributes to biodiversity conservation and ecological restoration, by effectively protecting a coral reef, which has at least 90% of the scleractinian coral species recorded in East Africa (Veron, personal communication). The Chumbe Reef Sanctuary has become a refuge for the critically endangered Hawksbill Turtles and a proactive Crown-of-Thorns Starfish (COTS) removal program is also undertaken whenever invasions occur, further reducing stress to the reef environment.

The sanctuary is a hot spot for seagrasses, as it hosts seven out of the thirteen species known from the region (Bandeira and Björk 2001; Knudby & Nordlund, 2010). In Zanzibar, invertebrate harvesters have observed or experienced a decline in animal abundance and seagrass cover over the last two decades, and this is a pressing problem in the Western Indian

Ocean (WIO), where many of the livelihoods of local people are immediately dependent on invertebrates collected from seagrass beds (Nordlund et al., 2010). The Chumbe no-take MPA is also a safe haven for less-known flora and fauna typical of inter-tidal reef flats that are little researched.

Promoting a Robust Forest Habitat in the Face of Climate Change

Threats to the Forest and Potential Impacts on Eco-tourism

Climate change also threatens terrestrial environments in a number of ways. Some areas may experience more frequent and more extreme storms, lower precipitation levels and increased evaporation rates. Such threats can impact forest habitats, create changes in soil (e.g., moisture levels, erosion and acidity) and cause loss of habitat cover through landslides and patch exposure from fallen trees. Such conditions also pose a threat to the health and safety of tourism operations in forest habitats (Cabrini et al., 2008).

Reducing Anthropogenic Threats by Creating a Forest Reserve

Like the marine environment, the forest environment is at its most robust when other threats are removed. Land clearing, for example, opens up forest areas to the impact of severe weather, causing loss of topsoil and soil integrity that is normally maintained by the complex root network of robust forests. Pollution, the use of agrochemicals and habitat fragmentation also lead to reduction of biodiversity and habitat strength in the face of climate pressures.

On Chumbe Island the coral rag forest habitat (i.e., tropical semi-arid forest on fossil coral) is fully protected as a gazetted forest reserve. The forest is dense with solid canopy cover and home to the world's largest known population of the red-listed Coconut Crab (*Birgus latro*; Hartnoll, personal communication). The forest is monitored, and its biodiversity maintained, whilst other coral rag habitats in Zanzibar suffer from felling and reduction in habitat cover and diversity.

Invasive species are another threat encouraged by anthropogenic factors, including climate change. Like many formerly inhabited islands, rats (*Rattus rattus*) were a serious pest on Chumbe Island at the start of operations – threatening bottom-breeding bird species, reptiles and other species, and leading to habitat degradation and biodiversity loss. CHICOP successfully eradicated rats in 1997 and since then closely monitors to prevent re-infestation. Through this active intervention, the pristine Chumbe forest biodiversity was restored and now remains in prime condition.

Such undertakings have encouraged further biodiversity conservation initiatives, including, in 1998, the translocation of the critically endangered mini-antelope Aders' Duiker (*Cephalophus adersi*), from its diminishing range on mainland Zanzibar to the secure sanctuary on Chumbe Island. Here it has become established and started breeding. The undisturbed Chumbe Island also supports the successful breeding of rare migrant birds, e.g., the red-listed Roseate Tern (*Sterna dougalli*).

Eco-architecture: Building for Sustainability

Threats From, and To, Coastal Development, and the Implications for Eco-tourism

The development of coastal tourism infrastructure can both increase the overall vulnerability of a coastline (and other coastal infrastructure) to climate change impacts and be vulnerable to conditions caused by climate change.

In tropical areas, threats from coastal tourist development include pollution caused by operations (such as sewage run-off, solid waste going into the sea, etc.) and these can all have negative effects on a coral reef environment. Consequently, the threats to coastal tourism development from climate change are exacerbated. Increased frequency and intensity of storms and wave surges created by climate change are already an anticipated threat to coastal infrastructure. Where a coral reef is damaged, the wave surge breaks closer to shore and can cause dramatic erosion of the beaches and coastal land areas.

Other threats to coastal tourism include the potential damage to infrastructure caused by increased intensity of storms, flooding and damage to historical, architectural and cultural assets, increased rainfall rates in some areas and, conversely, increased evaporation rates and reduced water availability in other areas. This latter issue can lead to competition over fresh water sources, both between tourism developments in any one area, and between tourist developments and local population needs.

Add to this the potential threat of sea-level rise (that is anticipated through climate change) and it becomes clear that the challenges of climate change need to be considered at all stages of a coastal tourism development process (from design, planning and management), including the importance of maintaining the integrity of the surrounding marine and coastal environment for the sustainability of any tourism investment.

Chumbe Island's Eco-architecture and Infrastructure Adaptations

To effectively ensure that the tourism development on Chumbe does not cause any threats to the surrounding marine and coastal environment, Chumbe has, from its outset, been committed to ecologically sustainable architecture and operations that have close to zero impact on the sensitive terrestrial and marine ecology of the island.

In addition to this, all buildings are situated at least four meters (vertically) above the high-tide mark (from the highest spring tides), and are set back from the beach to avoid potential damage from storm surges and coastal erosion. The buildings are also positioned to channel winds in line with the predominant seasonal wind directions, reducing stress on the bungalow architecture and providing natural air-conditioning (reducing any need for energy-consuming air conditioning).

Reducing Threats to the Surrounding Environment

Vegetative Gray-water Filtration. To avoid introducing pollutants from wastewater (or gray water) from showers, washbasins and the kitchen, all wastewater is passed through a particulate filter (and series of grease traps for the kitchen water) before entering a specialized vegetation bed that absorbs large amounts of phosphates and nitrates before clean water is released.

Composting Toilets. Sewage is totally avoided by using composting toilets that also avoid any fresh water consumption. The anaerobic composting process of these toilets turns human waste into nutrient-rich dry matter, which is reduced to a sixth of its original volume and can in turn be re-used as fertilizer and compost mulch.

Management of Other Waste. Organic kitchen waste is also recycled to provide the compost for the toilet systems. Fresh food products purchased on open markets avoid the acquisition of non-biodegradable products, in particular food packaging. Any other waste that cannot be recycled or composted is removed from the island.

Solar Power for Light, Communication and Water Heating. As climate change is caused by the increased CO_2 in the atmosphere through the burning of fossil fuels, photovoltaic panels generate all electrical energy on Chumbe Island, while solar water-heaters provide warm water for showers. These ecologically sensitive technologies effectively save energy and reduce all pollutant emissions from the eco-lodge.

Use of Local Building Technologies and Materials. All buildings on Chumbe Island reflect local architectural low-energy and low-input traditions, cultures and technologies, reducing the site's carbon footprint and, by giving preference to local products and technologies, also creating a thriving market for local producers.

Rainwater Catchments. With no groundwater available on the island, all buildings have been designed to make maximum use of rainwater that is collected by palm-thatched roofs and filtered through combined gravel and sand filters. The water is then stored in the large cisterns under the floor of each building to provide for water needs year round.

Promoting Social Resilience to Climate Change

Social resilience, much like biophysical resilience, is the ability of a system to recover after a stressful event. The system, in this instance, is society itself – the communities living in and around coastal areas who are vulnerable to the impacts of climate change. Threats to these societies from climate change include, for example, injury and loss of lives, properties and livelihoods from flooding, and from increased frequency and intensity of storms.

However, low resilience to these climate change factors can be socio-economically driven, such as absence of support systems in the face of disasters, low awareness of potential threats or necessary preparatory actions and no plans to mitigate impacts, low savings to help families through difficult times, and unreliable or lacking employment opportunities to enable self-dependence in planning and preparing for any potential event.

This can also be considered a population's "adaptive capacity" to cope with climate change effects (Berkes & Jolly, 2001) and in areas already afflicted by high rates of poverty, these added challenges can cause social unrest and conflict (Reck, 2008). When coupled with a degraded environment from other, non-climate change stressors, these threats are further exacerbated, such as reduced in-situ food security from reduced fisheries stock and a lack of awareness on methods to mitigate coastal erosion and manage sustainable fisheries continues to compound the challenges that will be faced.

Responses to such events can depend on many factors, including the social capital (or ability for collective action) of communities, and studies have shown that coastal communities with such social capital are more resilient to disasters (Adger, 2003).

Chumbe Island's mission to conserve the Reef Sanctuary and Forest Reserve for conservation and education purposes supports social resilience to climate change on a number of levels.

Environmental Education and Awareness Raising

Education programs on Chumbe Island are targeted at a range of sectors in the community – from fisher groups to schoolchildren to government personnel.

Community-education Programs. Since the start of operations on Chumbe Island, thousands of community members have participated in a range of activities to promote public awareness about the need for sustainable management of coastal resources and issues related to climate change. Through a community outreach program CHICOP has also trained 25 representative "peer educators" from ten villages around southern Zanzibar, who educate their fellow villagers on environmental issues, including measures to mitigate and be resilient to climate change.

Fisher-education Programs. Chumbe helps create awareness among fishermen in a range of ways. The Chumbe rangers manage the reef sanctuary through educative enforcement methods, sharing with fishermen information about the nature of coral reefs, the purpose of the MPA as a breeding ground for fish and the importance of marine conservation for sustainable coastal management. The park rangers, former fishermen themselves, have also trained fisher representatives, and this has resulted in fishers voluntarily complying with the regulations for the no-take reef sanctuary, to protect their fish stocks, and a far greater awareness of the impact of non-sustainable activities in reducing coastal resilience.

School-education Programs. School education in Zanzibar, as elsewhere in the region, is based on rote learning of an extremely academic syllabus having little relationship with the immediate surroundings. Though Zanzibar is a coral island, coral reef ecology and coastal management is insufficiently covered in school syllabi. Extra-curricular activities such as field excursions are rarely organized and very few children have a chance to visit reefs and tropical semi-arid (coral-rag) forests.

Chumbe's school program provides an excellent opportunity for hands-on learning for both students and teachers from Zanzibar and other parts of the world (Figure 16-3). CHICOP implements Education for Sustainable Development (ESD) through an Environmental Education (EE) program. Since 1995, this program has welcomed more than 5,000 students and 800 teachers for one-day school excursions to Chumbe Island.

Figure 16-3. Khamis Khalfan is introducing snorkeling to Zanzibari school children. This is a unique learning opportunity for girls since Zanzibar is a predominantly Islamic country where the dress code for women is very conservative, and as a result, women do not learn how to swim. The Environmental Education Program encourages schoolgirls to learn how to swim and snorkel across in coral reefs in appropriately conservative clothing. Photo credit: Lina Mtwana Nordlund.

Nature trails and educational materials (in Kiswahili and English) have been developed about the forest and the reef, and the excursions offer environmental hands-on education for schoolchildren, at the same time as giving teachers ideas for how to conduct field-based activities related to marine biology, forest ecology, biodiversity and climate change. Chumbe has also helped establish environment clubs in schools and communities across Zanzibar, who are addressing issues such as waste management, biodiversity loss and climate change mitigation projects such as tree and mangrove planting activities.

Institutional and Governmental Capacity Building

Chumbe contributes to capacity building of government staff. From the beginning in 1991, more than seven government departments were involved with important issues concerning the establishment and management of the reserve. An advisory committee with representatives from government departments, research institutions and local community leaders has been in place from 1995, and through this, the group receives considerable capacity building and awareness raising on the various programmatic issues.

Cross-institutional training has also been given on a range of projects, such as rat eradication programs (to maintain healthy island forest habitats), to training of rangers in other emerging MPAs and coastal conservation programs across the region.

In addition to this, all staff members on Chumbe, sourced predominantly from communities proximate to the MPA, receive training in all aspects of the work – no matter their position on the project. This includes training on basic marine and coastal ecology, eco-technologies, such as rainwater harvesting, and sustainable waste management practices, such as composting, recycling and handling solid and hazardous waste appropriately.

Promoting Employment and Entrepreneurial Opportunities

The promotion of social resilience and social capital is also supported through the employment of forty-two local people on Chumbe, each with an average of twelve dependents. As a fully managed nature reserve, and also due to the particular eco-technologies chosen, CHICOP is very labor-intensive. A third of the staff of forty-three employees are directly involved in conservation management and education.

With only seven rooms, CHICOP has probably the highest employee/room ratio of any tourism business in Tanzania, and, according to a study of the International Finance Corporation, three times the international average for eco-lodges (IFC, 2004). In addition, numerous entrepreneurial opportunities have been enabled as the park has received a steadily increasing flow of visitors. The considerable international publicity Chumbe has gained with the many prestigious international awards for environmentally and socially responsible tourism has helped to market Zanzibar as an attractive eco-tourism destination, and as a consequence created wider employment opportunities. These include makuti palm thatch producers, taxi drivers, boat handlers, and a women cooperative making biodegradable organic soaps and other products.

In addition to this, the project is yielding many insights for coastal management practices, and for solving the problem of financial sustainability in the management of protected areas in Zanzibar and elsewhere. The experiences are contributing to the development of a sustainable protected areas system in Zanzibar and the region.

Conclusion/Summary

This paper highlights the ways in which one eco-tourism destination, Chumbe Island Coral Park, in Zanzibar, Tanzania, has pro-actively responded to the challenges of climate change. It can be seen that considerations of climate change, and development of eco-tourism operations that promote long-term sustainability and achieve the mission of the project – to conserve the area for conservation and education purposes – have been considered at every step of the development process: from project conceptualization, to lodge design, to staff recruitment and training, to tailored education programs and the promotion of wider entrepreneurial endeavors.

This shows us the possibilities that exist for tourism developers to ensure their investments are conducted in such a way as to minimize (or at least avoid exacerbating) the threats to the coastal environment and, consequently, the threats to the communities dependent upon these coastal resources. It also shows us steps for preparedness that eco-tourism planners may consider – to ensure their investments are able to adapt to increasing pressures and threats from climate change events.

Finally, and of perhaps greatest importance, it highlights the inter-connectedness of the coastal environment – its people, its environment, its biodiversity – and the importance of understanding all of these elements in order to truly prepare for the challenge that climate change presents.

Acknowledgements

The authors would like to thank Louis D'Amore, IIPT Founder and President, for inviting us to write this paper. Further we would like to thank the referees for their insightful comments, and the Chumbe team.

References

Adger, N. W. (2003). Social capital, collective action, and adaptation to climate change. *Economic Geography* 79, no. 4 (Oct., 2003): 387–404 (Clark University).

Banderira, S. O., & M. Björk (2001). Seagrass research in the Eastern Africa region: Emphasis on diversity ecology and ecophysiology. *South African Journal of Botany* 67: 420–425. doi: 10.1579/.

Berkes, F., & D. Jolly (2001). Adapting to climate change: Social-ecological resilience in a Canadian western Arctic community. *Conservation Ecology* 5, no. 2: 18.

Keller, B. D., D. F. Gleason, E. McLeod, C. M. Woodley, S. Airame, B. D. Causey, A. M. Friedlander, … & R. S. Steneck (2009). *Climate Change, Coral Reef Ecosystems, and Management Options for Marine Protected Areas. Environmental Management.* Springer Publications. doi: 10.1007/s00267-009-9346-0.

Cabrini, L., G. Vereczi & S. Fotiou (2008). Climate Change and Tourism – Responding to Global Challenges. Published by the World Tourism Organization and the United Nations Environment Program. Printed by the World Tourism Organization, Madrid, Spain.

Grimsditch, Gabriel D., & Rodney V. Salm (2006). *Coral Reef Resilience and Resistance to Bleaching*. IUCN, Gland, Switzerland.

Hughes, T. P., D. Bellwood, C. Folke, J. Pandolfi & R. Steneck (2007). No-take areas, herbivory and coral reef resilience. *Trends in Ecological Evolution* 22, nos. 1–3. doi:10.1016/j.tree.2006.10.009.

International Finance Corporation. (2004). The International Ecotourism Society. Ecolodges: Exploring Opportunities for Sustainable Business. Washington/DC.

Knudby, A., & L. Nordlund (2011). Remote sensing of seagrasses in a patchy multi-species environment. *International Journal of Remote Sensing* 32, no. 8: 2227–2244. doi: 10.1080/01431161003692057.

Nordlund, L., J. Erlandsson, M. de la Torre-Castro & N. Jiddawi (2010). Changes in an East African social-ecological seagrass system: Invertebrate harvesting affecting species composition and local livelihood. *Aquatic Living Resource* 23:399–416. doi: 10.1051/alr/2011006.

Obura, D. O., & G. Grimsdith (2009). *Resilience Assessment of Coral Reefs. Assessment Protocol for Coral Reefs, Focusing on Coral Bleaching and Thermal Stress.* IUCN working group on Climate Change and Coral Reefs. IUCN, Gland, Switzerland.

Reck, R. A. (2008). Developing Social Resilience Amidst Climate Change and Global Insecurity: Finding Peaceful Pathways to the Future. The Forum on Public Policy.

Riedmiller, S. (2012). Can ecotourism support coral reef conservation? Experiences of Chumbe Island Coral Park Ltd in Zanzibar/Tanzania. In *Sustainable Hospitality as a Driver for Equitable Development*, ed. P. Sloan, C. Simons-Kaufmann, and W. Legrand. London: Taylor & Francis.

Salm, R. V., & J. M. West (2003). Enhancing coral reef survival in a changing climate: Additional marine protected area guidelines. In The Nature Conservancy and Partners (TNC) (2004) R2- Reef resilience: Building resilience into coral reef conservation; Additional tools for managers: Coral Bleaching. Volume 2.0. CD ROM Toolkit. The Nature Conservancy: Honolulu (Hawai'i).

Tyler, E. (2005). The effect of fully and partially protected marine reserves on coral reef fish populations in Zanzibar, Tanzania. PhD diss. University of Oxford.

CHAPTER SEVENTEEN

CONSERVING IN MISOOL, INDONESIA, THROUGH ECO-TOURISM

SHAWN HEINRICHS,[1] ANDREW MINERS[2] AND MARIT MINERS[3]

Misool Eco Resort (MER) is located in the remote southern part of Raja Ampat, Indonesia. The small resort is located on the island of Batbitim, deep in a vast archipelago of uninhabited islands, 240 kilometers from the nearest resort and half a day's journey from the nearest village. Misool Eco Resort is deeply committed to a policy of environmental and social responsibility. We seek to provide exceptional and enriching diving experiences in a sustainable environment. We aim to protect and revitalize both our natural surroundings and the community in which we operate. We are committed to demonstrating to our hosts, our guests and the local government that tourism can support a local economy on much more favorable terms than mining, logging, over fishing or shark finning.

[1] Founder, Blue Sphere Media (www.bluespheremedia.com).
[2] Managing Director, Misool Eco Resort.
[3] Director, Misool Eco Resort (www.misoolecoresort.com).

In doing so, MER entered into a lease agreement with the local community to establish a 425-square kilometer Marine Protected Area (no-take zone) surrounding Misool Eco Resort. Within this area, all fishing, shark finning, harvesting of turtle eggs and shellfish are strictly prohibited. In 2010, this no-take zone was expanded and now covers an area of 1,220 square kilometers. We also require all boats to practice reef-safe anchoring. We regularly patrol the area for illegal fishing and shark finning. In addition, Misool Eco Resort's charity arm, Misool Baseftin, conducts social projects such as supporting local schools, and is developing libraries in two of the local villages. Perhaps most importantly, MER is dedicated to safeguarding the local community in which we operate. Our labor force, drawn largely from the closest village, is offered favorable working conditions, health benefits, job training and English lessons.

Project Overview

Agreement Mechanism

A land-lease agreement, which includes a specified area of surrounding sea, was the formal agreement mechanism used to establish the Misool Eco Resort No-Take Zone (MER NTZ) in the southern Raja Ampat. The agreement was signed on the 28th of November 2005. Key provisions of the agreement include:

- Law: Indonesian law and Papuan law
- Duration: 25 years
- Payments: 5-year terms with payments due at the start of each term
- Lease Area: Approximately 425 sq. km. surrounding Batbitim, including two small Islands of about 1 sq. km. (Batbitim) and 1⁄4 sq. km. (Jef Galyu)
- Leaser: Heads of Bahale and Yelfom families
- Lessee: Andrew Miners and Misool Eco Resort
- Sign Date: November 28, 2005

The extension of the NTZ was concluded in October 23, 2010. This was through a second lease agreement for a NTZ surrounding the islands of the Daram archipelago to the east of the original NTZ.

- Law: Indonesian law and Papuan law
- Duration: 15 years
- Payments: 5-year terms with payments due at the start of each term
- Lease Area: Approximately 246 sq. km. surrounding the Daram island group
- Leaser: Heads of Macap, Soltief and Wainsaf families
- Lessee: Andrew Miners and Misool Eco Resort
- Sign Date: October 23, 2010

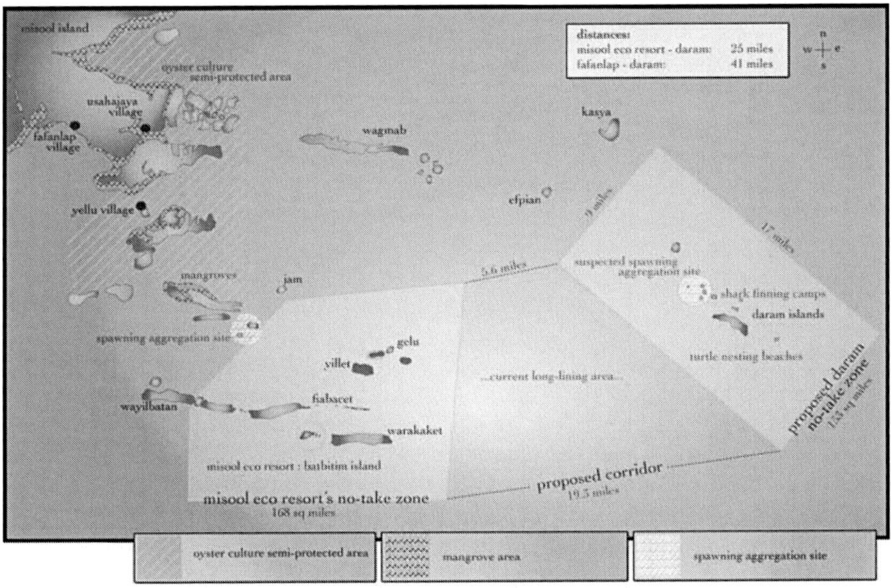

Governing Law

The contracts were founded on principles supported by Papuan law (or Hak Adat) and formalized under Indonesian law. The Indonesian Central Government recognizes Hak Adat as part of special autonomy status (Kursus Autonomy) whereby the state of Papua has a certain degree of autonomy from the rest of Indonesia. Under Papuan law, people can own not just islands but beaches, reefs and perhaps most importantly the fishing rights to the waters surrounding them.

Key Parties

- Andrew Miners, Founder and Managing Director, Misool Eco Resort
- Bapak Yasuf Salim, Camat (pronounced "chamut") District Head South East Misool
- Haji Jalil Bahale, Kapala Desa (Head of Yellu Village)
- Bapak Hadir Yelfom, Yellu Village Secretary
- Bapak Haji Yahya Bahale, Adat Leader Yellu Village (local traditions and culture leader)
- Bapak Rajak Soltief, Adat Leader Fafanlap Village
- Bapak Rasid Wainsaf, Head of the Wainsaf family, Fafanlap Village
- Bapak Mansuar Macap, Head of the Macap family, Usaha Jaya Village
- Bapak Marcus Wanmar, Bupati of Raja Ampat (Head of Raja Ampat Regency)
- Bapak Mambrasa, Head of Tourism for Raja Ampat

Lease Rights and Limitations

Under the provisions of the lease, MER secured exclusive rights to Batbitim and Jef Galyu Islands, including hills, forests, trees, water, animals and the surrounding lagoon. In addition, rights were secured to designate approximately 425 square kilometers of surrounding seas as a no-take zone (NTZ), including animals, coral reefs, turtles, sharks, rays and fish. Under the terms of the lease, anyone other than MER is prohibited from taking any marine products from the NTZ or granting permission to any other party to do the same.

The extension of the NTZ to cover the Daram Archipelago followed the same structure and offers the same protection to the ecosystems there. In addition to the lease payments, additional funding was sought to build two kindergartens in the two villages where the families that have the traditional rights to Daram come from. This was so that all the members of the community could immediately see a clear benefit from the lease.

Fees and Services Exchanged

The villages of Yellu, Fafanlap and Usaha Jaya benefit in a number of ways from their agreements with MER. The most obvious benefit is

MER's lease fee, payable every five years. However, both the communities and MER have a clear understanding that this payment is a nominal bonus in comparison to the significant long-term benefits of bringing sustainable tourism to their area. Tourism and the establishment of the NTZ offer several important economic benefits to the people of Yellu:

- New job opportunities
- Employee benefits
- Goods and services
- Language skills
- Thriving marine environment for future generations

Process and Issues

Decision-making and Implementation Process

Frequently there are no official documents that state who is the rightful owner of land/sea in Papua. In the case of the MER NTZ, the local Adat committee had to confirm who the rightful owners of the area were. A deep knowledge of the area and good relationships with the local people were critical to having the ability to form the agreement.

Initially, the idea of leasing the area from "owners" who lacked any formal documentation of their ownership rights was cause for great concern. There was a strong possibility that we might enter a lease with the wrong people or end up in a situation where counter claims undermined or invalidated the lease. To gain comfort, we went to great lengths to meet with the relevant folks in the local government and community leaders. Several factors worked in our favor:

- There are only a few villages bordering the proposed NTZ with only one obvious candidate for ownership.
- The Camat provided excellent guidance in helping locate the proposed NTZ in an area where no known counter claims existed.
- A large pearl farm operating about 10 kilometers from Batbitim and leased from the same family that MER leases from had operated for 12 years in this location without any counter claims.

These factors combined with our extensive due diligence provided the necessary assurance that a formal legal agreement would be sufficient to secure the lease.

Lands and Resources

Southern Misool offers some the best and most pristine diving in Raja Ampat. One dive site, Fiabecet, is considered by many to be one of the top ten dive sites in the world. In addition to obtaining permission to build MER on Batbitim Island, steps were taken to secure a NTZ reaching in all directions from the island. The goal was to include as many of the healthiest and most diverse reefs within the NTZ. Andrew Miners, having been a dive master in this region for years, was intimate with many of these sites.

At the same time, care was taken to not "overreach" and remove too many of the traditional fishing areas. It was critical that the area start with the support of the local villages and that support continued into the future. When forming the agreement, public discussions were held with the heads of the local families and community leaders to ensure that there was sufficient opportunity for everyone to ask questions and voice their opinion. This process was essential in deciding on the appropriate and acceptable boundaries for the no-take zone.

In the first lease agreement, MER secured exclusive rights to Batbitim and Jef Galyu Islands, including hills, forests, coconut trees, water, animals and the surrounding lagoon. In addition, rights were secured to designate 425 square kilometers of surrounding seas as a no-take zone (NTZ) including animals, coral reefs, turtles, sharks, rays and fish. Under the terms of the lease, MER is prohibited from taking any marine products from the NTZ or granting permission to any other party to do the same. Likewise, MER has the right to prevent any other parties for attempting to extract marine resources from the NTZ. There is one exception to this rule established to honor traditional fishing practices. Once every two years certain shellfish can be collected by member of Yellu village over a period of only two weeks without the use of air compressors or dive equipment. Even this is restricted from designated dive sites. In the second NTZ lease agreement, a similar system was developed with the same protection; however, this time it was agreed that the shellfish collection would be only once every five years.

The MER NTZ is located in the southern most boundary of Raja Ampat. South of the NTZ, open water stretches for seventy miles until you reach the islands of Seram and Ambon to the south. Historically, many of the outside fishing fleets entered Raja Ampat from these locations, passing into and through what would become the NTZ area. Many of these boats engaged in incredibly destructive fishing practices, including reef dynamiting, shark finning, turtle harvesting, turtle egg collecting and destructive trawling. These fleets were unwelcome by the local villages, but the locals were ill-equipped to do anything about it. Frequently these boats would anchor and dynamite or fin sharks on the southern sides of the islands to avoid detection. By locating MER and the NTZ on this boundary, a buffer was established that would significantly reduce the level of encroachment by these outside fishermen. No longer could they hide on southern aspects or enter this area of Raja Ampat undetected. You might say we decided to hit the problem head on rather than hope it didn't find us.

To further emphasize the above point, MER itself is located on an erstwhile dynamite-fishing camp. Many of Batbitim's south, east and shallow bay reefs were almost completely destroyed by dynamiting that occurred in the late 1990s/early 2000s. The northern reefs, by comparison, are stunning and virtually intact. The decision to locate MER on such an island was a deliberate one. The objective was to select a location where tourist traffic would not negatively impact the local reefs (the water cottages are situated primarily behind dynamited reef) and where active conservation/reef building efforts would significantly improve the health of the local reefs with each year that passed.

Threats

The primary threats to the NTZ are illegal fishing, shark finning, turtle/turtle egg harvesting, reef damage from net fishing, reef damage from boat anchors, reef life harvesting (hookah divers), pollution (flotillas of plastic debris that wash in with certain currents clogging bays and mangroves) and climate change.

The NTZ agreement expressly forbids any kind of fishing or animal (turtles/eggs) harvesting and it gives authority to MER to enforce these provisions. Since all reefs are protected within the NTZ, the agreement enables MER to administer and enforce boat anchorage policy. There is nothing that can be done locally to combat the outside pollution and climate change. These are serious realities that are affecting marine environments globally.

Organizational and Partner Capacity

A sound knowledge of the local area was essential to determining the appropriate location for the two NTZ's. Relationships with key local stakeholders (existing and new) proved critical to securing the land/water ownership rights. As described above, navigating ownership ambiguity, traditional fishing seasons (*sasi*) and establishing the necessary buy-off required strong relationships at the local, district and regional levels.

Villagers in Raja Ampat speak both Indonesian and local dialects in their daily lives. The ability to speak proficient Indonesian proved essential in getting to the heart of their needs and concerns, as well as establishing mutual trust and respect.

Established non-governmental organizations (NGOs) provided important guidance and resources when it came to the nuances or establishing and administering a NTZ. These benefits included access to their local and regional government contacts, research and reports, maps and surveys, fisheries data, interpretation of laws, insight on pending legislation and even transportation. These resources proved valuable, as we had never before established an NTZ.

Legal Framework

The leases were founded on principles supported by Papuan law (or Hak Adat) and formalized under Indonesian law. The Indonesian Central Government recognizes Hak Adat as part of special autonomy status

(Kursus Autonomy) whereby the state of Papua has a certain degree of autonomy from the rest of Indonesia. Under Papuan law, people can own not just islands but beaches, reefs, and fishing rights in the waters surrounding them. The region has a long-standing local tradition of fishing seasons (*sasi*), which the local villages "open" and "close" in their areas and surrounding villages generally respect.

The *sasi* proved to be a fundamental element to obtaining the leases. Because five families hold all rights to the area land, sea and creatures within the sea, they thereby also have the right to exclusively lease the area. Gaining the support of the village leaders, the district head, the regional head and the tourism department created the necessary buy-off to form the agreement.

Having secured all rights to control the land, water and creatures within the NTZ, MER was empowered to take the steps necessary to enforce the "no-take" provisions of the agreement. MER has decided to abide by the same provisions that prevent the leaser from extracting anything from the NTZ, thereby establishing an important check and balance. MER believes that if either party were to extract from the NTZ, it would be far more difficult to explain and regulate such activity.

Socio-economic Considerations

From the outset, the local community was keen to bring tourism to the area. Their positive experience with the local pearl farm helped foster an environment of receptiveness to outside business ventures in their area. It is important to note that MER (the resort) was of much greater interest to them than the NTZ. Whereas the NTZ required that they give up some of their traditional fishing areas, the resort offered the prospect of long-term employment/income for the villagers. In reality, the NTZ was formed on the back of the resort, with the favorable economic possibilities of resort driving their desire and willingness to grant the NTZ.

As described in the overview, the local communities benefit in a number of ways from their agreement with MER. The most obvious benefit is MER's lease fee, payable every five years. However the communities and MER have a clear understanding that this payment is a nominal bonus in comparison to the significant long-term benefits of bringing sustainable tourism to their area. Tourism and the establishment of the NTZ offer several important economic benefits:

- *New job opportunities.* Under the terms of the agreement, MER has agreed to give priority to local communities in the village of Yellu and surrounding areas when recruiting non-skilled staff. MER jobs include construction, maintenance, resort services, dive services and NTZ rangers. All staff is paid above the "going rate" set by the manpower department. MER currently employs 120 people with 75% coming from the local communities.
- *Employee Benefits.* All staff members, both full-time and temporary, receive health, accident and pension benefits. Full-time staff members receive benefits for their families. This is certainly not the "norm" for the area or required by the state.
- *Goods and services.* MER purchases fuel, fish, some dry goods and some vegetables from the local villagers. MER is encouraging them to grow more to sell to MER.
- *Language skills.* MER employs two professional language instructors and is teaching the village staff to become proficient in English. This is an important and valuable skill necessary for them to succeed in the tourism business.
- *Thriving marine environment.* Outside fishermen have increasingly been entering the traditional fishing grounds of the village and taking fish without securing permission or paying a permit fee. Even when a permit fee is paid, the amount received by the local community is relatively insignificant. For this reason, many in the village feel that their marine produce is being "stolen." Furthermore, the elders in the village fear that in the future, outsiders may severely damage their reefs and deplete their fish stocks. The NTZ provides for the long-term health of their reefs and fish stocks, with the anticipated spillover effect to surrounding waters offering sustainable and abundant fish catches for the villagers. Given that the concept of "closed" seasons and areas is fundamental to *sasi*, they are able to grasp the idea of "spillover" benefits from the NTZ. Furthermore, they appreciate the fact that

healthy reefs ensure the long-term success of the dive tourism in the Misool area, and hence their ongoing MER employment.

Outreach

Initial Lease Agreement – Key Stakeholders

- Bapak Yusuf Salim, Camat South East Misool: District Pak Camat is an elected Government Official and District Head of South East Misool. He is one level senior to the Kapala Desa.
- Haji Jalil Bahale, Kapala Desa: Kapala Desa is the Government Head of the village. Pak Desa has been Head of Yellu village for more than twenty years and is a powerful and respected man. He is also the younger brother of Pak Haji (Adat Leader).
- Bapak Hadir Yelfom, Yellu Village Secretary: Second in command to the Kapala Desa, Bapak Yelfom is the Head of the second largest family in Yellu and joint owners of the land and seas in the area.
- Bapak Haji Yahya Bahale, Adat Leader Yellu Village: Adat is the name given to local traditions and culture. Adat is well respected by local inhabitants of Raja Ampat. Pak Haji is an influential man and head of the Bahale family, which as the longest lineage in Yellu and holds traditional rights (Hak Adat) to all the islands around Batbitim and Yellu.
- Bapak Marcus Wanmar, Bupati of Raja Ampat: Bapak Wanmar is the Government Head of the Raja Ampat Regency. Although he was not directly involved in the lease he has lent his support to it.
- Bapak Mambrasa, Head of Tourism for Raja Ampat: A number of meetings were held with Bapak Mambrasa; however, it was his staff members that were more helpful. Although friendly, the tourism department of Raja Ampat is very young and quite inexperienced. Very little direction and assistance was provided in arranging the lease.

Each of the above players played an important role in the formation of the initial agreement. The process began in early 2004 with a year spent on site surveys. In March 2005, meetings began in earnest with the land planning department to better understand laws governing land leasing in Raja Ampat. In May, we were introduced to the Camat who identified the landowners and facilitated a meeting with the village. Scheduling conflicts and the monsoon season delayed further meetings until September.

During this delay period, time was spent identifying investors to provide seed funding. In September, the meetings resumed and we learned that the Camat had already shared the concept with the village leaders. We held meetings with the village and though they were very enthusiastic, both the Secretary and Kapala Desa were not present. Returning to Sorong, we met again with the Camat and the Village Secretary to draw up the lease agreement. The Village Secretary recommended a final meeting with the village to make certain they were very clear that they could not fish in the NTZ. This meeting was held, and on November 28, 2005, the lease was signed.

Funding

Because the grant of the NTZ was tied to the establishment of the resort, we needed to secure funding for both MER and the NTZ to get started. The initial NTZ lease payment was funded through a private loan. Two financing rounds were necessary to fund the construction the resort. Construction was completed in October 2008.

The charity Misool Baseftin plays a critical role in the management of the NTZ. The Conservation Center is responsible for oversight, research, projects, patrolling and regulations within the NTZ. Most importantly, it ensures that the no-take provisions are strictly adhered to and work to restore reefs that have been damaged by destructive fishing methods.

The Second Lease – Key Stakeholders

- Rajak Soltief, Kapala Adat and Head of the Soltief family: Bapak Rajak is the senior Adat figure and was key to organizing and agreement between the three families. He is retired now but was the Head of the Department of Tourism for the Province of West Papua, in which Raja Ampat is a Regency.
- Rasid Wainsaf, Adat Leader and Head of the Wainsaf family: Also an influential figure and key to bringing the Waisaf family to agreement. Theirs is one of the largest families in the Fafanlap village.
- Manuar Macap, Adat figure and nominated Head of the Macap family: Important figure as he represented the Macap family who are based in Usaha Jaya village.

Each of these members were very important in the extension of the NTZ to cover the Daram Archipelago. The lease agreement was more complicated as three families claim the rights to the Daram islands and the families hail from two different villages.

Discussions started in 2009 and it took until late 2010 before an agreement was reached. Funding was also raised during this period through Wild Aid, which provided money to set up the project and run it for the first year. Currently we are applying for an extension of funding for operational expenses for the next three years.

The goal is to develop self-funding initiatives during this period.

The driving motivation to secure this area was the fact that it was acting as a base for outside fishermen who set up illegal camps and destructively fish in the Misool region. They paid no fees to the villagers and plunder the reefs of fish, turtles and sharks. Establishing an NTZ along with ranger stations and a patrol vessel creates an incredible sanctuary within Raja Ampat and provides a sustained financial return and ranger jobs to the villages that own the area.

Conclusions

Perhaps the most important factor maintaining the sanctity of the NTZ is the full-time, long-term presence of MER. Every day the villagers are reminded that the NTZ plays a critical role in the success of MER and therefore their own personal incomes. Our model for a combined resort and NTZ can be replicated in other parts of Raja Ampat as well as abroad.

It is worth noting that the first NTZ was formed on the back of the resort, with the promise of employment more than justifying the sacrifice of a portion of the fishing grounds. This may be a good model for future NTZ development. The second NTZ was formed using the first one as an example of how communities and businesses can work together to secure their resources for future generations. The partnering of conservation groups with business enterprises can be a powerful force in marine conservation. The business enterprise can bring jobs and income while the conservation group brings programs, education and protection of resources. As time passes, the conservation group can transition its functions to the business and local community, thereby creating a completely self-sustaining model.

Scale

The MER NTZ is operating at the local level both ecologically and politically. Though it covers an incredible abundance and diversity of reefs (1,220 square kilometers), it is still relatively small in comparison to the larger MPAs in Raja Ampat. In addition, the area is leased from local villages and not a state or regional government.

If this model were embraced at the program level, NGOs could seek to group a number or private enterprises together in a given geography to support larger shared NTZs. This would require increased orchestration and collaboration between a larger number of stakeholders.

Benefits to expanding this strategy include:

- Larger NTZs would provide greater protection for species that move and migrate more.
- Economies of scale as more stakeholders share in the cost of the NTZ.
- Increased monitoring and patrol with more stakeholders.
- Larger-scale community impact thereby increasing longevity of NTZ.
- Creating a legal framework whereby future NTZs could be established more rapidly and be held to best practice standards.

Limitations to expanding this strategy include:

- Limited number of eco-resort operations can be sustained in any given area.

- Finding private organizations committed enough to take on the challenge.
- Identifying suitable locations where locals are willing to give up fishing rights.
- Patrolling and enforcing larger marine areas as poaching is rampant in MPAs.

Complementation

TNC, CI and WWF have been working together with the regional and local governments in Raja Ampat to establish a series of MPAs in the regency. These MPA became officially recognized in December 2006 when Raja Ampat was declared a Marine Regency (or Kabupatan Bahari). The charter of the Regency is to focus on marine economics, including marine tourism.

The broader MPAs will be zoned into different use areas, including traditional use, tourism and strict conservation. The MPAs are still awaiting zoning and with a few exceptions, there are no NTZs in place. One of the MPAs covers southeast Misool. As they look to zone this region, MER and TNC are working together to use the MER NTZs as the center of the larger NTZ for this MPA.

There is a great opportunity to form a complimentary strategy, which includes smaller private groups and larger NGOs such as TNC, CI and WWF. While these NGOs can work from the top down with central and local governments to change policies and establish broader conservation measures, the smaller private groups can work at the community level to implement them. In the case of MER, the NGOs broader efforts to establish MPAs in Raja Ampat combined with MER's grassroots work to establish its own NTZ made for a perfect complementary strategy. MER is on the ground 24 hours a day/7 days a week. The staff are primarily from the local villages, enabling closer and deeper relationships to form with the communities. The local leaders visit the resort and NTZ and see for themselves the conservation practice in action. Overtime they know from their own people's accounts that MER has adhered to the covenants of the NTZ lease and have been protecting it from outside intruders.

An NTZ is perhaps the strictest form of MPA and often the most controversial. As such, it necessitates real buy-off from the community for it to be granted and actually adhered to. MER offers the kind of community-integrated project necessary to secure this buy-off and create "real" NTZs.

Lessons Learned

What we did right and will continue to do:

- Selected an area of incredible biodiversity and abundant fish life.
- Chose an area where ownership rights could be established with reasonable confidence.
- Created an NTZ instead of an MPA (with managed fishing).
- Invested in relationships and secured buy-off at the local and regional levels.
- Created a very real and long-term economic opportunity for local communities.
- Took a serious and uncompromising "eco-approach" to the resort.
- Maintained a constant presence since day one.
- Kept our word when we made commitments to the local community.
- Engaged the community in the protection and improvement of the NTZ.
- Realized that constant communication face-to-face is essential in order for the NTZ to be correctly socialized and to avoid any misunderstandings.

What we did right but could do better:

- Constant socializing of the concept of the NTZ with the local community so they felt part of it (initially the younger members felt the elders had sold off a part of their heritage).
- Secured funding for the project prior to forming agreement.
- Understood the scope and complexity of the project in such a remote location.
- Provided for adequate transportation to and from MER.
- Accounted for the difficulty in regulating and patrolling the NTZ.
- Garnered local government support for implementing regulations for all the live-aboard and land-based operators in the area.
- Established clear understanding with live-aboard operators regarding the rules and regulations of the NTZ.

Things we didn't do but should have:

- Provided voice and data communications from the island.
- Secured funding for the establishment of a Conservation Center.

- Employed a full-time community relations officer.

Things we didn't do and are glad we didn't:

- Established the resort without first securing the NTZ.
- Compromised our commitment to resort eco-construction and operation.
- Compromised our commitment to provisions of the NTZ (i.e., allow the resort to fish but not villagers).

General Point

By not compromising on the "eco-ness" of our building practices, we offer a clear, visible message to the local community when they visit that we are really doing what we say we are – protecting the environment. They know we are doing good work and are there to preserve the environment.

Recommendations

There are significant benefits to having a private enterprise work at a local/regional level to establish MPAs/NTZs. NGOs frequently focus primarily on issues of law, regulation, zoning and management. These more global concerns are critical to the long-term sustainability of MPAs and must be addressed. However, they take time, resources and patience. Moreover, many NGOs frequently view their role as temporary (months to years) with the objective of creating a self-sustaining model. This also is an important objective.

In the meantime, we are all aware of how rapidly critical marine habitats are being stripped of their resources. At the current rate of loss for marine habitats and fish populations, every day without protection is significant. In addition, many regions lack the proper infrastructure to provide much or any economic alternatives to local peoples. This is where private conservation groups coupled with business enterprises can fill the gap. Eco-tourism can offer immediate jobs and income to the local community in exchange for the rights to establish MPA/NTZs. The community sees immediate benefits from their decision and become advocates and enforcers of the MPA regulations. And finally, because the community and private enterprise depend upon each other and form long-term relationships, there is a constant daily reminder of the importance of the MPA to the futures of both groups.

For more information, please visit Misool Eco Resort online at http://www.misoolecoresort.com/aboutMER.html.

CHAPTER EIGHTEEN

COMBATING CLIMATE CHANGE THROUGH STRATEGIC DESTINATION PLANNING: A QUADRUPLE-BOTTOM-LINE APPROACH[1]

JAMES MACGREGOR[2]

The Tata Destination Region is one of the hottest regions of Morocco. Summer temperatures frequently exceed 50°C, and temperatures seem to have been rising over the past two decades, in part because of the influence of climate change. Furthermore, the region's major resource, the oasis, has been rapidly degrading because of climate change, reduced available water and consequently decreased management. The reduced viability of agricultural production also has a significant impact on lifestyle. It is increasing the cost of urbanization, leaving the communities with fewer young people.

The region has not traditionally been a tourism destination, although it has been a significant pass-through territory for package tours and individual 4 x 4s traveling between Ouarzazate and the Atlantic coast of southern Morocco. While many tourists stop in the region, their length of stay is typically very short and there is little economic contribution from tourism.

The decision was made to establish the region as a destination that holds these pass-through travelers and attracts destination-bound tourists. Because of the interest in climate change from the major source markets to the region, including France, Germany, Spain and England, it was decided to highlight climate change as a theme.

---

[1] This article is based on the strategic sustainable tourism development plan for the Tata Destination Region in the Western Sahara prepared by Ecoplan:net sarl for the UNDP and the Program for the Oasis of the South and the Agency for the Social and Economic Development of the Provinces in the south of the Kingdom of Morocco. Project management was by Abdullah Hachimi. James MacGregor was the team leader with associate Hassan Aboutayeb, owner, Atlas Kasbah Eco-lodge (sustainable tourism consultant).

[2] President, ecoplan:net sarl, Morocco.

Figure 18-1. Hiking the hills of Tata.

Two sub-themes were also developed:

(a) Living with a hotter climate: the people of the region are familiar with adapting to and coping with 50°C-plus temperatures. Their architecture, lifestyle and customs reflect these conditions and provide important lessons for their western European market.
(b) Visiting a low-carbon destination: This implies converting the region from one of relatively high carbon consumption, from both the residents and tourists, to one that conscientiously seeks to reduce carbon emissions.

The tourism development criteria and project proposals were based on these two themes. In practical terms, all current and planned tourism facilities and services will adopt climate-friendly infrastructure, equipment and practices. As well as this, tourism activities will focus on those that are particularly low or neutral with respect to carbon emissions. Tourists will be encouraged to park their 4 x 4 vehicles and take any number of hiking, cycling, photography, etc. tours.

The supply chain itself has slowly adopted a number of technologies and practices that are meant to reduce the level of greenhouse gases.

The region has been keen to adopt this new quadruple-bottom-line approach based on social, economic, environmental and climate considerations. First of all, the president of the region, M. Mehdi Elhabibi, who has been elected to the position consistently for the past thirty-six years, possibly the longest political career in Africa, is known as one of

Morocco's strongest environmentalists. His support for the proposed environmentally responsible and carbon-reducing initiative has been instrumental in promoting the concept of a carbon-neutral development plan.

Figure 18-2. Map of Western Sahara region.

1. Overview of the Tata and Western Sahara Region

The Tata Tourism Destination Region is located approximately 200 kilometers south of the popular Moroccan beach destination of Agadir on the northern limit of the Western Sahara. The area is also strategically positioned between the interior Moroccan city of Ouarzazate and the Atlantic Coast. The region has traditionally relied upon oasis agricultural production, including dates and other cash crops as well as growing food for local consumption. The oases themselves have been an excellent source of natural resources used in the production of a variety of products that have sustained the "oasis-ian" lifestyle and economy for generations.

Water is truly the source of life in this region. It is channeled from the higher altitudes through a complex series of canals that bring water to the individual plots of land in the oasis of those families living adjacent to the oasis. This system of irrigation has been effective for centuries in part because it assures equal distribution of the precious water resources to each family. This in itself provides a powerful message to the modern traveler as well as future generations who must ensure that future water resources must be shared to ensure survival.

Figure 18-3. The largest oasis in Morocco.

Figure 18-4. Measuring water level in the basin for equal distribution.

Global warming, including desertification, is influencing the health of the oasis. Increased temperatures and reduced water supply have begun to discourage the continuation of management practices that have maintained the oasis for generations. Decreased economic opportunity combined with changing interests of the younger generations has already caused the first wave of exodus from the region to major cities such as Casablanca and Marrakesh or to Europe.

2. Response to Changing Economic and Social Conditions

The president, of the region, Mr. Mehdi Elhabibi, recognized the need for introducing and integrating new economies into the communities if they were to survive as a viable region within Morocco and Western Sahara. Tourism was identified as the leading economic opportunity despite relatively little infrastructure and the complete lack of recognition in the domestic and international tourism marketplace. Despite these factors, about 40,000 tourists have been coming to the region, particularly from December to April, and they seemed to enjoy their contact, albeit limited, with the people and the various natural and cultural resources. But for the most part they were passing through and rarely spent one night in the region.

The challenge, then, was to increase the length of stay as well as reinforce the recognition of the region as an appealing Moroccan tourism destination.

The decision was made by the president and counsel to aggressively pursue the tourism market. UNDP and the regional economic development agency were approached to contribute partnership funding and expertise.

A proposal call was issued and the Moroccan sustainable-tourism consulting firm, ecoplan:net sarl was hired to prepare a long-term tourism development strategy.

3. A Classic Tourism Strategic Planning Process … Well, Not Exactly! (Phase 1)

Despite the wealth of natural and heritage resources throughout the region, the consultants immediately recognized the need to go beyond the traditional strategic-tourism-planning process if the region was to achieve its development goals. However, this did not preclude undertaking certain typical exercises that are associated with standard sustainable-tourism planning.

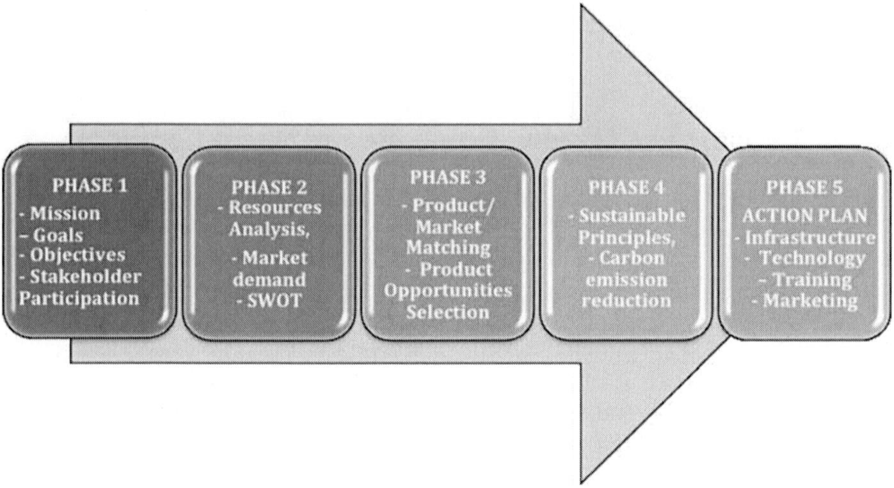

Figure 18-5. The sustainable tourism planning process.

Figure 18-5 demonstrates it was necessary to establish a clear mission and set of goals and objectives (phase 1). This was achieved by engaging a variety of stakeholders from the communities. Their demands were rather conventional and included increasing tourism volumes, increasing the length of stay and encouraging higher expenditures. Furthermore, there was an understanding that the impressive mix of cultural, heritage and natural resources were sufficiently attractive to support a viable tourism sector.

However, it was the stakeholders' strong sense of protecting their resources that most impressed consultants. Perhaps it was because these people live in such a harsh environment that the combination of sharing, intelligent use of and conservation of the limited resources are an integral part of their vocabulary and behavior. The physical size limitations of the oasis suggests that every square meter must be protected and managed with great care, because beyond the edge of the oasis there is only desert; which, of course, is limited for agricultural production and even sheep grazing. In other words, the oasis is the lifeline to the survival, shelter and culture of the people.

Despite this encouraging respect for their local resources, the challenge of positioning the region as a significant destination still remained. Essentially, it was necessary to develop a strategy that could attract substantially more tourism to a relatively unknown region of north Africa and Morocco.

4. Resources and Market Analysis (Phase 2)

The resource analysis identified the typical physical attractions that one would expect to find in a desert environment and its communities, including:

- A vibrant, nomadic culture throughout the region with their traditional herding and gathering activities;
- Traditional villages and Kasbahs, still fully occupied and supporting cultural and lifestyle activities dating back many generations;
- Dramatic desert and mountainous landscapes and an abundance of petroglyphs;
- Lush oasis with a still abundant date production as well as other agricultural (and fully organic) products;

Figure 18-6. The stunning environment beyond the oasis.

- Unique structural features, including the irrigation system that brought water from higher elevations and distributed it to the villages and agricultural areas in the oasis; and
- Some of the world's oldest archeological sites.

The tourism market profile was equally impressive. They were almost exclusively western European, well educated, well traveled, sophisticated travelers with a sense of adventure, curiosity and interest in Moroccan culture and lifestyle and, given the opportunity, demonstrated a strong preference toward environmental responsibility.

5. Product/Market Matching Process (Phase 3)

The product/market matching process was prepared in collaboration with local stakeholders and matched the identified long-haul travel consumer interests with the available resources of the region. It was this dynamic process that identified the concept of introducing the educated and environmentally aware market to the local practices associated with living in this hot environment. There was a strong message that connected the local population's ability to survive and flourish in these extremely hot conditions with the travel markets concern about global warming.

What was initially considered a "negative" – the increased regional temperatures due to climate change – became a "positive" feature. It was possible to exhibit to tourists:

(a) The impact of carbon emissions and subsequent global warming on the destination (and in particular, a desert environment); and
(b) Local, traditional knowledge practices that can demonstrate how to effectively live in a hot environment (i.e., with a warming planet).

These two messages are very compelling and, when tested, proved to appeal to the current (and presumably future) visitor market.

6. Product Development Opportunities

The results of this product/marketing process and the identification of potential visitor activities resulted in the selection of numerous product opportunities associated with the global warming theme. They were grouped under two sub-themes:

Sub-theme 1: Living with Climate Change and Increased Temperatures

(1) Climate change interpretation visitor center
(2) Interpretation and demonstration projects including
    a. Traditional architectural building techniques

b. Irrigation and water conservation and sharing methods
c. Local food production and preparing traditional meals
(3) Traditional transportation methods

Sub-theme 2: Establishing a Carbon-Responsible Destination

(1) Minimizing fuel-based transportation and excursions for both visitors and residents; providing options for the visitor to move within the destination by hiking, cycling and on horseback or camel.
(2) Reducing energy consumption for visitor accommodation, food services and tour operations.
(3) Increased local food production and decreased imports.
(4) Environmental education and improved visitor awareness with a specific emphasis on carbon-reduction activities.
(5) Restoration of traditional irrigation techniques and revitalization of the oasis food production.

7. An Environmentally Responsible Tourism Destination Brand

The combination of responding to the above sub-themes, as well as the government support and political will from the local authorities, and the expressed interests of the current market, suggested branding the destination as "carbon responsible" with a conscientious focus on reducing carbon emissions.

The concerted efforts were targeted at four levels, including:

(1) Visitors
(2) Residents
(3) Tourism supply chains, particularly accommodation, food services and receptive tour operations
(4) Local business not necessarily associated with tourism

The strategy outlined how each of these components can work collectively to ensure that greenhouse gases (GHG) are kept to an absolute minimum, thus projecting a strong environmentally responsible image to the national and international marketplace.

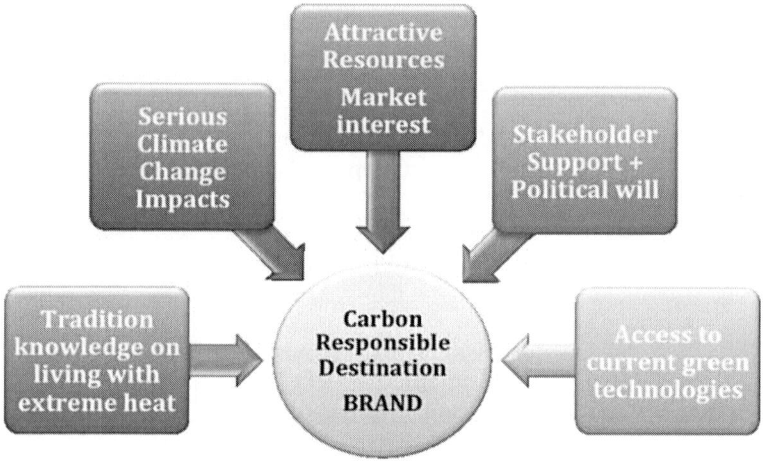

Figure 18-7. The brand: a carbon-responsible destination.

8. Establishing Carbon-Reduction Criteria and Practices (Phase 4)

In phase 4, the strategic tourism destination planning process deviates from the standard approach. In order to ensure that there is a conscientious and proficient implementation of the quadruple-bottom-line approach, and in particular carbon-reduction initiatives, it is necessary to provide government, residents and the industry with a comprehensive set of criteria that guarantees the proper planning and development of the various tourism recommendations.

Four categories of implementation criteria were established, including:

(1) Tourism facilities and infrastructure: accommodation, food services, attractions, visitor centers, interpretation kiosks, etc.
(2) Tours and tour packages: hiking trips, cycling, horseback, etc.
(3) Purchasing procedures: local purchasing, bulk buying.
(4) Training programs and educational materials: accommodation suppliers, tour operators and government authorities.

For each of these categories, general standards were provided that would enable any stakeholder within the tourism supply chain to immediately adopt carbon-friendly initiatives. Of course the use of these standards did not preclude other areas of environmental, social and economic responsibility.

9. Tourism Facilities and Services Grading Program

The application of carbon-friendly practices will be reinforced by the launch of an international level-grading scheme. This would initially be developed for the accommodation sector, and then followed by restaurants, attractions and tour operations. It would be a comprehensive grading program including quality assurance, responsible tourism and universal accessibility. The responsible tourism standards in particular would ensure that all available carbon-reduction standards are integrated into the overall tourism operations.

10. Training Programs, Educational Materials and Awareness Campaigns

The offering of sustainable tourism training programs, with a specific emphasis on carbon-reduction practices was an important component to advancing the destination towards "low-carbon-emission status."

Carbon emission reduction training includes introducing concepts of:

- Transportation efficiencies and reduced-travel planning
- Energy-saving technologies
- Integrating local purchasing practices and related operations
- Sourcing and purchasing energy-efficient equipment

Furthermore, well-designed manuals support all training programs, and technical documents that when circulated throughout the destination make a significant contribution to understanding social and environmental responsibility and identifying carbon-reduction practices.

These training efforts are further enhanced by a destination-level environmental awareness campaign.

The following diagram demonstrates the scope of the training materials.

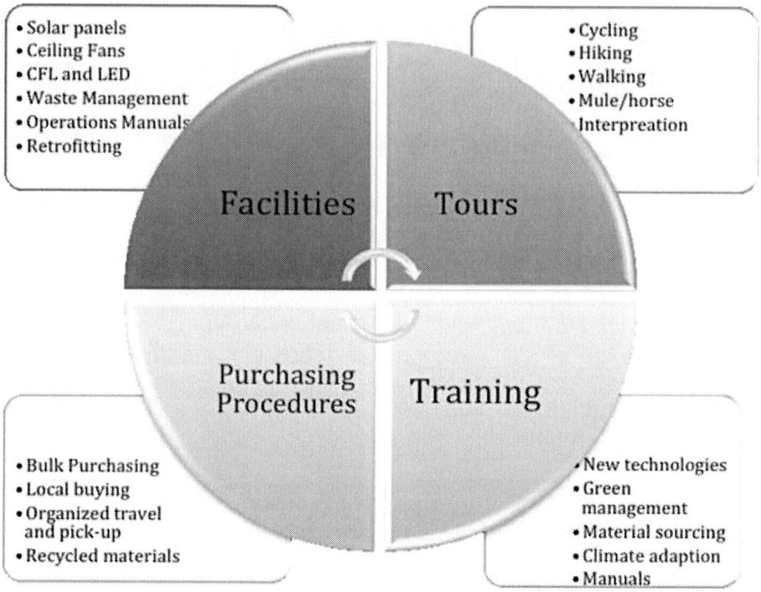

Figure 18-8. Developing carbon-reduction criteria.

11. Preparing an Aggressive Action Plan and Attaining Carbon-neutral Status (Phase 5)

Moving the Tata Destination from a conventional southern Morocco region to one that could eventually attain carbon-neutral status requires a comprehensive and integrated action plan. Furthermore, the various stakeholders had to be committed to the overall plan as well as their individual responsibilities and within the prescribed time frame. The various partners were also committed to the costs of implementation.

Parallel priority streams characterized the first five years including:

Stream 1: Establishing the tourism products that identified the region as a carbon-reduction-responsible destination, including:

(a) Creating a variety of hiking trails within the oasis with visits to women's cooperatives, interpretation exhibits and enjoying local cuisine.

(b) Inaugurating multiple-day hiking and cycling trips throughout the destination with particular emphasis on identifying select responsible communities as staging areas.

(c) Promoting interpretive tours of heritage Kasbahs to demonstrate how architectural materials and layout provided an ambient environment in a hot climate.

Stream 2: Integrating energy-conservation technologies and building practices throughout the tourism supply chain. A few examples include:

(a) Retrofitting hotels and guesthouses, including the introduction of natural ventilation techniques, improved insulation, etc.
(b) Promotion and distribution of solar panels for water heating.
(c) Establishing individual, local purchasing policies and contracts for accommodation and food-service suppliers.
(d) Respecting traditional architecture that combines both aesthetics and functionality in terms of ventilation.

12. A Phased Planning Process

The local stakeholders validated the following three-phase implementation process and action plan.

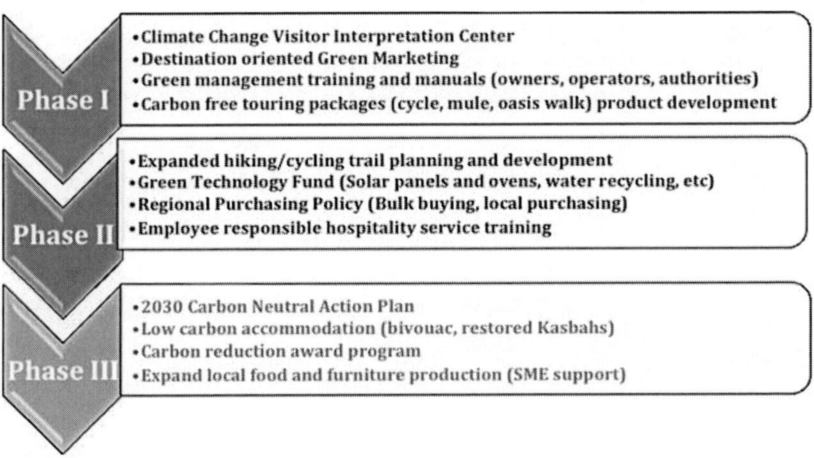

Figure 18-9. Tata destination carbon-reduction action plan.

13. Conclusion

The Tata Destination Region and the Western Sahara are positioned to become one of the first carbon-responsible and possibly carbon-neutral tourism destinations in North Africa and even all of Africa. The combination of strong political will, an active social and economic development agency and a clear mandate by the regional stakeholders suggest that the region could be carbon neutral by say 2020. This date corresponds with the Vision 2020 plan of the Morocco Tourism Ministry that envisions sustainable development at the core of its tourism strategy.

Positioning the destination as "carbon responsible" is proactive on carbon-emission reduction and also enables the crafting of a unique identity and brand that will generate the necessary level of interest that can attract visitors to a relatively remote and unknown area of Morocco. By identifying climate change as a development theme, the destination will continue to build new environmentally responsible products for a wider audience that will expand as concerns for global warming and the reduction of greenhouse gases increases.

This then becomes a win-win situation. The destination benefits from increased exposure and consequently more tourists, Morocco benefits by supporting the sustainable tourism initiatives in the destination, the travelers are educated on how to live in a warming world and the planet benefits from a decrease in carbon emissions.

14. News for Other Tourism Destinations

Other destinations can also benefit as well from introducing similar carbon-responsible initiatives. The concepts and techniques identified in this Western Sahara sustainable-tourism plan can be voluntarily applied elsewhere. For the most part, they are relatively inexpensive, are readily adopted by most stakeholders across the tourism supply chain, use technologies that are currently available, and, for the most part, are cost-saving.

In fact, much of the effort in the first two years of a carbon-reduction plan can focus on changing local attitudes and behaviors. Green management workshops (particularly for the accommodation sector) and training seminars for the travel trade in general as well as environmental awareness campaigns for residents and business can contribute to significant and immediate reductions in carbon emissions. Even these modest changes send a positive message to the travel consumer about how your destination acts in a responsible and professional manner.

It is hoped, by this author, that eventually there can be a network of tourism destinations whose members are designated because of their special emphasis on implementing initiatives to reduce greenhouse gases. Collectively, they can send a message to the marketplace that the tourism industry is aware of its contribution to global warming and yet there are select destinations that are committed to reducing the impact.

CHAPTER NINETEEN

ECO-AGRITOURISM AS A MEANS TO PRESERVE CULTURE AND THE ENVIRONMENT

NIKKI ROSE[1]

Abstract

The benefits of eco-agritourism are discussed as a way for rural communities to help preserve their distinctive cultural and natural heritage. Since 1997, Nikki Rose has supported the efforts of more than forty small businesses and individuals working on action programs to preserve Crete's heritage. Rose formed a network of organic farmers, artisan producers, chefs, lodge owners, historians, nature conservationists and many other residents to collaborate on a range of dynamic educational programs. By rekindling resident and visitor interest in culture and nature, eco-agritourism can simultaneously help sustain communities, encourage expansion of organic agriculture and provide an extraordinary visitor experience.

[1] Founder of Crete's Culinary Sanctuaries.

Introduction

Crete is blessed with a fascinating history spanning 4,000-plus years, natural beauty and an abundance of healthy food choices, both wild and cultivated. There is much to discover and enjoy. There is also much to protect. Beyond the seaside resorts and imposing limestone cliffs are people preserving their heritage (sustainable organic farmers, artisan bread bakers, cheese makers, beekeepers, chefs and many others). They are maintaining what most of us have lost touch with – a connection between themselves, their community and nature. Their knowledge of sustainable living practices is beneficial to the global community. Many people around the world are striving to "return to the land," while many people in rural Crete have never left the land. But modern society beckons and rural communities are abandoned or developed.

Methods

Crete's Culinary Sanctuaries (CCS) introduces visitors to residents like Yiorgos, who maintains his small family farm much like his ancestors, using sustainable organic methods. He refuses to buy food from outside sources and even collects salt from a rocky beach nearby. "The chicken I eat must first dine at my house," he says. When we travel to foreign lands, it seems so natural to connect with residents outside of the tourism industry, to spend time in their villages and enjoy a leisurely meal together. CCS offers this opportunity. Before we eat bread, we meet farmers to see how grains are cultivated. Before we sample olive oil, we meet the farmers at their groves and factories. Before we sample cheese, we meet shepherds to discover how his or her flocks live – roaming the slopes and eating wild plants. Many wild plants are used in cooking and natural medicine, which we discover during hikes in the countryside and cooking demonstrations.

The renowned traditional healthy cuisine of Crete is not a phenomenon; it's a matter of respecting the land and the bounties it provides. Every great chef will tell you that they are only as good as their ingredients. The foundation of the Mediterranean diet concept is fresh, local, organic food and a clean environment. Both residents and visitors benefit from community-based preservation programs. Careful consideration and strong alliances are required to sustain these programs. CCS celebrates Crete's heritage in an ethical and professional manner. Residents are investing their time and money to share valuable knowledge.

Our seminars offer visitors a rare opportunity to discover the heart of Crete and obtain information that can enrich their lives.

Tourism and agriculture are primary industries in Crete. The majority of tourism planners support generic services, such as large beach resorts and continental food. Most operations are of little benefit to local communities in terms of providing financial stability or protecting Crete's cultural and natural heritage. More than 70% of Crete's residents are still involved in agriculture – primarily olive oil and wine production on a part-time basis. Both industries compete for increasingly scarce natural resources, commonly referred to as "the war on water." CCS promotes the benefits of educational travel via the window of traditional food ways. As agriculture is an integral part of Crete's culture, our programs center on the work of organic farmers and interrelated preservation work to serve several purposes at once.

Industrial agriculture in Crete is comprised of cooperatives, some of which represent hundreds of families cultivating a variety of produce on small plots of land. Slowly, some cooperatives and individuals are converting to organic production. Yet they face challenges if their neighbors are growing conventionally and public entities are not supporting their work. A growing number of specialized organic farmers are implementing projects. Most producers have incorporated agritourism into their work to promote their products, share their knowledge and generate supplemental income. Not all producers or stakeholders are collaborating with residents involved in interrelated preservation work that can enhance the quality of agritourism and the quality of life for residents.

Implementing Responsible Travel in the Case of Eco-agritourism

The fact that a destination has excellent cuisine and nature reserves does not mean that communities are able to open up the area to visitors. Before residents invite visitors to their villages and farms, they need to consider how they will present and sustain their programs and protect their communities.

Since most of us are not farmers, we don't know what to expect from an agritourism experience. Agritourism is meant to support farmers and their communities. In the case of organic agriculture, it also helps to protect our environment. Agritourism can make a difference as part of the bigger picture of preservation and responsible travel. Organic farming is a lifelong commitment, not a "nine-to-five" position. Supplying the world with excellent food and wine is more challenging than we might ever know or appreciate. While it's a lovely notion that farmers or fishers might take the day off to entertain us, in fact we are asking for the world. The time they devote to sharing their knowledge with us is a rare privilege.

Responsible travel covers a lot of ground today. There are interrelated categories – eco, sustainable, ethical, green, geo, etc. The bottom line is that people are actively working to support and protect the communities in which they live, work or visit. Responsible travel is part of the solution to a growing number of global issues relating to social inequality, exploitation, cultural preservation, food safety and environmental protection. It's a fine theory, embraced the world over, yet the practice is challenging. While the number of responsible-travel practitioners increases, the number of travelers partaking in such beneficial programs is minimal. Many people say they prefer "real" or "ethical" travel experiences, yet they are unaware that local communities must justly benefit from the services they provide.

It is the responsible-travel practitioner's job to develop valuable programs and create awareness of that value, which will be offered at a fair price. Preservation work is a partnership between providers and beneficiaries. Enjoying an action-packed series of cultural activities requires the participation of many local people. It must be financially rewarding, otherwise the arrangement is not mutually beneficial or ethical.

Sharing the Benefits of Responsible Travel

By tailoring small-group seminars in harmony with the seasons, cultural and environmental impacts, availability of residents and interests of attendees, CCS has the advantage of flexibility. In consideration of our collective and individual projects, CCS seminars serve a dual purpose (to support local preservation work and share those benefits with visitors). We link organic farmers with other members of the community through:

(1) our educational programs
(2) staying in locally owned lodges and eco-lodges
(3) providing free training and referral services
(4) global promotion of projects
(5) sharing seminar revenues (in contrast with standard tourism practices)

Many travelers benefit from community-based preservation work. We stress that communities must benefit the most if we expect to enjoy such services, including excellent fresh or local organic cuisine. CCS acts as a gateway for communities to expand or create their own programs.

There are established Women's Agricultural Cooperatives in Crete, producing and selling traditional products within their communities or beyond, depending on their resources and support for their work. There are also excellent home cooks, gardeners and artisan producers that do not have the resources to join cooperatives or host visitors. CCS invites residents from varied backgrounds to participate in our programs. By collaborating with the community, eco-agritourism stretches beyond the boundaries of a single farm. Communities maintain their way of life as they choose, rather than altering it to suit an outsider's vision of foreign travel.

Eco-agritourism must be advantageous for communities. Otherwise there is no incentive to provide these valuable services or expand organic production. Our approach can be implemented anywhere in the world. Eco-agritourism is a business run by accomplished professionals. All beneficiaries – the public and private sectors, the media and travelers – need to be aware of the value of these programs. CCS is an internationally acclaimed responsible-travel program. We also organize workshops for colleagues and our success has encouraged others to follow our lead.

Promoting Responsible Travel

As with all market trends, travel providers offering alternative forms of tourism in an unethical manner is damaging to authentic programs. As long as the majority of media and travel agencies instill the message that "cheap travel deals are the way of the world," people will not be aware of the true cost of travel, including the environmental and social impact that travel has on communities or the true cost of safe food. The majority of travel agencies are not yet trained in ethical travel practices. It requires that they modify their business practices and form mutually beneficial partnerships with local communities. More travelers are asking agencies questions such as, "How does your company contribute to the well being of local people and their environment?"

While responsible-travel options are increasing, practitioners are competing with volume-discount travel and powerful public-relations campaigns that overshadow their distinctive work. Mass media rarely covers small-scale programs because there is no apparent benefit to them. Discount deals cut in mega-travel cannot be cut in rural communities. Most media outlets expect free food and lodging, consultancy and entertainment. Small-scale practitioners cannot afford this form of advertising, nor are they certain to benefit from it, as the coverage is rarely specific to their work and could be subjective.

The first step to launching a small-scale program is to work from the ground up in collaboration with other local businesses, advocacy groups and public agencies, if possible. With the rapid growth of responsible-travel advocacy groups, entrepreneurs' participation in programs is increasing. Responsible travel is not a trend; it is a necessity.

Conclusion

Responsible travel, in the case of eco-agritourism, provides a wide range of benefits to communities and visitors. It provides an opportunity for entrepreneurs and travelers to be active participants in sustainable development programs. It requires direct collaboration with preservationists and tangible returns on their investments. Responsible travel can help preserve our world's sanctuaries and even reverse some damage caused by unsustainable development. It can build meaningful careers for future generations and protect the very reason why people visit countries like Greece – to discover her significant cultural legacy and natural beauty.

For more information on Nikki Rose and Crete's Culinary Sanctuaries seminars, projects, awards and press, go to www.cookingincrete.com.

References

Adams, B. (2008). *The New Agritourism: Hosting Community and Tourists on Your Farm*. Auburn, CA: New World.

Blangy, S. (2006). *Le Guide Des Destinations Indigenes*. Indigène Editions: Collection Indigène Esprit.

Kyriakopoulos, V. (2007). *Lonely Planet Guidebook to Crete*. Lonely Planet.

Miller, D. (2008). *The Jungle Effect*. Harper Collins.

National Geographic Center for Sustainable Destinations: www.nationalgeographic.com/travel/sustainable.

Transitions Abroad, Responsible Travel Handbook. (2006). www.transitionsabroad.com/listings/travel/responsible/responsible_travel_handbook.pdf.

World Tourism Forum for Peace and Sustainable Development. (2006). Case Study, Crete's Culinary Sanctuaries. www.destinations.net/cases/plonearticle. 2007-10-31.0051745210.

United Nations Convention on Biological Diversity Success Stories: http://www.cbd.int/2010/stories/.

CHAPTER TWENTY

MEETING CHALLENGES IN A PROTECTED LANDSCAPE: JURASSIC COAST WORLD HERITAGE SITE, SOUTHERN ENGLAND

SALLY KING[1]

Abstract

The Dorset and East Devon Coast World Heritage Site is England's only natural World Heritage Site. It is popularly known as the Jurassic Coast. The site is located on the south coast of England and covers 95 miles (155 kilometers) of exceptional coastline from East Devon to Dorset, with rocks recording 185 million years of the Earth's history. UNESCO declared the site a World Heritage Site in 2001. World Heritage status was achieved because of the site's unique insight into the earth sciences: it clearly depicts a geological "walk through time" spanning the Triassic, Jurassic and Cretaceous periods that make up the Mesozoic Era of geological time, between 250 and 65 million years ago.

The Jurassic Coast has a high visitor numbers and the careful management of the site, including sustainable tourism and access, is of paramount importance. Conserving the integrity of the site involves allowing the natural erosion processes, which formed the basis for World Heritage Site designation, to continue. This can create conflict. A national project called Coastal Change Pathfinder has sought to help communities, including tourism businesses, adapt and become more resilient to change, based on sound science and local knowledge of climate change and its consequential impact on coastal change.

[1] World Heritage Visitor Manager, Jurassic Coast Team.

1. Our Global Connections

The Dorset and East Devon Coast World Heritage Site (also known as the Jurassic Coast), stretches for 95 miles (155 km) along the southern English coastline, encompassing one of the most spectacular geological sequences in the world. The internationally renowned coastal exposures of the Jurassic Coast were awarded World Heritage Site status in 2001 based on a near complete sequence of Mesozoic rocks, which record the development of early reptiles through to the age of the dinosaurs. For a site to obtain World Heritage Site status, it must exhibit cultural or natural features that are of "outstanding universal value" and must be protected for present and future generations for all humanity. Protecting the integrity (or condition) of the site is essential for maintaining the qualities that led to inscription.

Figure 20-1. Durdle Door. Photo credit: Jurassic Coast Team.

The World Heritage Committee, on behalf of UNESCO, designates World Heritage Sites. Around the world, there are currently 936 World Heritage Sites in total; 725 of these are cultural sites, 183 natural and 28 are mixed sites (2011 figures). The Jurassic Coast is a natural site, which are designated under Criterion (viii) "to be outstanding examples representing major stages of earth's history, including the record of life, significant on-going geological processes in the development of landforms, or significant geomorphic or physiographic features." Put simply, the rocks and fossils along the Jurassic Coast represent 185 million years of the Earth's history (a third of the record of life) in just 95 miles of coastline. This is the basis for the site's Outstanding Universal Value.

In the United Kingdom there are 28 World Heritage Sites (2011), of which just four are natural sites. The Jurassic Coast is England's only natural World Heritage Site and this is likely to remain the case into the future.

2. Management and Coastal Change along the Jurassic Coast

The Jurassic Coast World Heritage Site Management Plan 2009–2014 (Jurassic Coast Team, 2009) is structured around eight long-term aims that are supported by a number of targets and indicators against which progress is measured.

The Management Plan guides the work of the Jurassic Coast Steering Group and partners in managing the site now and into the future. The Steering Group consists of broad stakeholder involvement and is the main body responsible for the delivery of the World Heritage Site Management Plan. The Jurassic Coast team and associated partners deliver the work program. It is, however, the duty of all stakeholders involved to help manage and protect the integrity of World Heritage Sites.

An important management policy for the Jurassic Coast is ensuring the natural erosion processes that form the basis for its designation are allowed to continue. Erosion is what gives the site its value; without it, the Jurassic Coast would not have received World Heritage Site status. Articulating this to local and global audiences and stakeholders, and changing the common perception that natural processes (including erosion) must be stopped, is an important part of the management of the Jurassic Coast.

In 2010, a national project called Coastal Change Pathfinder, funded by DEFRA (Department for Environment, Food and Rural Affairs), sought to ensure that coastal communities are well equipped to understand, debate and take part in decisions about coastal change, through meaningful engagement and participation. The aim of the project has been to help communities, including tourism businesses, adapt and become more resilient to change, based on sound science and local knowledge of climate change and its consequential impact on coastal change.

Along the Jurassic Coast, the project focused on six case study sites, using scenario planning to engage with communities to identify risks and opportunities that might arise from coastal change in the future, and consider options for adaptation.

The outputs of the project to date have included a research project into how spatial planning can best support sustainable adaptation to coastal change, as well as an education project to embed coastal change into the geographical curriculum locally in Dorset and Devon and equip future generations with the knowledge and skills to address coastal change issues.

The project, and the communities it has engaged with, has recognized that there are potentially opportunities in coastal change as well as threats, and these can be best drawn out through a well-designed and facilitated process of discussion. Approaches must be inclusive and engage all key stakeholders, including statutory bodies. Particularly strenuous efforts should be made to engage those with most to lose from coastal change (i.e., those with properties or businesses at risk), but all coastal communities and businesses should be involved in risk management and discussions on the need to adapt to change.

The project used computer-generated visualizations to illustrate how the coast in the six case study communities could evolve in future and to demonstrate change. This was found to be a powerful means of communicating coastal change impacts to communities and businesses, as well as statutory agencies, local authorities and government. Involvement in the project of public bodies has highlighted that they are equally in need of adapting their policy and regulatory frameworks to take account of the issues raised by coastal change and support communities in the process of adapting to change.

The key findings and outcomes from the Pathfinder project are available on www.jurassiccoast.com/pathfinder.

3. Sustainable Tourism and Access

3.1 Engaging the Audience

Two of the aims of the Jurassic Coast Management Plan relate specifically to sustainable tourism.

The coastline has always been a popular holiday destination. The responsibility of the Jurassic Coast team is not so much to encourage more visitors to the coast, but to promote the interests in respect to the site's geological credentials. In other words, rather than marketing the location, the team markets the geological and associated elements of the product offer. This approach poses numerous challenges mainly because geology and the earth sciences can often be an inaccessible and abstract subject area for non-specialists.

A first step is to examine ways in which the geological stories can be interpreted to engage the audience on a variety of levels. In 2003, the Jurassic Coast World Heritage Steering Group commissioned the Natural History Museum in London to investigate how this could be done. The scoping study revealed that written interpretation should be targeted to an audience aged between seven to fourteen years and that there were a variety of geological stories that could be told with site-specific elements (Natural History Museum, 2003).

Bearing this in mind, the interpretation provided by the Jurassic Coast team is designed to be accessible to a broad audience, often without specialist scientific knowledge. Examples include the Jurassic Coast World Heritage Site "Miniguide" (Jurassic Coast Team, 2011) that gives an overview of the whole site, while encouraging people to explore the area by public transport, walking or cycling and incorporates information on responsible fossil collecting and other activities.

In addition, a series of publications for sale include an official guide to the Jurassic Coast (Brunsden et al., 2005) and a series of more detailed site-specific guides. Income generated from these sales is managed by the Jurassic Coast World Heritage Site Trust and goes towards reprints, distribution and supporting conservation and education projects.

As a World Heritage Site, the Jurassic Coast attracts interest on a global scale. This has the potential to increase significantly when the Jurassic Coast becomes the backdrop for the London 2012 Olympic and Paralympic Games sailing events that will be held in Weymouth and Portland. The Jurassic Coast website (www.jurassiccoast.com, Jurassic Coast Team, 2006) contains a wealth of information to inspire, inform and engage potential audiences of all ages and nationalities (key pages are translated and podcasts are available in a variety of languages). As new technology continues to emerge, the scope for providing visitors with current, up-to-date and personalized information is constantly evolving. It is important to recognize that not all visitors will have access to these new and emerging technologies, so high-quality, physical on-site interpretation will always be needed.

Along the coast, outdoor interpretation panels have been located in appropriate locations (such as car parks) but care has been taken to avoid imposing on the tranquility of the natural environment. In many cases, natural, local materials (e.g., setting a panel in a local stone plinth) have been used to complement the surrounding environment.

Recently there has been a move towards using more creative means for conveying messages about the Jurassic Coast to the general public. The Jurassic Coast Arts Program has been specially created to deliver this objective. Recently commissioned work includes a site-specific arts project called "Universal Value," which explored the inter-relationship between people and their environment. The artist recorded interviews with local residents along the Jurassic Coast and then projected the edited film with no sound and in slow motion onto a rugged cliff face. The result was a stunning visual portrayal of people's reaction to their natural heritage, played out on the very environment that stimulated their reactions in the first place.

Arguably the best way to truly engage with any audience is through first-person interpretation. Experiencing a guided walk and learning how to find fossils with an expert is much more evocative and memorable than reading about it in a book or online. The Visitor Center and museums showcase some of the key interests of the World Heritage Site and most of them offer visitors the chance to experience guided walks, evening talks and seminars.

3.2 Visitor Management

The movement and geographical spread of visitors across the area and throughout the year is greatly significant to the condition of the site's setting. Efforts are made to alleviate the pressure of large numbers of visitors at "honey pot" sites i.e., those locations that are so popular they are potentially close to or at their capacity. Usually, locations along the coast are under most visitor pressure in the summer months – July through the end of August. However, there is now a growing trend towards taking short trips during half-term school holidays (particularly October, when the weather on the south coast of the United Kingdom can still be very favorable) and during Easter.

Capacity is notoriously difficult to measure. Perception often interferes with real issues of capacity; for example, what to one person might constitute unacceptable levels of congestion, may well seem perfectly acceptable to another. The question can arise as to whether one is monitoring capacity or managing the perception of capacity. Both are important, but neither is easily resolved.

What is clear is that some parts of the Jurassic Coast are much busier than others. Often this is due to the accessibility of a site and the provision of facilities (car park, toilets and cafes), but it can also be down to historical factors (e.g., Lulworth Cove has been a popular holiday destination for centuries), media coverage and a destination's marketing strategy. For example, if every brochure, website and newspaper article on the Jurassic Coast uses Durdle Door as the accompanying image, that location will (and has!) become a heavily visited site.

Sometimes extreme action is required; for example, Avebury World Heritage Site, also in the south-west of the United Kingdom, has a non-marketing policy, as the site managers, together with the community, took the decision that the sensitive landscape setting was nearing capacity. To further complicate matters, while some parts of the Jurassic Coast may well have the capacity to receive more visitors, it must be remembered that tranquil zones along the coast are an integral part of its appeal and indeed, integrity. But should tranquil zones remain as they are so that visitors are encouraged to descend on the traditionally busy parts of the coast?

Careful marketing clearly has a role to play, and for the Jurassic Coast team this means influencing and providing advice for other marketers who wish to use the Jurassic Coast in promotional material, be they accommodation providers, local authority tourism officers or businesses seeking involvement with the brand.

Figure 20-2. Ladram Bay, East Devon. Photo credit: Jurassic Coast Team.

A Jurassic Coast Quality Business Scheme has been set up, which encourages business involvement with the coast and aims to support businesses to be in part responsible for the management of what is, essentially, "their" World Heritage Site. Many of the businesses involved are accommodation providers, recognizing that they are often the first point of contact with a visitor to the area. Nearly 100 businesses have been accredited and have thereby gained a number of benefits including use of a logo closely associated with the Jurassic Coast. Of those businesses accredited, 69% felt the Jurassic Coast was "very important" to their business and 30% felt it was "quite important." It is hoped the numbers involved will continue to increase and create Jurassic Coast ambassadors who will be advocates for the World Heritage Site and all that it stands for.

Engagement with the wider community has always been at the core of the principles of the Jurassic Coast. The team endeavors to work with local communities to generate a sense of pride and ownership about their environment. Local communities are increasingly gaining confidence with their role in the management of the site, including adaptation to coastal change.

3.3 Managing Access

Access to the Jurassic Coast is complex due to various factors, including erosion, climate change, the physical geomorphology and the fact the coast has multiple landowners. To preserve the integrity of the site, it is important that access is given due regard within the remit of visitor management.

Eighty percent of visitors to the Jurassic Coast come by car and the coastal roads have seen a significant increase in traffic. For example, in Dorset as a whole, traffic grew by 5.5% between 1999 to 2004; whereas, on roads servicing the Jurassic Coast, it grew by 8.6% in the same time frame. Traffic growth is undoubtedly part of a national trend in the United Kingdom, but that does not make it any less concerning. Changing attitudes and behavior towards travel is a huge challenge but one that those individuals involved in the management of the Jurassic Coast are attempting to tackle.

A starting point has been to try to get people to leave their cars behind once they are on holiday in the area. This can be marketed as an appealing option (avoid traffic jams, no need to pay for parking, more relaxing, etc.) and has been especially successful where the alternative to the car is seen as an "experience" or "attraction" in its own right.

In 2002, an existing bus service was the beneficiary of a successful three-year Rural Bus Challenge grant submitted by Dorset and Devon County Councils and First Bus company. First Bus increased the frequency to a two-hourly service between Exeter and Poole and six new, low-floor, double-decker buses were purchased and branded in an engaging and distinctive "Jurassic" livery. Local people, through a competition, have named the buses and the drivers have taken part in training courses that aim to raise their awareness and understanding of the Jurassic Coast. Passenger usage has increased fourfold.

The potential for waterborne transport along the Jurassic Coast has only recently begun to be explored. While boat trips do operate out of several harbors, they are currently seen more as an excursion rather than a means of traveling from one place to another. They have an important role to play in demonstrating the coastline to visitors – probably the best way of seeing the Jurassic Coast's unique geology is from the sea – and local operators have benefited from Jurassic Coast training. Work is currently under way to investigate the potential for waterborne transport as a means of transport along the coast, as opposed to purely recreational boat trips. This is attracting significant interest from potential investors and the Jurassic Coast hopes to be successful in attracting funding and investment to develop services and infrastructure, working together with a number of potential partners.

Other than from the sea, the second best way to see the World Heritage Site is to walk along it. The South West Coast Path National Trail runs along the entire length of the site. Recreational counters to monitor path use are in place at key locations along the route and provide valuable information for monitoring and managing the trail. A challenge is the expense associated with the on-going erosion of the coast path and the necessity to create diversions to maintain the accessibility of the route. Management of coastal defenses is directed by Shoreline Management Plans prepared for South Devon and Dorset Coastal Authorities Group (Halcrow Group, 2008) and for the Poole and Christchurch Bays Coastal Group (Royal Haskoning, 2008), which encourage a policy of "no active intervention" along the majority of the Jurassic Coast. Since erosion is a key process that maintains the natural beauty and integrity of the World Heritage Site, conflicts can arise with the need to maintain and promote the coastal footpaths.

4. Conservation and Conflicts

The geological sequences along the Jurassic Coast chart monumental changes encompassing hot deserts, deep oceans, shallow warm seas, dense forests, swamps, lagoons and salt lakes. The fossils revealed along the site are crucial to our understanding not only of past environments but also how life adapted and evolved over time.

Coastal erosion is an ongoing, natural process and once fossils are released onto the beach, they are at the mercy of the ocean that inevitably damages them and washes them out to sea. Once this happens, potentially scientifically valuable specimens could be lost forever. For this reason, responsible fossil collecting is acceptable along certain parts of the

Jurassic Coast. While it may seem surprising that people are allowed to take away a part of our natural heritage, the history of fossil collecting along this coast has always followed the premise that collectors are only collecting that which would otherwise disappear forever. Fossil collecting has been an important part of the culture of the Jurassic Coast for more than 200 years. Collectors have rescued numerous specimens that are important to science with open access to the site.

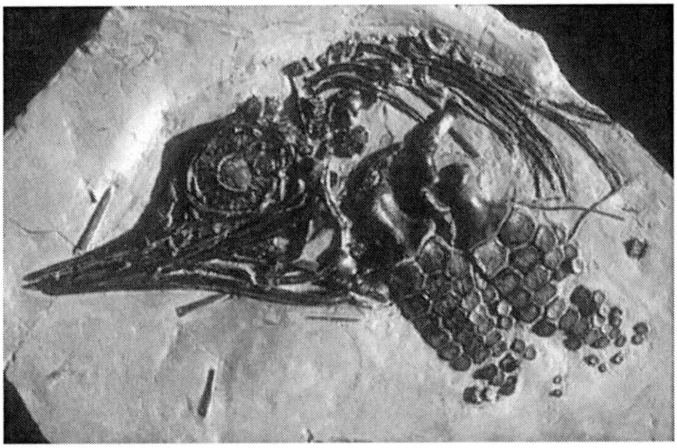

Figure 20-3. Ichthyosaur Fossil. Photo credit: Jurassic Coast Team.

In fact, some of the specimens found by renowned fossil hunter Mary Anning from Lyme Regis, 200 years ago, still help to answer key scientific questions today. Precisely when scientifically significant fossils emerge along the Jurassic Coast is impossible to predict. We do not know which cliffs or geological sections will yield a specimen that will be new to science. For visitors, ammonite and belemnite fossils are easily found on the beaches, particularly in winter and early spring. Having an activity that is appealing to visitors in the off-peak seasons is a gift in terms of sustainable tourism. The Jurassic Coast is lucky that one of its key interests is far more readily accessible in winter than in the already extremely busy summer months and gives us a golden opportunity to promote the coast out of season.

Along the West Dorset coast there is a Fossil Collecting Code of Conduct (Appendix 4 World Heritage Site Management Plan (Jurassic Coast Team, 2009)) by which landowners agree to transfer ownership of fossils to the finder provided they are collected according to the terms of the code. Collectors who do not abide by the code may be regarded as stealing the fossils. The terms of the code state that "scientifically important finds" should be registered. The fossil is photographed, a record is taken and then the specimen is handed back to the collector. Collectors and visitors alike should not collect fossils in situ (stable cliff exposures) and advice is given on safe and sustainable methods of collecting.

Throughout the last twelve years, fossil collectors have helped to discover several new species of Ichthyosaur that have contributed to our understanding of the Jurassic Period (Charmouth Heritage Coast Centre, 2006). Another collector has saved possibly one of the largest and best-preserved skulls of a Pliosaur (not in code area), which could be a species new to science, over a five-year period, as it eroded out of the cliffs. This particular find attracted significant international media coverage both when it was initially uncovered and two years later in 2011 when the preparatory work was complete and the fossil went on display at the Dorset County Museum in Dorchester for all to see.

Fossil collecting is one of the major aspects of the World Heritage Site that really bring it to life for visitors, communities, site managers and, perhaps especially, young people. It is vital that this engaging and historically important activity continues. Rather than damaging the site in any way, if carefully managed, fossil collecting can only enhance the site's integrity as new specimens come to light and as more people recognize the value of heritage, in all its forms, whether cultural or natural.

Through stimulating this shared sense of pride and ownership about the environment, and continuing to open up dialogue with communities and stakeholders about adaptation to climate change in all its forms, it is hoped that the Jurassic Coast World Heritage Site will continue to be available and valued for generations of people who come to enjoy its beauty and its amazing geology, fossils and geomorphology.

References

Brunsden, D. et al. (2008). Official Guide to the Jurassic Coast 3rd ed. Wareham: Coastal Publishing.

Charmouth Heritage Coast Centre (2006). Fossil Database, www.charmouth.org.

Fossil Collecting Code of Conduct, Appendix 4 World Heritage Site Management Plan (in Jurassic Coast Team, 2009). rev).

Jurassic Coast Team (2011). www.jurassiccoast.com/pathfinder.

—. (2009). World Heritage Site Management Plan (2nd Jurassic Coast Team (2006). www.jurassiccoast.com.

Natural History Museum (2003). Jurassic Coast World Heritage Site: Scoping Study on Interpretation Facilities, report prepared for World Heritage Steering Group.

Royal Haskoning (2008). Two Bays Shoreline Management Plan, report prepared for Poole & Christchurch Bays Coastal Group.

CHAPTER TWENTY-ONE

MEETING THE CHALLENGES OF CLIMATE CHANGE TO TOURISM IN AUSTRIA

KLAUS RADUNSKY[1]

Abstract

This article provides some basic facts about Austria and the relevance of tourism for the Austrian economy; the impacts of climate change in the past decades as identified in Austria; the ongoing efforts of Austria to prepare an adaptation strategy; the vulnerability of tourism in general and winter tourism in particular; and adaptation actions that have been already taken or are being considered. The lessons learned from adaptation activities in the winter-tourism sector conclude this article.

Some Facts About Austria

Austria is a land-locked country of 83,858 km^2 located in central Europe. About 40% of the total area lies at more than 1,000 meters above sea level (m asl). Austria has 8.32 million inhabitants (2007) who live in a Federation with nine federal provinces (Länder), each of which has its own government and parliament. It has been a member state of the European Union since 1995.

Tourism is a very significant economic activity in Austria: 124.9 million overnight stays from 33.4 million guests – about one third from abroad – are served in about 60,000 businesses with an overall capacity of one million beds. Winter tourism has become increasingly important for business despite the impacts of climate change already being felt. The importance of tourism in Austria can be seen in the fact that it provides about 10% of the jobs there and contributes about 4% to the gross domestic product (GDP) (with indirect effects of about 9%), delivering a surplus of balance of payments in the range of €5 billion per year.

---

[1] Umweltbundesamt Wien.

Climate Change in Austria

In the last fifty years the temperature has increased between 1–2°C; and in the last forty years the precipitation has increased by 10–15%, with the exception of southeast Austria which has become drier. The retreat of glaciers and permafrost is significant. The loss of glacier area between 1969 and 2006 is in the range of 30%. This has consequences for the infrastructure, i.e., skiing on glaciers. In the lowlands, snow cover has been reduced by three to four weeks per year within the last twenty years and this trend is getting even stronger. Since 1970 the snowfall has decreased by 10%, even at an altitude of 3,100m. This is a clear indication of global warming. Looking at the future, based on the A2 emission scenario, the snow duration in the Alps in altitudes of 2,000m is likely to be reduced by 50% and by 95% at levels below 1,000m (IPCC AR4). This will impact winter tourism in Austria, as most of the skiing regions are located in altitudes between 1,000 and 2,000m asl.

Adaptation Framework in Austria

The development of the framework was triggered in 2007 by a request of the federal provinces. This trigger was informed by the development of the Green (June 2007) and White Papers on Adaptation to Climate Change (April 2009) prepared by the European Commission. The milestones of the development of the Austrian framework on adaptation have so far been as follows:

- Kick off 2007
- Assessment of already ongoing activities
- Five workshops (09/2007; 03 and 11/2008; 06/2010; 11/2010)
- Preliminary recommendations (agriculture, forest management, water management, tourism, electricity sector, biodiversity, transport infrastructure, health, settlements and natural hazards. For more information see: http://www.klimawandelanpassung .at/newsletter-registrierung/newsletterarchiv/.)

In addition, the Umweltbundesamt has started a participatory process. Within the government, the lead for developing a National Adaptation Strategy (NAS) in 2007 was taken by the Austrian Federal Ministry of Agriculture, Forestry, Environment and Water Management, which is supported by a group of experts representing other ministries and the provinces, called Kyoto-Forum. The NAS is planned to be finalized by

2012. The government decided to set a timescale of five years to come forward with this strategy. This iterative process will result in a final strategy in 2012. Some of the future activities will include workshops addressing health, ecosystems/biodiversity, transport infrastructure, buildings and natural disasters, with the goal being to elaborate further recommendations. The participatory process, engaging eighty-four organizations, should lead to a common understanding on how to address adaptation in Austria as well as to the identification of synergies and conflicts of interest and of needs for further research.

Vulnerability of Tourism

In studying the vulnerability of tourism, a distinction has been made with respect to four tourism categories (winter tourism in the alpine region, summer tourism in the alpine region, summer tourism at lakes, and urban tourism). The vulnerability of tourism was considered in general and a strong dependence on climate change was found in winter tourism in the alpine region, but also with respect to the summer tourism at lakes. This has high importance for the economy in general and in particular for those regions for which winter tourism is the main economic activity and which thus are very sensitive to the impacts of climate change. The regional variability of vulnerability has been assessed and is significant.

An additional point, with a linkage to Africa, is tourism at lakes, in particular Lake Neusiedel. This lake has shallow water that is a home to many bird species that come in spring from east Africa to Austria and go back later in autumn. This lake might disappear due to evaporation and lower precipitation, triggered by climate change.

Winter tourism is, however, the most vulnerable tourism category in Austria. Autonomous adaptation had already started many years ago with the introduction of artificial-snow production. This technology makes a large demand on water and energy resources and therefore artificial snow production is quite expensive. A basic snow cover of 30cm depth requires about 2,500m^3 artificial snow for 1 hectare of ski slope. The water consumption is about 600 to 1,500 m^3 per hectare and the energy demand 5,000 to 27,000 kWh per hectare. The costs amount to about €10,000 per hectare (data from 2005). Yearly investment in Austria in artificial snow production is over €100 million. A lack of sufficient snow cover raises the risks of skiing accidents.

Whereas currently 85% of the skiing regions would have a fair snow cover without artificial snow production, this number is expected to decline to about 70% by 2025, 50% by 2050 and only 20% by 2100 (see table 21-1).

Table 21-1. Snow cover.

Area of slopes	with snow making equipment	%
25,400 ha	16,760 ha	66

Number of skiing regions	Fair snow cover (without snow making equipment)	+ 1°C (about 2025)	+ 2°C (about 2050)	+ 4°C (about 2100)
228	199	153	115	47

Possible Adaptation Measures for Skiing Regions

- Regional co-ordination of investments
- Broadening of the basis for financing beyond cable-car businesses (e.g., co-operation with communities) – not very successful until now
- Shift to slopes in less climate-sensitive regions (e.g., higher elevations)
- Close inefficient slopes
- Improvements in artificial-snow production – has been significant in the past
- Diversification into other tourism activities then skiing that are independent from the availability of snow
- Diversification into other economic activities than tourism

Conclusions

It is evident that climate change is a significant challenge for winter tourism in the alpine region in Austria; however, the vulnerability depends very much on the specific local circumstances. Therefore, every region has to identify its appropriate adaptation strategy. The private sector is quite responsive to the need to meet the challenge and has sought to compensate for climate change mainly by technical solutions. More holistic approaches that are imbedded in the sustainable development strategies of a region/a community are lacking; therefore, the development of the NAS is very important to prepare the ground for a more holistic perspective. Currently the approach is mainly driven by economic considerations from the perspective of the cable-car business, based on the need to achieve a return on investment within four years.

CHAPTER TWENTY-TWO

QUEENSTOWN TOP 10 HOLIDAY PARK CREEKSYDE

ERNA SPIJKERBOSCH[1]

Introduction

Queenstown Top 10 Holiday Park "Creeksyde" was established in 1987 in the pristine mountain and alpine area of Queenstown, New Zealand. The park was originally a site for camper vans, and has grown into a popular holiday accommodation park that offers a wide range of facilities and accommodation types. The site provides a more up-market facility for caravan users, providing luxury facilities that are not found in most other campsites.

Queenstown's award-winning holiday park is only a 5-minute walk from the town center, located in a sheltered garden setting alongside a peaceful stream. A variety of accommodation types are nestled amongst

[1] CEO, Queenstown Holiday Park.

the trees and the site with shops, including a large supermarket nearby. Queenstown is a pristine mountain and alpine four-season resort with a great appreciation of outstanding natural landscapes.

A unique quality that draws tourists to this destination is the distinctive pole-and-beam construction cantered around a 12-meter-high, 10-ton boiler tube. This graffiti pole has become a local landmark due to the thousands of signatures from previous guests covering its surface.

Environmental Considerations

Creeksyde has had environmental considerations in mind since their establishment more than twenty years ago. The park joined the EarthCheck program in 2001 and has benchmarked its environmental practices nine times, achieving commendable results each year.

The park received Gold Certification due to its efforts in the program and the sustainability achievements they have made. Key achievements in 10/11 include:

- Energy consumption was 17.3MJ per Berth Occupancy Day, 69.1% better than Best Practice
- Greenhouse gas emissions were 1 kg CO_2-e per Berth Occupancy Day, 51.5% better than Best Practice
- Potable water consumption was 120.8L per Berth Occupancy Day, 42.5% better than Best Practice
- Waste sent to landfill was 1.7L per Berth Occupancy Day, 40.8% better than Best Practice
- Waste recycling rating was 91.7 points, 11.7 points better than Best Practice
- Community contributions rating was 100 points, 20 points better than Best Practice
- Paper, cleaning, and pesticide products rating was 100 points, 20 points better than Best Practice
- Renewable energy consumption was 100% (hydro and solar hot water pre-heating)

Key Performance Areas

Reducing the Use of Freshwater and Wastewater Generation

Queenstown Holiday Park is an urban property that runs on town water reticulation. Sites in the park are mostly of permeable make up so that rainwater returns directly back into the land and the adjoining fish-spawning creek, rather than into the storm-water system, which includes road run off. The park has undergone a number of initiatives to ensure they are not consuming excessive water throughout their property.

- To reduce unnecessary water loss, Creeksyde ensures they check for leaks on a weekly basis. The reading of water meters allows leaks to be identified and repaired.
- Low/dual flush toilets were installed in a number of bathroom facilities across the property and all taps and showers were upgraded with low flow fittings. Additionally, motion sensors were installed on urinals to limit flushing. These initiatives helped achieve a water savings rating of 86.9 points, 6.9 points better than Best Practice.
- Native plants are used throughout the park to reduce the amount of required water for irrigation. Additionally, irrigations systems are used at night where possible and function on timers to ensure only the necessary amount of water is used.
- Staff members are trained to turn off taps while cleaning and undertaking other activities.
- All services are overlooked daily and readings undergone monthly. As a result of these initiatives, Queenstown Holiday Park measured potable water consumption of 120.8L per Berth Occupancy Day, 42.5% better than Best Practice.

Energy Efficiency and Reducing Greenhouse Gas Emissions

Queenstown Top 10 Holiday Park Creeksyde has performed extremely well in terms of energy consumption and management over the past 10 years. Several initiatives have been implemented in order to reduce their energy use and consequently, their environmental impact.

- Queenstown sources all of its electric energy from hydro sources, resulting in a renewable energy consumption of 100%.

- Hot water supplies are preheated from captured heat from laundry dryers across the property as well as solar panels.
- Guests and staff are encouraged to turn off lights and equipment when not in use and lighting runs on a light or motion sensor where possible.
- Double-glazing, heavy curtains and good insulation is used throughout the property to reduce heat loss, and therefore reducing the pressure on the heating system. These initiatives helped reduce energy consumption to 17.3MJ per Berth Occupancy Day, 69.1% better than Best Practice.
- As a result, the Holiday Park's greenhouse gas emissions are notably low, at 1 kg CO_2-e per Berth Occupancy Day, 51.5% better than Best Practice.

Solid Waste Reduction

The Holiday Park is performing at Best Practice in many aspects of waste management. Recycling systems and guest and staff education have helped the park achieve an effective waste disposal system that significantly reduces the amount of waste sent to landfill.

- To communicate the park's commitment to waste management, recycling messages are located in all accommodation compendiums and on signs in the kitchen area. The park's guest map asks for help with the 'reduce, recycle and reuse' ethos.

- Well-signed recycling stations that separate waste according to material are also located around the park to encourage staff and guests to dispose of their waste responsibly. This waste-management system achieved the park a waste recycling rating of 91.7 points, 11.7 points better than Best Practice.
- Grounds staff members check the sorting bins and remove as much as possible, which guests have not sorted. All garden waste, and as much food waste as possible, is composted using a Worm Farm.
- Additionally, food scrap bags are provided in all accommodation units and in all communal kitchens to encourage guests to separate out food scraps for composting. The waste sent to landfill by the park is significantly low due to these initiatives, measuring at 1.7L per Berth Occupancy Day, 40.8% better than Best Practice.
- Unwanted materials are reused where possible. For example, crushed glass is used in the floors and paths of new buildings, and for the floor and tabletop of the BBQ Kitchen. As a result, 40.7% of the site's waste is recycled/reused/composted.

Sustainable Purchasing to Reduce Waste

Queenstown Top 10 Holiday Park has established green procurement techniques that ensure their supply chain is operating in an environmentally friendly manner, and waste sent to landfill is minimized.

- Suppliers are made aware of the park's ethos via direct and media information. All purchasing is based on a criteria checklist that staff members refer to ensure supplier's environmental policies reflect those of the park. The park's purchasing policy includes waste minimization strategies for all items and reuse policies are also in place. As a result, 20% of the operators and suppliers dealt with by the park are environmentally accredited.
- Full training on cleaning methods is given to all staff to ensure environmentally friendly cleaning products are used and employed responsibly. Internal training is ongoing through a Housekeeping-monitoring program that monitors the quality and output of cleaning and grounds staff.
- The Executive Housekeeper is also compiling a booklet of cleaning tips for use in all departments including housekeeping, grounds and laundry. These cleaning practices achieved the park a

'cleaning products' rating of 100 points, 20 points better than Best Practice.

- No pesticides are used on the property, resulting in a pesticides products rating of 100 points.
- Additionally, recycled and environmentally friendly paper products are prioritized by the park, which also achieved them a paper products rating of 100 points, 20 points better than Best Practice.

Community Empowerment and Support

Creeksyde recognizes that the social welfare of its community is a key priority to their business and has implemented a number of programs to ensure they support their local community in a variety of ways.

- Creeksyde supports its local community and economy by providing employment to local community members; 95% of staff at the park lives within a 20km radius of the property, resulting in a community commitment rating of 95%, 35% better than Baseline Level.
- Additionally, the park sources 50% of its consumable goods from within their own country (supporting local suppliers).
- The park contributes to the good of the community by taking part in local government forums and involving themselves with various local, regional and national organizations to help spread their ethos and form partnerships. Regular events such as tourism functions and S.T.A.R steering group (Sustainable Advisors in Regions) meetings are attended to help promote sustainability practices within the region.
- Frequent media releases as well as posters and framed copies of the park's sustainability code located throughout the park reinforce and promote the park's sustainable ethos and their partnership with EarthCheck.
- 100% of staff members are formally trained in a variety of practices depending on departments in order to ensure operational practices are carried out effectively and with a minimal ecological/social impact. Senior staff members are developing cultural training for all staff on the nationalities of guests; this will promote and support the wide range of visitor's cultures.

- Additionally, employment agreements include a clause that is signed by staff in acknowledgement of the park's sustainability ethos.
- The park hosts and provides lectures for tourism students approximately four to five times a year.
- Surplus items and items left behind by guests are redistributed by the Salvation Army. As a result of these community initiatives, the park received a community contributions rating of 100 points, 20 points better than Best Practice.

PART IV

TOUR OPERATOR SUCCESS STORIES

CHAPTER TWENTY-THREE

INTREPID TRAVEL'S CARBON MANAGEMENT PLAN: A CASE STUDY

JELINA MITROVIC[1] AND JANE CROUCH[2]

Background

Intrepid Travel operates small, group adventure trips worldwide. Commencing in 1989, Intrepid has grown to take around 100,000 travelers on a range of trips to more than 100 countries. We employ more than 1,000 staff members and have offices around the globe. As things have changed and grown, Intrepid's philosophy has stayed the same – to provide the best value small, group adventures in the world.

[1] Responsible Business Manager, Intrepid Travel.
[2] Responsible Travel Manager, Intrepid Travel.

Intrepid Travel has a long track record of innovation in sustainability. We are renowned for our responsible travel approach in our operations – managing our environmental, social, cultural and economic impacts. We aim to have the most effective systems and good governance practices in place to help us understand and manage our impacts on society and the environment, and we are committed to working against corruption in all its forms. We see adherence to our sustainable development policy as essential to the way we operate our business, live out our core values, achieve our purpose and work toward our long-term vision.

In 2006, Intrepid Co-Founder and Director Darrell Wade had his "moment of truth" around facing climate change realities. In recognizing that as a business, we encouraged people to get on planes and fly across the world to take Intrepid trips, Darrell felt that we had to make some bold and effective plans to take responsibility for our emissions and make reductions wherever possible. The decision was made for Intrepid Travel to become carbon neutral, company-wide, by the end of 2010. The groundwork for how we would get there came later!

Carbon Management at Intrepid

Intrepid recognizes that climate change is one of the most urgent problems facing our world today and that the tourism sector is a growing contributor to the problem. As a travel company that creates and promotes holidays within the tourism sector, we see it as our responsibility to ensure that the negative impact we have on global warming is minimized and that we work toward sustaining our environment; therefore, as a business we made a commitment to tackle climate change through the development of our Carbon Management Plan.

An extraordinary amount of thought and effort went into the roll-out of our Carbon Management Plan within a four-year period. In 2008, we became a signatory to the United Nations Global Compact and part of our commitments included working toward environmental sustainability by addressing climate change. In late 2010, we were very proud to announce that we'd become a carbon-neutral company!

Our Carbon Management Plan supports the practice of our company's core values, including operating our business in a sustainable manner. It also contributes toward addressing our environmental commitments under the United Nations Global Compact, which in turn works toward the seventh Millennium Development Goal of "ensuring environmental sustainability."

Carbon Management Plan Outline

The first step in our journey toward carbon neutrality was to measure our impacts. Intrepid undertook an extensive Life Cycle Assessment (LCA) on the business to understand where our environmental impacts were. We found that our main carbon emissions came from:

1. Our corporate/office sites
2. Our trips
3. The flights that passengers were taking to meet our trips

We chose to include the sources of the following six gases (where appropriate) covered by the Kyoto Protocol:

- Carbon dioxide (CO_2)
- Methane (CH_4)
- Nitrous oxide (N_20)
- Hydro fluorocarbons (HFCs)
- Per fluorocarbons (PFCs)
- Sulphur hexafluoride (SF_6)

We report emissions of these gases in units of carbon-dioxide equivalents (CO_2-e).

We then created a plan to help us manage our impact on climate change. Here are some of the actions outlined in our plan:

1. Corporate/Office

We have adopted the following principles to manage the environmental footprint of our offices and stores globally:

Measure: We measure our emissions – everything from electricity, gas, waste and business travel to paper usage – from ALL our global offices (20 international offices and 10 domestic offices/sites). See the Corporate Greenhouse Gas Inventory below.

Avoid: We adopt energy-efficiency measures, such as automatic computer shut off at 8 p.m. for head office staff and Skype conferencing rather than traveling for face-to-face meetings.

Reduce: We're reducing the number of business trips we take, paper we use and waste we generate.

Offset: Whatever we can't avoid, we offset by investing in high-quality renewable energy projects, which benefit the local community.

Intrepid-owned properties and leased properties, where Intrepid has operational control over light and power, are incorporated; however, Intrepid is not able to claim full operational control of base power of a leased property (such as power used for elevators, air conditioning and central heating) unless Intrepid is the sole tenant.

Table 23-1. Intrepid travel's GHG inventory.

Corporate Greenhouse Gas Inventory
Scope 1 (Direct Emissions)
Gas consumption, often used for heating, water heating and cooking
Fuel consumption, as used in owned vehicles
Scope 2 (Indirect Emissions)
Electricity, as consumed by the offices directly, which is usually used for lighting, HVAC (heating, ventilation and cooling), office equipment, appliances and sometimes water heating
Scope 3 (Optional Material Emissions)
Fuel consumption from the use of taxis, public transport and aircraft
Electricity consumed in hotels during business travel
Waste generation and disposal
Paper use

After following the above actions to reduce our corporate footprint, Intrepid Travel globally produced a total of 1988.84 tons CO_2-e in the 2010 Australian financial year, which we have subsequently offset. The pie chart below shows the sources of our office emissions in the 2010 financial year.

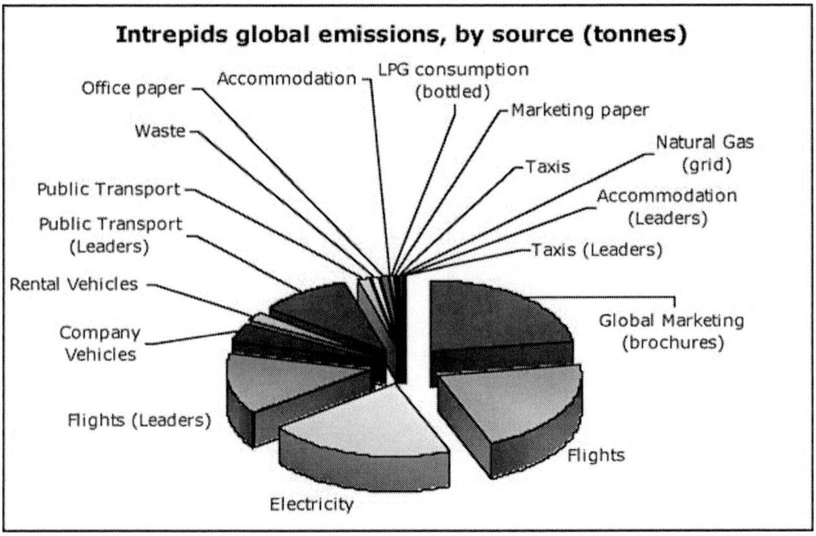

Figure 23-1. Intrepid's GHG emissions, 2010.

Although our emissions from business flights, brochures and company vehicles increased from the previous year, we managed to reduce our overall emissions from electricity and waste, as well as divert more waste from landfills by recycling. We are hoping to achieve further reductions in the future by changing our everyday behaviors in the office when it comes to energy use, business travel and waste disposal.

2. Trips

All trips directly operated by Intrepid Travel are carbon offset – that's almost 500 trips. We have measured and offset the main sources of the emissions created on our trips by our passengers. Where measurement was not possible, we carefully extrapolated this information. The trip components assessed were:

- Transport
- Accommodation
- Waste

Our trips are generally low-impact by design. We use public transport where possible, stay in locally owned and simpler styles of accommodation and eat at locally owned eateries where the food has been locally sourced (therefore reducing food-mile emissions).

In the 2010, we offset 25,000 tons of CO_2-e emissions of our passengers' footprint – that's equivalent to taking 5,000 passenger vehicles off the road for one whole year.

3. Flights

We offer customers an opportunity to offset the most carbon-intensive portion of their trip – their flight. When customers book their flight through Intrepid, we offer an offset component to their airfare. From 2007, Intrepid has managed to offset more than 45,000 tons of emissions through our flight-offset program. That's investing more than AUD$500,000 in renewable offset projects from this program alone.

Methodology

Intrepid has used methods based on the World Resource Institute and the National Greenhouse Accounts Factors (from the Australian Government's Department of Climate Change, 2008). Flights are measured as per 2008 Guidelines of the UK Government's Department for Environment, Food and Rural Affairs (DEFRA) GHG Conversion Factors: Methodology Paper for Transport Emission Factors, 2008. The flights are measured region to region and take into consideration Great Circle Distance (GCD) and Radiative Forcing Index.

Staff Resource Commitment

1. Responsible business manager role

- Overall facilitation and management
- Definition of scope and boundaries
- Updating management plan and carbon factors
- Contracting of offset provider
- Distribution and calculation of carbon worksheet
- Providing training and information on how to reduce carbon
- Website maintenance
- Reporting

2. General managers (covering 30 offices)

- Responsible for overall carbon management at their offices
- Drive carbon-reduction activity
- Ensure finance teams have filled out carbon worksheets

3. Finance teams globally

- Fill in and submit carbon worksheets globally

4. Operations and destination managers

- Provide trip information that allows carbon to be measured on each trip
- Seek ways to reduce carbon on trips through design of the product
- Ensure that environmental audits (Annual Responsible Travel Evaluations) are completed on each trip

5. Responsible travel manager

- Prepare policies and training material that supports the embedding of environmental sustainability practices into Intrepid trips

6. Leaders and sales staff

- Relay our commitment to reduce our impact on climate change to customers and potential customers
- Encourage customers to offset flights and also minimize their footprint

Some Outcomes

The process undertaken to become carbon-neutral was the result of careful planning and execution. There were significant monetary contributions (more than AUD$970,000 across three years) on behalf of our sector, travelers, suppliers and staff members to seven different internationally certified carbon-abatement projects. This in turn, contributed to:

- The development of local communities through infrastructure development around the carbon-abatement projects;
- Improvement of the environment of the local communities by providing an alternative to fossil fuels and therefore producing less pollution;
- Generation of employment for local people; and
- Increased energy availability and accessibility to communities with improvements to quality of life.

Our carbon credits are obtained by investing in internationally accredited Voluntary Carbon Standard (VCS)-compliant projects that are based in some of Intrepid's most popular destinations:

- Bargaran Mini Hydro and Jamnagar Wind Power projects, India
- Macahoe and Sichuan Province Hydro and Quzhai Waste recovery projects, China
- Rice Husk Biomass project, Thailand
- Callahuanca Hydro Efficiency project, Peru

We are adding two other projects to our list in 2011:

Biomass project (Brazil) will address the significant deforestation problem in Brazil, saving 24,000 tons of native wood simply by swapping wood for renewable biomass alternatives such as peanut shells and sawdust to fire the ceramic plants across the country.

Hydro project (Vietnam) will construct a new 3km road and directly supply clean electricity to 50 households living near the project, and will devote part of its revenue to a local poverty-reduction program.

The Carbon Management Plan also created awareness of the importance of addressing climate change amongst our stakeholders, including staff members, passengers and suppliers, by engaging them in addressing the issue through the Intrepid business.

Some Challenges

There were a number of challenges in rolling out our Carbon Management Plan across the company, but the two biggest hurdles were:

- Finding a way to measure the footprint of thirty global offices, our entire trip portfolio and all of our passenger flights was a huge and lengthy undertaking. Our processes needed to work for a wide variety of locations. It took much trial and error before we found a simple, effective solution.
- Because of the global nature of our business, we experience language difficulties communicating the concept of climate change (and consequently our Carbon Management Plan) to our various stakeholders. To address this issue, we developed illustrated training materials to help us effectively communicate, regardless of spoken language.

We are constantly reviewing and improving our Carbon Management Plan to simplify the process and improve efficiencies. We make changes to address any issues as they become apparent.

The most important factors contributing to the overall success of our Carbon Management Plan are:

- Support for the project from the top. Even during the global economic slowdown, there was still unwavering support from Intrepid's co-founders and senior management group.
- Staff support from all areas of the company, including finance, product and operations, sales and marketing, who devote their time and energy to seeing that Intrepid ensures environmental sustainability at our operating destinations.

Ensuring Sustainability is Ingrained in Our Culture

At Intrepid, we're genuine about engaging in responsible, sustainable operations and management at all levels.

We've integrated respect for the environment into our company culture in a variety of ways, most notably through the company's core values – staff members are rated on their contribution to environmental and social sustainability in their annual performance reviews.

Also, the carbon-neutral objective and Carbon Management Plan (outlined above) were put forward by our passionate oo-founder Darrell Wade and were announced as a company-wide goal. This ensured there was wide support from all staff to get behind it.

Other smaller initiatives to engage staff include:

- A global office competition for the smallest environmental footprint;
- Annual sustainability awards for staff; and
- Monthly incentives for staff to offset client flights.

We've also integrated our environmental commitment into our website, newsletter, trip notes for passengers and instructions for group leaders.

All these little initiatives help to build a culture around respecting the environment and working toward its preservation.

Working with Suppliers

Through our Carbon Management Plan, we try to engage our suppliers to also address the issue of climate change. One of the ways we are working with suppliers is through our partnership with Rainforest Alliance. Rainforest Alliance helps train our suppliers (mainly accommodation providers) in South America to operate in a more sustainable manner, which aims to benefit the environment, communities and businesses.

Independent Review

To ensure that our Carbon Management Plan adequately addressed our carbon footprint as a business, we sought an independent review of the plan from Price Waterhouse Coopers (PwC). PwC's findings were positive overall, which sees our methodology closely following the tough new National Carbon Offset Standards set by the Australian Government.

The Bigger Picture

We hope that we have set a high benchmark for our industry, one that we expect others will meet and exceed for the sake of our environment. We also see our voluntary actions in tackling climate change, alongside other Australian businesses, as contributing to the progression of climate-change talks and actions at the federal level of our government.

As a company, we will continue to live out our values, which include operating in a sustainable way, address our United Nations Global Compact commitments and be accountable for the impacts we have on our people and environment. The Carbon Management Plan is just one of the initiatives we have embarked on to help us successfully realize these goals.

CHAPTER TWENTY-FOUR

CLIMATE CHANGE AND RESPONSIBLE TOURISM: TRAVEL ANOTHER INDIA – THE VILLAGE EXPERIENCE

GOUTHAMI[1]

The tourism industry is growing worldwide, and especially so in India. As disposable incomes increase, more people are traveling on more occasions. People are curious about other cultures, cuisines and places. A couple of decades ago, Indians traveled either to pilgrimage centers – a church, temple, mosque, gurudwara – or on an annual voyage back to their village or "native place." Typically, travel involved many forms of transport – train, bus, bullock cart, rickshaw, ferry, etc. Travelers carried with them all that they would need during the journey, as they were not sure of availability or quality en route – water, a lot of food in typical Indian Tiffin carriers, bedding, personal belongings and, of course, gifts for those back home. The whole process of travel involved a lot of fun for the younger ones and a lot of activity for the older people. Very rarely would a family or group go just to enjoy the beauty of a place or admire a culture different from theirs or "travel for travel's sake." Every journey had a purpose – to see God, to see family, to go for "higher" studies and so on.

In the last twenty years, this has changed rapidly in India. Transport facilities – air, road and rail – have improved dramatically while costs have actually come down in real terms. An air ticket from Chennai to Delhi cost me 5,000 Indian rupee in 1994. It didn't matter if I booked the ticket the day before or a year before – the cost was the same. Today, I can get the same ticket for 3,000 rupee if I plan just a month earlier! With the growth of digital technology, I can book tickets, rooms, taxis and guides at the click of a button. And this has translated into more than 600 million Indians traveling around India in 2009 and 2010.

With its population of more than 1.2 billion, the numbers that India throws up for any sort of activity always leads to a groan. Given that this

[1] CEO, Travel Another India.

trend towards tourism is only going to increase, responsible tourism offers a way forward for travelers and service providers to reduce their impact on the planet.

Responsible tourism means that we keep the triple bottom line of "People, Planet and Profit" very clearly on an equal priority as we set up tourism ventures. A typical business venture would focus only on the profits. However, for long-term sustainability, it is equally important to focus on the environmental impact as well as the social impact on the community where the venture is located.

We translate this focus on the triple bottom line in the way we plan and set up responsible tourism ventures in Travel Another India.

While all three – People, Planet and Profit – are equally important, (alphabetically) "People" gets our first consideration. This would include the communities who host the ventures as well as the guests who come to enjoy the experience.

People

- The primary stakeholder is the local community or their representative(s).
- They need to get a tangible economic benefit out of the venture while ensuring minimal negative social and economic impact on the community.
- They need to be involved in all the decisions involved in setting up the venture, as it will impact their lives directly.
- The guests need to have an enjoyable experience.
- The venture should be accessible for all – people with disabilities, single women, older people, families with children, etc.
- Finally, the experience should be based on mutual respect between the host community and the guests and should enhance the pride of the host community.

All our ventures are owned, managed and controlled by the local community or their representatives. As far as possible, local people are employed by the venture or provide services to the venture. If it is not possible to find trained staff in the surrounding villages, then we work on ensuring that with training and exposure, a trained team is built up over a three-year period.

Accessibility is a key issue while planning the venture, especially for persons with disabilities. We start by identifying itineraries for our guests. An access audit is conducted to make sure that the itinerary is made accessible for guests using wheelchairs, those with a visual or hearing impairment, guests with autism, etc. The access audit document details the steps that need to be taken to make each of the places accessible – place of stay, monuments, places of worship, picnic spots, restaurants and, most importantly, toilets.

If a place is impossible to make accessible, such as the steeper monasteries in Ladakh, it is then dropped from the itinerary and replaced with another place of touristic importance. All the suggestions are made keeping in mind physical feasibility, financial viability and practicality. If we are constructing a guest house, we make sure that the doorways to the rooms, bathrooms, dining area, lounge, etc. are wide enough to take a wheelchair; supports are given in the bathroom; switch boards are at a lower height; furniture is widely spaced; switches are clearly marked; natural light is enhanced, etc.

Helpers, drivers and all staff are trained in handling special needs with dignity.

Through all our communications material, we highlight the fact that the customer is a guest in the host community to obtain a unique experience and not merely making a payment for a room. We ensure structured interaction with local craftspeople and artisans to highlight the skills of the host community.

Planet

- We focus on minimizing the environmental impact through the venture.
- We enhance the natural and cultural heritage through the venture.

All our guesthouses are built using local materials, reusing old material and in local style (vernacular architecture). Traveling across India one sees a variety of construction in rural areas. On closer examination it is explained that housing has evolved, keeping in mind the material available nearby as well as the climate of the area. So along the west coast where rainfall is typically high, one finds sloping roofs with a higher incline than in the interior where the rainfall is less. One also sees that overhangs are much greater in high-rainfall areas to protect the main building from the direct impact of the rain. In Kutch where it is hot during the day with bright sunshine and cool at night, windows are small and placed so as to ensure maximum cross ventilation.

Typically, the construction is such that it has an inbuilt temperature control system that minimizes extreme hot or cold conditions. This reduces the need to have artificial temperature control and leads to a huge savings in energy. There is a belief that it is cheaper to build using local material. While there is a marginal difference, it is important to ensure quality of construction so as to ensure maximum benefits. It is also important to bring in some modifications – to bring in modern conveniences and make the structures suitable for guests with disabilities.

Most old houses will have a threshold at the doorway to ensure that bugs are kept out. This however, poses a problem for guests using wheelchairs. Doorways are typically narrow and low to keep rooms cool or retain their warmth. There is a need to modify intelligently to allow for greater access while ensuring the building retains its traditional properties.

One area where there is little compromise is in the construction of the bathroom and toilet. Again, rather than designing for international standards, one needs to take into account the local water table in deciding if we provide a bath tub or a bucket bath and a shower.

It is important to explain to guests that it is not a mere frugality on our part to request them to reuse towels and use less water, but rather that the local water situation that demands we do so. This generates an interest in the guest to find out more about the place they have come to and perhaps count their blessings in living in a place with abundant water! In our destinations, guests are given the option of reusing sheets and linen and using thin towels that use less water. Guests are encouraged to go swimming in clean and safe natural water bodies if they are really keen to swim.

The influx of urban guests often leads to a problem of waste creation and disposal. We work on a principle of zero-waste and ensure that plastic use is kept to a minimum. (Unfortunately the best makers of water tanks in India are Sintex – plastics!) Organic material is composted while plastics are sent off to the nearest town for recycling. The whole village is involved in this, including the Gram Panchayat (local governance body) so that the village itself is transformed.

Guests are encouraged to carry back plastic waste, batteries, etc. Again this often gets them, especially the younger ones, active on the whole issue of waste management in their cities, leading to a larger impact on climate change.

We are also in the process of designing shopping bags made of cloth as souvenirs for our guests.

Guests are given the option of clean drinking water that is not in plastic bottles and a chance to refill their own plastic bottles. It is important, especially in India, that guests keep hydrated through drinking water. So special efforts need to go into ensuring that they carry water, but with minimal environmental impact. A leading hotel chain in India has taken the positive step of setting up their own bottling plant and using glass bottles in their hotels. This is a step in the right direction and needs to be widely followed.

We highlight the value of local fauna and flora – bird sanctuary, lake, forest, etc. Local guides are trained in understanding the importance of the ecosystem. A guidebook is prepared for each destination that we set up. This helps guests to better understand and appreciate both the natural and cultural heritage of the region.

Finally, we use only CFL lamps where there is electricity. We are increasingly switching over to solar lights since power cuts are so frequent in the areas that we work. We are also replacing our battery-operated torches with solar-powered torches. We are considering pedal pumps for lifting water to the overhead tanks and solar-powered fans. We will order and use a "fridge made of mud" designed by a potter in western India. Two of our partners are completely off the electricity grid. Using alternate energy serves to reduce our impact on the planet and also meets the very practical need of erratic supply on the grid.

Many of us from urban areas ignore climate change and its impact since we do not understand how it could change our lives. In rural areas, it is easier to explain the interconnectedness of the ecosystem using live examples. It is also easier to see the physical signs of climate change, the changing weather patterns, the need to sow different crops because of that, etc. The stories of the farmers across India immediately illustrate a complex global phenomenon in very simple language.

We work alongside the Gram Panchayat and the Village Tourism Committee to ensure conservation activities at the local level. It is important at every level to explain the changes, their roots and impact.

Profit

Activist groups working in the area of environment often disdain profit. Perhaps it is not possible to be "green" and profitable if profit is the only goal of a business. However, if we are clear that we are pursuing the triple bottom line of "People, Planet and Profit," then it becomes much easier to take decisions that have a minimal negative impact on the environment. The smallest decision in a tourism venture has a cost impact – the decision to request guests to reuse towels, for example. The decision options are many – we can decide to insist that guests have to reuse towels or we can request them or we can simply replace towels everyday. Which decision we opt for will be decided based on our product and the company's goals. We have decided that we will request, not insist, that guests reuse towels. We are walking the thin line between focusing purely on profits and purely on the environmental impact.

Equally, local people need to make profits out of the inflow of guests to ensure a greater buy-in into the tourism initiative. In a rural setting, it would be easy for one disgruntled person to make tourism a negative experience. However, if there are more local people getting revenue from tourism, they are likely to come together to face that one person.

Tourism highlights the natural and cultural heritage of the area. When we are trying to conserve a way of life that naturally conserves its environment (as in much of India), tourism can be a powerful tool to support the initiative. Just like any tool, it can work positively as well as negatively; therefore, it is important to convey the positives of going in for responsible tourism. This becomes easier if local people also benefit economically from the initiative.

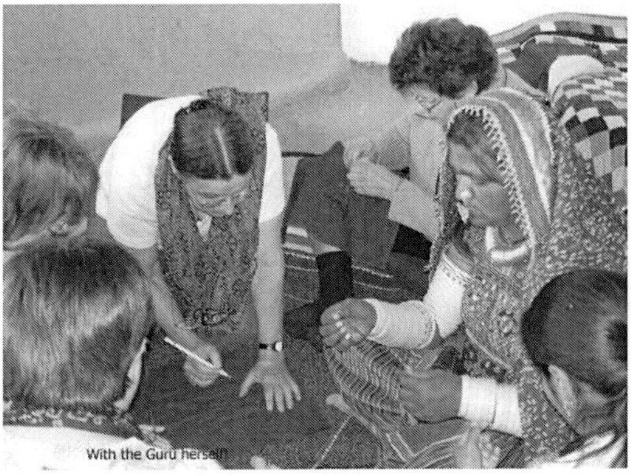
With the Guru herself!

We ensure that local people are able to set up and benefit from businesses, such as accommodation, food, craft, transport, etc.

Most villages in India have different caste groups and while the cuisine is similar based on what is grown, there are variations between caste groups and between different religious groups. So it is possible even within one village to have different families profiting from sharing their food with guests.

Most rural areas have a variety of crafts. While one source of revenue is the sale of souvenirs, another could be through workshops in that craft. Of course, a three-hour workshop in a craft, such as weaving or pottery, does not make one an expert. But it does help the urban guests to understand the complexity of a craft that is often taken for granted.

Providing bicycles gives the guests a green and simple way of getting around the area. It also revives an interest in what is rapidly being forgotten in urban areas in India – the joys of cycling.

Of course, the best people to show you around an area are the local youth. They are trained to highlight the attractions of the village as well as share local myths, ghost stories, etc.

We ensure that in each destination we encourage local businesses based on the principles of people and planet. It is easy enough to fall into the trap of focusing only on profit and ignoring the other two. For example, the food provider may find it less messy to use plastic disposable plates. So it is important to explain to them during training the advantages of using plates that a family would normally use for a special occasion and the disadvantages of using disposable plates (or the option to serve on leaf plates, as is common in many parts of southern India).

It is not the easiest option to work with local communities who often do not understand what responsible tourism is all about; however, in the long term, it is the only sustainable way – economically, socially and environmentally.

Conclusion

As a company, we are in the business of setting up tourism ventures and selling holidays. The difference is that we request our guests to consider a different type of holiday. As a starting point, I would suggest that everyone take at least one holiday a year responsibly. In addition, on every holiday, try to go beyond the routine and understand the life of the people in the area you visit. Today it is easy to travel around the world and yet stay in the same kind of rooms, eat the same kind of food and even buy similar souvenirs. Make a conscious decision that on each holiday you will view at least one aspect of the local people and their lives – be it food, craft, architecture, the local school or simply having a ten-minute conversation with someone other than your service providers (waiter, taxi driver, masseur, etc.)

I am not going to wax eloquent about the charm, warm hospitality natural beauty and exquisite cuisine of our villages – I will leave it to you to try out!

Keep the principles of responsible tourism in mind – People, Planet and Profit – and you can't go wrong!

Travel Another India is in the business of providing a unique experience to discerning clients exploring another India, rich in diversity, of cultures, cuisines and comforts, while enriching lives along roads less traveled.

We promote Responsible Tourism with travelers and hosts through supporting village communities and interested individuals set up Responsible Tourism Ventures (RTV). We provide support on sensitizing people to Responsible Tourism, planning the experience, bringing in technical and financial resources, reaching out to guests, ensuring appropriate capacity building, facilitating learning across RTVs and identifying allied livelihoods that can be enhanced.

CHAPTER TWENTY-FIVE

TOURISM, PEACE AND SUSTAINABILITY: THE STORY OF THREE SISTERS ADVENTURE TREKKING AND EMPOWERING WOMEN OF NEPAL

LUCKY CHHETRI[1]

Background

In 1993, the Chhetri sisters, Lucky, Dicky and Nicky, were running a restaurant and a lodge in Pokhara, where they had the opportunity to meet women from all over the world. They met some unhappy and frightened solo female travelers who had been harassed by their male guides, and the idea for their own trekking business, by women for women, was born. They summoned all their courage and started their Female Trekking Guide and Porter Services for Lady Trekkers in 1994. Lucky Chhetri notes three major factors that inspired them to start the business:

- Requests from solo women trekkers.
- The mountaineering training that the eldest of the three sisters had received.
- The difficult life of rural women from west Nepal.

[1] Co-founder, 3 Sisters Adventure Trekking.

Although the law states that Nepali men and women receive equal rights, the reality often does not reflect the regulation. Most women are uneducated and underprivileged, working grueling hours in the household and fields. The Female Trekking Guide and Porter Services gave local women an alternative to domestic labor and traveling tourists the comfort of a female guide. It was a natural match, but there were plenty of challenges to face. They were working in a male-dominated profession where women were made to feel unwelcome and education opportunities were limited. Our aim has been, and continues to be, to empower and develop women through tourism and to encourage sustainable tourism in remote areas where there is little hope for the future.

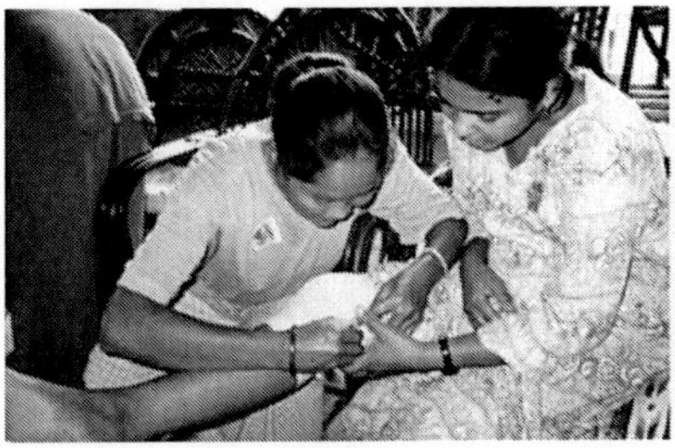

To combat these challenges, the Chhetri sisters pushed hard for instruction at the Council for Technical Education and Vocational Training Institute. This was an eye-opening experience, because they were now able to design their own introductory training program designed for disadvantaged and rural mountain women. The course was tailored to their level of understanding and aims, giving them confidence to join this profession and an eagerness to learn and earn.

The first training took place in 1996 with ten participants. The trekking company 3 Sisters Adventure Trekking was officially registered in 1998. To continue the training program, they founded Empowering Women of Nepal in 1999. Since then, 3 Sisters and EWN have been working closely together to empower women in the trekking industry and to make them independent, confident and self-sufficient.

Between 1999 and 2008 EWN conducted eighteen basic female trekking guide-training programs with more than 600 participants. The women who entered the tourism business through the program came from all parts of the country; some of them low-caste women whom society looked down upon, others socially disadvantaged and facing challenges in life.

Trekking training offers a unique combination of very practical life skills and a broader global perspective, improving cognitive skills and fostering independent thinking. There is a large demand for female trekking staff in the market, but a limited supply of skilled women. EWN and 3 Sisters Adventure Trekking has filled this void in the male-dominated trekking industry.

A guide participant says: "I learned to become an ambassador for my country." In just one month, women who had no means of earning a living – without additional help – are set on the path to self-sufficiency. At the end of the training, the women enter an apprenticeship program with 3 Sisters Adventure Trekking where they earn wages equal to those of men and acquire field experience working as trainee guides. From their apprenticeship, they gain immediate economic benefits and develop the skills they need to emerge as independent entrepreneurs.

Since 1998, the three sisters have managed to bridge a gap in the market and contribute successfully to women empowerment. Their story has been the topic of many international publications, documentaries films and other media coverage. Thanks to this exposure, more and more women are finding their way into the trekking apprenticeship. The project has received global recognition and awards from different organizations, such as Travel Plus Leisure and National Geographic.

Through the three sisters' work and activities, it is now acceptable for women to work in the Nepali tourism industry. 3 Sisters Adventure Trekking offers its employees not only fair wages (they earn USD $1,200 annually, whilst the Nepali average is USD $240) but also benefits such as insurance, tuition fees for their children and a savings and credit program. The improvement of the women's social and economic situation also directly influences their families and communities. Through contact with travelers from all over the world a vital cultural exchange takes place, which benefits both sides.

Their wages allow them to pay for university, family, medical expenses, siblings' education and their own businesses, resulting in boosted self-confidence, independence and professional skills. The three sisters have created a platform for women to experience their capacities and interests, which enables them to make their own choices about their future. The program gives the women a chance to expand social boundaries and, in the long run, to contribute to social change and a better quality of life for all Nepali women.

Tourism, Peace and Sustainability

Tourism activities range from leisure to the extreme adventure. In Nepal, adventure-tourism trekking and mountaineering are the main attraction. Tourism plays a leading role in the country's economy. It provides huge employment opportunities.

Tourism consists of knowledge, opportunities and development. It provides automatic knowledge of the places, people and the communities within each tourist market. It also allows people to experience different cultures and their communities. Various international development projects have been initiated by tourists. The late Sir Edmond Hillary started a health and education project in the Everest region after reaching the summit of Mount Everest. The result of this project is that the now-famous Sherpa people are educated, have businesses in Nepal and abroad, and hold different world records on mountaineering. They are happy, sound and peaceful.

Tourism encourages travelers to be involved in humanitarian work, such as volunteering and other social welfare work. It also helps to build peace and prosperity. Modern travelers are socially responsible, often traveling in order to support the local economy and many other issues. Every year, trekkers fund raise for breast cancer by climbing Mount Kilimanjaro.

Peacemakers are working in many ways and tourism is a factor in creating peace. Where tourism is developed, travelers visit, local peoples become involved and engaged in creating employment, building infrastructure, culture exchange and in sharing experiences – all helping to develop and maintain peace and harmony.

Peace is a product of people's emotions, activities, cooperation and good energy; sustainable peace is connected with the local, emotional atmosphere.

Likewise, the 3 Sisters Adventure Trekking Company and Empowering Women of Nepal (EWN) are organizing women treks for peace and development for the arena of community development. These treks are in the remote mountain regions of Nepal, badly affected by poverty and the domestic Maoist revolution. Every year the locals suffer from food shortages, along with many different minor diseases. There is no other source of income other than farming on the barren land.

The EWN team decided to visit the most affected area of west Nepal, Karnali, in 2003. They walked all the way from Humla to Jumla, via Rara Lake, for fifteen days. During the visit, they met community leaders, individuals, officials and political leaders. They talked about the current situation of those areas. All individuals and community members expressed the same emotions with a heavy heart, and anger. It was understandable, "hungry people become angry people – and angry people are more likely to become violent people."

The team felt sad to see all the facts and the harsh situation. The entire area was highly affected by the domestic Maoist revolution and generalized poverty. The local people were suffering from food shortages, different health issues and a lack of income sources. Men were traveling to the cities and India for employment. All local youth were forced to support the Maoists.

In the two years since the west Nepal visit, EWN developed a training package on eco-tourism, consisting of food (local food preparation), health (sanitation and hygiene) and tourism as an alternative income source. This was to meet local requirements for the development of community tourism in the remote mountain areas of west Nepal.

Prior to this program, EWN were working to address a patriarchal society through women's empowerment by the means of adventure tourism, such as trekking, rock-and-ice climbing training and providing employment opportunities.

The Maoist revolution was in process when EWN traveled and the area was not safe. They interacted with Maoist rebels along the way and had intense dialogues with them. They saw a mixture of fear and need among the people. They thought deeply about the areas and the possibilities of doing any productive programming there. But it was not easy. The main reason to extend the program to those areas was to support and maintain peace and prosperity in the country.

There were several reasons for violence in the region, including unrest due to the struggling local economy. EWN found everyone was very desperate and hopeless for his or her future. There was a huge gap between urban and rural economies and other facilities. The training package helped put ointment on the people's wounds by raising hope with the tourism-development program.

The people are now excited and hopeful regarding their future within tourism development. Now the younger generations are engaged to develop and promote tourism rather than just frustrated with their economic situation. Slowly, local people are secure and optimistic about investing locally and settling down. Peace can be maintained or created. The possibility of building peace is an important phenomenon for development, and tourism.

CHAPTER TWENTY-SIX

VOLUNTEER TRAVEL FOR WOMEN IN RESPONSE TO THE CHALLENGE OF CLIMATE CHANGE TO WOMEN IN RURAL AFRICA

LINDA RIVERO[1]

Abstract

Scientific data presented at conferences over the past few years has made it abundantly clear that climate change has progressed from being a grave concern to an urgent reality. The 5th IIPT African Conference (May 2011) provided an in-depth forum for examining the impact of the dramatic global change on the tourism industry worldwide. This paper focuses on the effects of climate change on women farmers in rural Africa, particularly in Senegal, and the specific challenges faced by these women in terms of their everyday lives, as well as their ability to participate in the societal change necessary to cope with these climatic shifts. The paper then presents one approach to coping with these problems through a program of responsible, community-based, volunteer travel for women that aims to address, on a grassroots level, some of these challenges resulting from climate change that are faced by rural women in Senegal every day of their lives.

Introduction

The volunteer travel programs of Women Travel For Peace have evolved from Peace Through Travel for the purpose of developing volunteer travel

[1] Linda Rivero, Founder and CEO of Global Action Network of Entrepreneurial Women LLC, President and Founder of Women Travel For Peace and Director and Founder of Peace Through Travel LLC.

Correspondence concerning this article should be addressed to Linda Rivero, Founder & CEO, Global Action Network of Entrepreneurial Women, P.O Box 8104, Alexandria VA 22306. Contact: lrivero@ganew-connect.com.

programs for women that support and empower women's economic development in emerging societies.

Today, the Travel Division of Global Action Network of Entrepreneurial Women LLC is made up of both Women Travel For Peace and Peace Through Travel. This Travel Division is known as "Adventure Quest." The Global Action Network of Entrepreneurial Women LLC (also known as "ga'NEW") is focused on taking action to connect, support and engage entrepreneurial women worldwide, in both the industrialized and the developing worlds, for the development and empowerment of all women.

Our travel programs bring western women to African villages to work side-by-side with rural women to strengthen their economic standing. Village women determine the community program that will best advance their education and economic development and lead them to greater economic stability and self-sufficiency. Women Travel For Peace, also known as Adventure Quest Women, works to make this happen. The result of our travel programs is one concrete step forward toward greater empowerment for the local women, as well as empowerment for the western women, though of a different nature.

In our travel work in rural Senegal, we have found that the project choices made by local women inevitably are those most immediately needed and which also, perhaps not surprisingly, relate to the demands of changing climate conditions and the social repercussions of these transitions with which they must cope on a daily basis.

With each passing day, the challenge of climate change becomes a greater concern on our planet. Yet we must bear in mind that while climate change is a risk to our entire planet,[1] those countries without the economic, governmental and municipal infrastructure to cope with sudden and dramatic changes caused by climate shifts are more vulnerable to its devastating impact. We have all witnessed the severity of these problems during the aftermath of hurricanes in Haiti 2010 and the catastrophic tsunami in Asia in 2004.[2]

Tourism, the largest service industry in the world with international tourism having reached US$ 919 billion worldwide,[3] has not only the possibility, but also the responsibility to contribute to the amelioration and, as much as possible, the solution to these serious threats.

Given the vast expanse of the industry's reach, tourism is positioned to contribute constructively to these solutions in various ways, as was discussed during the recent 5th IIPT African Conference in Zambia. This paper will focus on one approach to tourism that addresses several issues at once, climate change being one of them.

Coping with Climate Change Requires Participation of All Members of Society

First, let us consider the impact of climate change and the need for all segments of society to participate in successfully coping with changing conditions. As has been discussed in previous global conferences, including the recent UN Climate Change Conference in Cancun, adaptation to climate change requires the full participation of all members of society.[4]

"The official Report of the Conference of the Parties on its sixteenth session, held in Cancun from 29 November to 10 December 2010, Part Two" outlines the decisions adopted by the Conference of the Parties.

> Recognizes the need to engage a broad range of stakeholders at the global, regional, national and local levels, be they government, including sub-national and local government, private business or civil society, including youth and persons with disability, and that gender equality and the effective participation of women and indigenous peoples are important for effective action on all aspects of climate change;[5]
>
> […]
>
> Affirm[s] that responses to climate change should be coordinated with social and economic development in an integrated manner, with a view to avoiding adverse impacts on the latter, taking fully into account the legitimate priority needs of developing country parties for the achievement of sustained economic growth and the eradication of poverty, and the consequences for vulnerable groups, in particular women and children.[6]

Impact of Climate Change on Agricultural Production and Rural Women in Africa

It is not surprising that shifts in climate systems can be expected to impact agricultural procedures and production. According to Gerdien Meijerink and Pim Roza, authors of "The Role of Agriculture in Economic Development," a paper published in 2007 as "Markets, Chains and Sustainable Development Strategy and Policy Paper No. 4":[7]

> It is especially the people who live in ecologically and economically marginal and poor areas who suffer most by a decreasing availability of natural resources, as their livelihoods directly depend on them. Climate change, characterized by more extreme and unpredictable weather, such as prolonged droughts affects these people disproportional[ly].[8]

The fact is that women are the main agricultural producers on the African continent. Journalist Christopher Connell states in his article on "African Women Farmers, An Untapped Goldmine," appearing in Afrik-News.com, October 2010:

> In Africa, women grow most of the crops and perform most of the farm labor, as they do in much of the developing world, proving that the hand that rocks the cradle also tills the field.

It is hard work, and they do it with distinct disadvantages, because, as U.S. Secretary of State Hillary Rodham Clinton noted in a recent speech, women farmers "are very often denied access to the best seeds and fertilizer and other assets that would yield bigger crops."

David Kauck of the International Center for Research on Women and a senior gender and agriculture specialist states:

> Women often farm with rudimentary tools and without the advice of government extension agents, who can show farmers how to get more from the land. Often the family farm is not in the woman's name, so she cannot mortgage the land or get crop insurance. And often they don't control the sale of their produce. They don't see what it sells for. They don't capture the gains of their labor.

With eradication of extreme hunger and poverty topmost on the list of the Millennium Development Goals, the United States, the UN World Food Program (WFP), and major foundations and development agencies all have made securing equitable help for women farmers a cornerstone of their anti-hunger strategies. It is embedded in Feed the Future, the Obama administration's $3.5 billion effort to help poor countries provide enough food for their populations, principally by investing in agricultural development.

"Women farmers are the untapped solution to this problem," said William Garvelink, a senior official at the U.S. Agency for International Development (USAID) who helps coordinate Feed the Future. As their crops and profits grow, women are "far more likely to spend those gains improving their family's access to health, education and nutrition," said Garvelink, the former U.S. ambassador to the Democratic Republic of the Congo.[9]

The realities and statistics of rural African farming women as outlined by Kasham Shawanma Keltuma, CEO of Mother Nation Initiative, support these statements:

> In most African countries, rural women are the food farmers, and carry the burden of life. Africa's 100 million rural women grow almost 80% of Africa's food, including food for subsistence and food for markets. This amounts to food production of 3 metric tons each year per woman. Rural women do almost 80% of the work to provide the proper transport and storage of Africa's food. They do almost 90% of the work to process Africa's food, including the tasks such as threshing, drying, winnowing, peeling, grating, sieving and pounding. They also do almost 60% of the work related to marketing Africa's food.[10]

Table 26-1. A summary of the role of women in agriculture in some African countries.[11]

Benin: 70% of the female population lives in rural areas, where they carry out 60% to 80% of the agricultural work and furnish up to 44% of the work necessary for household subsistence.
Burkina Faso: Women constitute 48% of the laborers in the agricultural sector.
Congo: Women account for 73% of those economically active in agriculture and produce more than 80% of the food crops.
Mauritania: Despite data gaps, it is estimated that women cover 45% of the needs in rural areas (further details not specified).
Morocco: Approximately 57% of the female population participates in agricultural activities, with greater involvement in animal (68%) as opposed to vegetable production (46%). Studies have indicated that the proportion of agricultural work carried out by men, women and children is 42%, 45% and 14% respectively.
Namibia: Data from the 1991 census reveals that women account for 59% of those engaged in skilled and subsistence agriculture work, and that women continue to shoulder the primary responsibility for food production and preparation.
Sudan: In the traditional sector, women constitute 80% of the farmers. Women farmers represent approximately 49% of the farmers in the irrigated sector and 57% in the traditional sector; women produce 30% of the food in the country.
Tanzania: 98% of the rural women defined as economically active are engaged in agriculture and produce a substantial share of the food crops for both household consumption and for export.
Zimbabwe: Women constitute 61% of the farmers in the communal areas and comprise at least 70% of the labor force in these areas.

These threats to the farmers of Africa are particularly significant to Women Travel For Peace (also known as Adventure Quest Women), as our dedication is to improving living conditions – and life – for entrepreneurial women worldwide.

Note that at the Global Action Network of Entrepreneurial Women, we interpret the term "entrepreneurial" to describe any self-generated, profit-yielding initiative. While the entrepreneurial style of women farmers in Africa may be quite different from that of the entrepreneurial women in the industrialized west, the African women are, nevertheless, earning their living through their own, independent, commercial efforts. Thus, we consider all of us to be in the same category of entrepreneurial women around the world.

Direct Effect of Climate Change on African Farming Women

Having established the fact that rural African women contribute an enormous amount to agricultural production on the African continent, let us consider the direct effect of climate change on the productivity of these women and, therefore, the effect on their lives and economic standing and that of their families and communities.

In her article entitled "Climate Change Policy Ignores Women Farmers" appearing in IPS News.net, August 2010, writer Kristin Palitza reports that:

> "The issues of climate change, poverty, environment and gender are tightly interwoven and cannot be separated," explained WWF South Africa national climate-change policy officer Louise Naudé during the meeting. Women farmers are particularly affected by climate change, food insecurity and disaster, so we have to drive gender equality and decrease women's vulnerability in the sector. Research has shown that women are more likely to feel the effects of climate change because they have less access to resources. Changing weather patterns increase poor women's work burden on gathering water and firewood. Girls may be forced to forgo school in order to contribute to the increased household work. Where traditional land tenure is practiced, women may lose land normally reserved for growing crops for household consumption to give way for commercial crops. "We need people-centered solutions that are context-specific, participatory and use local knowledge," she stressed. "Ultimately, we want to create environmental circumstances where women are in control and don't depend on others."[12]

African Women: Poverty, Illiteracy and Social Activism

African women are deciding that enough is enough. The United Nations Department of Public Information, in an article entitled, "Women Farmers, The 'Invisible' Producers," highlights the fact that African Women are Campaigning for More Official Support:

> "Women farmers in Africa may be poor and illiterate," says Ms. Celina Cossa, president of the National Farmers Union in Mozambique, "but at the same time we are the principal force in the struggle against misery,

backwardness and dependency." In many countries across the continent, rural women are pressing for a higher profile, to match their preponderant role in the cultivation and processing of the continent's food. Yet despite some progress in recent years, state agricultural programs and facilities in most African countries do not yet reflect this reality, and rural women generally remain the continent's "invisible" producers. The FAO estimates that [en bloc] in sub-Saharan Africa, 31% of rural households are headed by women, mainly because of the tendency of men to migrate to cities in search of wage labor … Despite this substantial role, observes the FAO, "women have less access to land than men; when women do own land, the land holding tends to be smaller and located in more marginal areas. Rural women also have less access than men to credit, which limits their ability to purchase seeds, fertilizers and other inputs needed to adopt new farming techniques." Only 5% of the resources provided through extension services in Africa are available to women, notes Ms. Marie Randriamamonjy, director of the FAO's Women in Development Service, "although, in some cases, particularly in food production, African women handled 80% of the work. Of total extension agents at work in Africa today, only 17% are women."

Meanwhile, rural women have been organizing themselves to a much greater extent. In most of the countries surveyed by the FAO, there has been some growth in the number of non-governmental organizations and women's associations involving or working with rural women. Sometimes these are mixed organizations, but frequently, rural women prefer to belong to groups run by women … According to Ms. Comfort Olayiwole, principal of Nigeria's Samaru College of Agriculture, "women's groups and projects are no longer isolated ventures, easily ignored by government or community members. Women are organizing themselves into a formidable political and social movement."[13]

Adventure Quest Women: A Travel Solution to the Climate Change-Induced Hardships of Women Farmers in Africa

Clearly, the inequities that exist between the education and access to information of policy makers and that of the poor, rural women who are most directly affected by said policies are profound. While it is interesting to ponder to what degree the travel industry can affect these hierarchical discrepancies and support those most affected by climate change, the fact is that with the enormous economic clout that the world tourism industry wields, surely our industry is capable of affecting significant progress. Intelligent leverage is always powerful.

The Travel Division of the Global Action Network of Entrepreneurial Women, known as Adventure Quest and formerly Women Travel For Peace, is dedicated to building international peace through intimate, volunteer travel programs that join together western women with rural African women through hands-on support of local women's grassroots community projects.

In concrete terms, our programs help rural African women cope with the alarming threats of climate change to their everyday lives – their incessant toil, their lack of education, their exclusion from the domestic and political systems that have the power to improve the quality of their lives and the lives of their children. The fact that our intimate, women-helping-women involvement results in mutual growth and enrichment of both groups is an additional and precious blessing.

Because our mission is to connect, support and engage entrepreneurial women worldwide for the evolution and empowerment of all women, we are committed to improving the lives of our sister entrepreneurs who face dramatically challenging conditions in their struggle to become profitable entrepreneurs. Through our travel program, we directly assist and work side by side with African women who are predominantly farmers, enabling them to expand their efforts through a variety of projects.

Adventure Quest Women travel programs exist to:

Support women-determined, grassroots projects that support the education and empowerment of those most vulnerable to shifts in climate change, namely the agricultural producers, i.e., women in our targeted area of participation, sub-Saharan Africa through volunteer programs that lead to concrete results;

Bridge boundaries of culture, race, religion, language and nationality through intimate, small-group, women-helping-women travel;

Provide the opportunity for the travelers involved to interact closely and consistently with the host population, experiencing the personal enrichment and joy that comes with this new cultural exposure, which for women means cooking, singing, dancing, childcare and other universally feminine activities;

Provide the host community the opportunity to interact closely with travelers who have chosen to make this community project a priority in their own lives and contribute not only their material resources, but also of themselves through physical work and their willingness to be personally involved and transformed by this experience; and

All for the purpose of creating more peaceful world through travel – one traveler at a time.

Specific Volunteer Projects of Adventure Quest Women

Our first project, the Senegal Women's Well-Building Project,[14] resulted in the first-ever, durable, concrete well for a group of village women in the Casamance region of Senegal. The trip itself:

- Provided the funding – through the travel fees of individual travelers – for the well construction.
- Included the labor of the travelers themselves for the completion of construction.

Note that the recipients of our project, the village women, had been farming all their lives with the grossly inadequate and highly unhygienic and unsafe conditions of a collection of hand-dug wells. These crude wells present serious problems, as they:

- Wash away during the annual summer rains, requiring that they be re-dug every year.
- Present a serious safety hazard to children and animals, as both easily fall into the wells.
- Yield water that is shallow and of poor quality.
- Yield water inconsistently because they are shallow. During the dry season, because the wells provide no water at all, the women must walk many miles to fetch water, which, given the limits of what they can carry, is typically not enough to provide adequate irrigation for their crops, in addition to the water needed for family use.

By the end of our project, the group of sixty women farmers who were the recipients of our efforts had an enduring concrete well, seven meters deep, that now yields water for them all year round.

This well has dramatically improved the lives of these women as it has:

- Increased their agricultural productivity.
- Enabled the women to have greatly improved access to sufficient water for both their crops and their families' personal use.

- Eliminated the need of the women to walk long distances during the dry, scorching hot months of the year to haul insufficient amounts of water from afar.
- Reduced the hours of grueling toil under an unrelenting sun, as they are able to now access a steady water supply relatively easily.
- Reduced the safety concern of finding children or animals stranded in the hand-dug wells.

In January 2012, Adventure Quest Women/Women Travel For Peace will return to Senegal to work with women of a different village to provide them with their greatest dream – a Women's Literacy Center.[15] Unfortunately, many adult women today are unable to read. In fact, according to a United Nations Children's Fund statement, it is not only mature women who are either illiterate or barely literate. "In Senegal itself, only 15% of girls are able to go to secondary school – and later in life there are only 6 literate adult women for every 10 literate men."

While literacy may not correlate directly with climate change, it is, in fact, a critical factor in coping with environmental shifts. The problems presented by climate change are profound and require the skill, resources and participation of all members of society. Furthermore, on the African continent, which is largely dependent on agriculture for economic growth, it is imperative that the women, who contribute the vast majority of farming labor and food processing, be equipped to take their rightful place in the policy decisions that affect them and their daily lives most directly.

In January 2012, Adventure Quest Women/Travel For Peace, will enable village women of the Casamance area of southern Senegal to move their lives forward by learning to read. This reality begins with step one – providing a place for them to study. This is the mission of our next volunteer travel program.

The travelers' fees will fund the construction, and the travelers will work together with the local women, side by side, during the final days of construction.

Our work will conclude with a glorious celebration for which African communities are so well known – a day of music, dance, song and the shared joy of accomplishment.

An added benefit of a travel program such as ours is that the western women – rich in resources but likely limited in exposure to the catastrophic effects of global poverty and the impact of climate change on local life – experience the transformational effects of getting to know, working side by side with, and supporting women who are poor in economic resources but rich in spiritual and community strength.

Lasting Benefits of Volunteer Travel

Conclusions of a study prepared by Cross-Cultural Solutions in Collaboration with the Center for Social Development at Washington University in St. Louis (May 2009)[16] indicates that:

> International volunteer service positively affected volunteers' cross-cultural understanding and career path, in many cases transforming their lives. The majority of volunteers also believed they made a significant contribution to the host organizations and communities, including transferring a specific skill or providing money, time or other resources. The vast majority of volunteers did not believe that their presence in the community caused problems, and nearly all believed the community desired their services.

Intercultural Understanding

One of the most frequently reported outcomes of international volunteering is an increase in intercultural understanding and competence. Consistent with this claim, more than 95% of the alumni reported that international volunteering exposed them to communities different than the ones they grew up in, exposed them to new ideas and ways of seeing the world, increased their appreciation of other cultures, and helped them gain a better understanding of the community where they worked."[17]

My own travelers who joined me for the Women's Well-Building Project in Senegal noted a similar transformation:

> This was a wonderful experience that touched and opened a deeper level of humanity within me. I felt so very connected with some of the women there and their rootedness to what I would call spirit and what really matters. We contributed to the women in terms of the well and they were so very thankful in so many ways. … Without a doubt, I was so very touched by many of the people of Senegal that I long to go back. (Patricia DiVecchio)

> I left Senegal with a new appreciation for my time, my views and my life. I discovered that in essence we are all the same, looking to help our village. (Chhayal Mehta)

> In theory, the experience sounded altruistic, humanitarian, benevolent and kind. In reality, the experience gave me more than I gave and filled me with such gratitude that I never could have hoped for yet dreamed for. (Maria Sanchez)

Conclusion

The travel industry, with its massive economic influence and steadily increasing potential, provides a powerful vehicle for supporting vulnerable destinations in coping with the changes of our modern world by focusing on profit in terms of the improved quality of life for all, both those in the destination and the travelers, as well. This combination of social responsibility, political and economic awareness, and sensitivity to the evolving demands of our world and our planet are more than likely to produce profitable rewards, both in human and financial terms. It is up to us, as socially responsible and politically and economically savvy travel providers to do our part to help those most economically vulnerable and also culturally rich to move forward, as we build our businesses and support our honorable industry.

Notes

[1] Additional information at http://www.nature.org/ourinitiatives/urgentissues/climatechange/threatsimpacts/index.htm.

[2] Additional information at http://www.wsws.org/articles/2004/dec2004/tsun-d27.shtml.

[3] According to the forecast prepared by UNWTO at the beginning of the year, international tourist arrivals are projected to increase by some 4% to 5% in 2011. More information at http://media.unwto.org/en/press-release/2011-05-11/ international-tourism-first-results-2011-confirm-consolidation-g.

[4] Decisions adopted by the Conference of the Parties. Retrieved from http://unfccc.int/resource/docs/2010/cop16/eng/07a01.pdf.

[5] A shared vision for long-term cooperative action, page 3: http://unfccc.int/resource/docs/2010/cop16/eng/07a01.pdf.

[6] Additional information at http://unfccc.int/resource/docs/2010/cop16/eng/07a01.pdf.

[7] Additional information at http://www.boci.wur.nl/NR/rdonlyres/98CCE2E3-0FA2-4274-BCA0-20713CA1E125/62608/Fullreport4_Meijerink_Roza.pdf.

[8] Climate change's adverse effect on farming in Africa leads to its diminished economic value. Additional information at http://www.boci.wur.nl/NR/rdonlyres/98CCE2E3-0FA2-4274-BCA0-20713CA1E125/62608/Fullreport4_Meijerink_Roza.pdf.

[9] Women often farm with rudimentary tools and without the advice of government extension agents, who can show farmers how to get more from the land. Additional information at http://www.afrik-news.com/article18377.html.

[10] The rural woman in this century is a sharp contrast with her urban counterpart: unlike the urban woman, the rural woman in most African communities is saddled with enormous and even hazardous responsibilities. Additional information at

http://www.mothernation.org/index.php?option=com_content&view=article&id=75&Itemid=78.

[11] This estimate is considered low as the 1991 census included the subsistence-farming sector for the first time. Additional information at http://www.fao.org/docrep/X0250E/x0250e03.htm.

[12] Research has shown that women are more likely to feel the effects of climate change because they have less access to resources. Retrieved from http://ipsnews.net/news.asp?idnews=52593.

[13] According to official labor force statistics issued by the World Bank and other institutions, 42% of the economically active population involved in agriculture in Africa is female. Retrieved from http://www.un.org/ecosocdev/geninfo/afrec/vol11no2/women.htm.

[14] Watch us in action at http://www.youtube.com/watch?v=8Pisc_R-bzs.

[15] An unforgettable experience – Come to Senegal. https://www.ganew-connect.com/adventurequest-senegal-winter-2012/.

[16] This is an edited version of CSD Research Report 09-10 which used a cross-sectional design to sample volunteer alumni who served with two different volunteer-sending organizations: Cross-Cultural Solutions (CCS) and World Teach. This version of the report presents the findings based on the responses from CCS volunteers.

[17] These sentiments were reflected in many statements from alumni that described how the international experience exposed them to new ideas and peoples, altered their worldview, and increased their intercultural understanding.

References

Adventure Quest, Aka Global Action Network of Entrepreneurial Women (2012). Women's literacy center project (Senegal), https://www.ganew-connect.com/adventurequest-senegal-winter-2012/.

Connell, Christopher (2010). African women farmers, an untapped goldmine (October 14), http://www.afrik-news.com/article18377.html.

Cross-Cultural Solutions & The Center for Social Development at Washington University in St Louis (2009). Perceived effects of international volunteering: Reports from CCS Alumni (May), http://docs.google.com/crossculturalsolutions.org/Washington-University-Report-1-Perceived-Effects-on-Alumni-09.

FAO Corporate Document Repository- Economic and Social Development Department (1995). Women, agriculture and rural development: A synthesis report of the Africa region. http://www.fao.org/docrep/X0250E/x0250e03.htm.

Keltuma, S. K. (2011) The twenty-first century rural woman in Africa (January 9), http://www.mothernation.org/index.php?option=com_content&view=article&id=75&Itemid=78.

Meijerink, Gerdien, & Pim Roza (2007). Markets, chains and sustainable development, strategy & policy paper 4. The Role of agriculture in economic development (April), http://www.boci.wur.nl/NR/rdonlyres/98CCE2E3-0FA2-4274-BCA0-20713CA1E125/62608/Fullreport4_Meijerink_Roza.pdf.

Palitza, Kristin (2010). Climate change policy ignores women farmers (August 24), http://ipsnews.net/news.asp?idnews=52593.

Symonds, Peter (2004). Devastating tidal wave kills more than 13,000 in southern Asia (December 2), http://www.wsws.org/articles/2004/dec2004/tsun-d27.shtml.

The Nature of Conservancy (n.d.). Climate Change: Threats and Impacts, http://www.nature.org/ourinitiatives/urgentissues/climatechange/threatsimpacts/index.htm.

United Nations Children's Fund (2005). Childhood under threat: The state of the world's children 2005 (October 29), http://www.unicef.org/sowc05/english/fullreport.html.

United Nations Framework Convention on Climate Change (2011). Report of the Conference of the Parties on its Sixteenth Session, held in Cancun from 29 November to 10 December 2010. Part Two: Action taken by the Conference of Parties at its Sixteenth Session. Decisions Adopted by the Conference of the Parties (March 15), http://unfccc.int/resource/docs/2010/cop16/eng/07a01.pdf.

United Nations Department of Public Information (1997). Women farmers, the 'invisible' producers: African women are campaigning for more official support. *Africa Recovery* 11, no. 2 (October), www.un.org/ecosocdev/geninfo/afrec/vol11no2/women.htm

World Tourism Organization (UNWTO) Press Release (2011). International tourism: First results of 2011 confirm consolidation of growth (Madrid), http://media.unwto.org/en/press-release/2011-05-11/international-tourism-first-results-2011-confirm-consolidation-growth.

CHAPTER TWENTY-SEVEN

SUSTAINABLE PRODUCT DEVELOPMENT AND MARKETING

MARIKA MANN[1]

> There is a lot of talk about developing sustainable tourism entrepreneurs. But we have to teach customers as well how to use sustainable tourism. If we want to have everything for free, we are not sustainable anymore. It is nice to contribute to and support the local communities. It shows appreciation. Cheap is not always best for those whose home you visit. With no income, people cannot live here and then everyone is forced to move to the city, which would be a tragedy. Therefore, I encourage everyone to use local services … (Marika Mann, Lusaka, Zambia, May 2011)

Introduction

Estonia is a land of great natural beauty. With its Siberian-like forests, Scandinavian peat bogs, water meadows full of geese and waders and a fascinating coastline, together with a low human population, empty roads and excellent infrastructure, Estonia has rapidly become a favorite wildlife destination in Europe. Even though Estonia now takes its place as part of the modern world, Estonians have always enjoyed a close affinity with the natural world, a relationship that is evident in the traditions, culture and language proudly passed from one generation to the next. This respect for the elements of nature and adaptation to living in a climate of seasonal extremes, has given Estonians a unique perspective on the natural world, which is demonstrated by an enlightened attitude toward the all-important issues of conservation in the twenty-first century.

---

[1] Executive Manager, Estonian Nature Tours.

Figure 27-1. Männikjärve bog in Endla Nature Reserve. Photo credit: Sven Zacek.

Facts about Estonia

- Estonia stretches 350 km from east to west and 240 km from north to south.
- Sea islands form one tenth and lakes about one fifth of Estonia's territory.
- Forests and woodlands cover almost half of the Estonian territory.
- The total area of Estonian fens and bogs is about 140,000 ha.
- An admirable 17% of the country is afforded protection within nature reserves that vary in size from the large national parks to the smaller, locally protected reserves.
- With an area of 45,000 km^2, Estonia is larger than, for example, Slovenia, Holland, Denmark or Switzerland.
- Conversely, Estonia's population ranks among the smallest in the world; as of January 2000, an estimated 1,361,242 people live in Estonia – a population density of only 30.2 people per square kilometer.

Estonian Nature Tours

Estonian Nature Tours (ENT) is a tour organizer specializing in bird-watching, botanical and mammal-watching tours. ENT's individual and specialized approach to tourism was developed over a number of years, and by organizing local travel services in Matsalu National Park, before broadening operations in the spring of 2004 to include visits to a wide variety of nature reserves throughout Estonia. The principal aim of these trips is to introduce visitors to the unique natural heritage and wildlife of Estonia, which have fortunately been preserved in many places thanks to enlightened attitudes toward conservation, and a deep-rooted tradition of respect for the natural world. A skilled Estonian leader with the necessary local expertise escorts each trip. Tour leaders are often professional biologists or nature-reserve employees. Services can also include the reservation of accommodation, meals and transport or an all-inclusive tour in accordance with the specifications of clients.

Mission: Their mission is to introduce Estonia's natural heritage, preserved by moderate human activity; to promote ecological thinking and awareness of nature; and to improve local tourist services.

Philosophy: Estonian Nature Tours considers cooperation with the local community an essential part of organizing nature trips. Lunch at a gourmet restaurant cannot compare with the direct experience that guests receive having conversations with locals at a farm picnic or buying handicrafts and farm products. When a trip tour guide neglects to count minutes or hours, the guest comprehends that the guide was able to give something back by making contact with locals, thereby valuing them and leaving warm feelings in the hearts of all the guests… And that gratitude toward the tour organizer is experienced over and over again.

Products: Bird watching; botanical tours; mammal watching holidays; butterfly and dragonfly holidays; nature study tours; wildlife photography.

Matsalu Ticket

The Matsalu Ticket is a very new and environmentally friendly brand meant for anyone who wants to travel in a sustainable way and knows how to respect nature as well as local people. The project teaches about wildlife, and coordinates visits to the Matsalu National Park, helping to avoid excess disturbance of both animal and plant species, as well as the local population. Bus tours and canoe trips are lead by local Estonian- and English-speaking guides.

The Matsalu Ticket benefits several companies, including about twenty different service suppliers – tour guides, transportation companies, boatmen, tourist farms, caterers, museums, nature centers and many local craftsmen.

Figure 27-2. Common Cranes. According to data from the Environmental Board, the peak number of Common Cranes occurred here in 2006, when 23,000 birds were counted in Matsalu area, with a total of 55,000 in Estonia. Photo credit: Remo Savisaar.

Figure 27-3. Orchids. Thirty-six species of orchids (including the Lady's Slipper shown here) have been found in Estonia. Photo credit: Anneli Palo.

Sustainable Tourism Development

Estonian Nature Tours has been a pioneer in developing nature and bird tourism throughout Estonia, by mentoring and encouraging tourism entrepreneur beginners (local networks) and organizing hiking days in Matsalu National Park (2000–2004) as well as national seminars such as "Does Estonia need bird-watching tourism?" (2005, 2007). These are just some examples of ENT's contribution to the development of sustainable tourism in Estonia.

Natural-history tourism, and especially bird watching, is a highly specialized area of tourism that demands a high level of professionalism and an intimate knowledge of the subject matter from the ground agent. Estonia enjoys increasing recognition as an attractive bird-watching destination, which each year brings more birdwatchers from all over Europe in search of its many interesting bird species. Estonia (and ENT) is proud of this interest in its natural attributes, but also aware of the potential dangers that uncontrolled tourism can pose to the fragile ecosystems supporting the birdlife. The organization of "Responsible Tourism" through liaisons with the management of national nature reserves is the ethos that dictates actions at Estonian Nature Tours.

ENT Product Development

Estonian Nature Tours has long-term experience in organizing botanical and bird-watching tours and other wildlife holidays, and the feedback from clients has always been good to excellent. Every year, ENT develops one or two new tour packages.

- In the autumn of 2007, ENT introduced a new offering, "Estonia in Early Spring," which has proved very popular among clients.
- In autumn of 2009, ENT came out with totally new opportunities in Estonia, including "Estonia in Summer" (wild flowers, birds and butterflies), and, in spring 2010, "Mammal-watching in Estonia" (mammals and birds).
- In autumn 2011, ENT came out with "Butterfly Holiday" and "Dragonfly Holiday," the first such holidays in Estonia.
- In spring 2008, the oldest tour, "Estonia in Spring," and, in autumn 2011, the "Matsalu Ticket" were accredited with the brand EHE (Quality System Pure and Interesting Estonia).

Marketing Project (2008–2011)

Estonian Nature Tours was allocated a grant for marketing by Enterprise Estonia in August 2008. The marketing project (2008–2011) totals were about €82,600. The result of the marketing project during three years has been a huge increase in visitors to the ENT website. There is also more and more interest being shown in Estonia by foreign naturalists and birdwatchers. At the moment, ENT has eighteen agents all over the world who sell tours to their clients.

Why Estonia?

What is special in Estonia? Why should nature enthusiasts from around the world visit Estonia? Do we have something that others do not? Who is our target group? Where are the target markets?

Estonia is a small and wonderful place to see wildlife at any time of year, with fantastic bird watching including Steller's eiders, Ural owl, pygmy owl, great snipe, nutcracker, citrine wagtail, lesser spotted eagle, eight species of woodpeckers, sea ducks, hazel grouse and massive numbers of birds on migration. Estonia is locked between the Finnish Gulf, the eastern coast of the Baltic Sea and Lake Peipsi near the Russian border. In this respect, geographically, the Estonian waters and coastline are the natural stepping-stones, the most natural flyway, between breeding and wintering areas for millions of Arctic water birds, making bird watching in Estonia fabulous.

Figure 27-4. Brown bear cubs. Estonian forests are renowned in Europe for their healthy populations of mammals, with around 700 to 800 lynx, more than 150 wolves, 500 to 600 brown bears and almost 20,000 beavers – tremendous numbers for such a small country. Photo credit: Valeri Štšerbatõh.

Butterflies, dragonflies and flowers are equally exciting, and the country is renowned for its large number of mammals, with more than 150 wolves, 500–600 brown bears and almost 20,000 beavers plus flying squirrels and an incredible number (700–800) of lynx, one of the strongest populations in Europe!

ENT clients are bird and nature clubs, mammal watchers, nature photographers, nature lovers, private groups, tour operators (agents) and individual travelers. Most of ENT's travelers are 50 to 80 years old, who prefer to buy all-inclusive holidays.

Project Activities

Within three years, ENT has developed a website, created a client database, compiled and distributed newsletters (four to five times a year) and bought advertisements to target destination bird-watching and nature interest (in all 55 times). ENT have organized three promotional tours to European tour agencies and journalists and attended the biggest bird and eco-tourism fair in the world, British Birdfair (2008–2011).

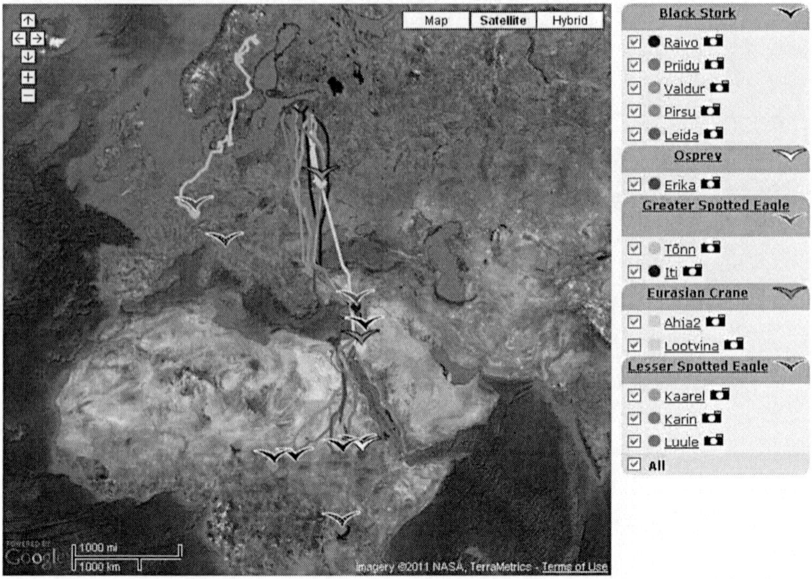

Figure 27-5. Migration routes. Since 2005, the Estonian Ornithological Society, www.eoy.ee, and the Eagle Club, www.kotkas.ee, have placed satellite and GPS receivers on seven adult birds and one young bird. Receivers are placed on the birds to research the species and organize its protection. Birds that have received valuable "backpacks" include black storks, greater spotted eagles and ospreys.

The migration route of osprey, black storks, lesser spotted and greater spotted eagles and Eurasian cranes can be observed at the following site, http://birdmap.5dvision.ee/en.

Climate Change: Peace through Tourism and Birds

Birds do not care much about frontiers. Many bird species that breed in Europe, winter in Africa. To achieve that, the passerines, many of which weigh barely ten grams, must traverse the Sahara Desert without any stopover.

Only the strongest ones survive this migration route – thousands of kilometers. Human activities have caused desert expansion in many countries of Africa and that is the reason why more and more migratory birds perish before reaching their wintering areas. A mere couple of decades ago, songbirds like the wood warbler, spotted flycatcher, barn swallow, red-backed shrike, turtle dove and many others were common breeders in Europe, but now their abundance has abruptly decreased.

Due to the changes in agriculture caused by human impact, many bird species are endangered. The number of barn swallows has declined remarkably, due to a decrease in animal husbandry; corncrake and skylark are about to disappear due to intensive agriculture and the cultivation of monocultures.

The most endangered birds are spotted eagles and black storks. In the breeding areas of Europe, their main threat is over-use of chemicals for fertilizers and insecticides and overgrowing of flooded areas, and in wintering areas of Africa, the diminishing of wetlands. On their migration, the birds are forced to traverse areas where hunters shoot them down as a sport. Massive illegal bird hunting on the island of Malta in spring threatens spotted eagles and black storks and affects the abundance of tens of other species. All the countries of the world must contribute to decelerate the processes that are causing declines in biological diversity.

CHAPTER TWENTY-EIGHT

ADVENTURE TRAVEL'S RESPONSE TO CLIMATE CHANGE: A SUCCESS STORY FROM INDIA[1]

SHANNON STOWELL AND CHRISTINA HEYNIGER[2]

Abstract

The field of adventure tourism is experiencing the effects of climate change, and businesses across industries operating in sensitive regions around the world are beginning to develop ways of adapting to it. In the tourism sector, the broader tourism industry has mainly focused on adaptation strategies for two major travel sectors: coastal island and ski tourism. In the adventure travel space, tour operators working in remote environments where climate-change effects are more pronounced are also taking steps to modify operations.

The article focuses specifically on some of the climate-change challenges faced by India. In a series of detailed interviews with a tour operator from India's Himalaya mountain region – Snow Leopard Adventures – a portrait of a business in transition, with lessons for other tourism companies, emerges.

[1] This article is based on two previously published reports by Xola Consulting with the permission of Christina Heyniger. The original paper titles are: "Climate Change, Water Scarcity, Sustainability and Adventure Travel in Brazil, Russia, India, China and Mexico," written by Henrietta de Veer and Christina Heyniger; and "Adventure Tourism Companies and Climate Change: Observations from the Himalaya, Amazon, and Polar Regions Spark Adaptation Strategies for Business," written by Ryan Piotrowski and Christina Heyniger.

[2] Adventure Travel Trade Association.

Adventure Travel's Response to Climate Change: A Story from India

In contrast to leisure tourism, where more emphasis may be placed on man-made settings, adventure tourism's business model is thoroughly exposed to the slightest environmental changes. Adventure tour operators, with products and services that depend on healthy, natural environments, have a regular and direct connection to important environmental issues such as climate change.

India is a leading international tourism destination and was also named one of the top ten climate-affected nations by the UN World Meteorological Organization (WMO). Six of the other nations were in Asia, and include the Philippines, South Korea, Indonesia, Vietnam, China and Afghanistan. With climate changes threatening agriculture in Asia in particular, WMO is recommending increasing investments in urban and indoor agriculture, sustainable farming practices and seasonal prediction and early-warning systems, particularly to guide farmers on when, where and what crop is best to grow.

India is experiencing changes in climate and the impacts of climate change, including water stress, heat waves and drought, severe storms and flooding, and associated negative consequences on health and livelihoods. All of these events, in addition to their effects on agriculture and day-to-day life also have important effects on impact tourism – particularly adventure tourism – with its emphasis on rural and outdoor activities.

The Indian government is focusing on promoting a variety of new tourism products, including caravan tourism, heliport tourism, medical and health and wellness tourism and adventure tourism in addition to its traditional products. The Indian healthcare and medical-tourism market is expected to reach USD $2 billion by 2012. With respect to adventure tourism, the country is promoting itself as one of the most exciting adventure tourism destinations in the world and intends to focus marketing campaigns in the future on the innumerable adventure and eco-tourism opportunities in the country. The government's focus has been on Ladakh, the Sariska Wildlife Sanctuary, Keoladeo Ghana National Park and Corbett National Park, among others. Areas such as Kaziranga National Park, Cir National Park and Kanha National Park have also received attention.

In a series of detailed interviews with a tour operator from India's Himalaya mountain region, Snow Leopard Adventures, a portrait of a business in transition emerges. This company is implementing operational and management focus toward climate-change adaptation.

This article, first published as part of a larger study by Cambridge University student Ryan Piotrowski in collaboration with Xola Consulting, discusses and recommends management and business practices for sustaining adventure travel businesses in sensitive environments affected by climate change. An adventure travel company's climate-adaptation model is provided in the final section.

Hypothesis

Under the hypothesis that adventure tour operators experience the effects of climate change sooner and to a greater degree than most other tourism businesses, a case-study approach was used to examine operators in three different countries and environmental regions – polar, mountain and tropical forest. The results of the Indian company operating in the mountain environment of the Himalaya are presented here.

The overall trend that we observe is that climate change has not yet had a significant effect on the polar or tropical region adventure tour operators interviewed, but has affected tour operations in India's Himalaya mountain region. All of the companies interviewed are implementing operational and management focus towards climate-change adaptation.

Significant Findings

- There is a high level of emphasis by each company to consider future climate-change effects and begin basing strategy for adaptation around it.
- Implementing itinerary changes, education, training and preparedness programs, and good environmental practices, such as reduced emissions, are adaptive strategies used by all the tour operators.
- Documentation and study of environmental and weather conditions varied in degree, but overall none of the companies maintained a database for reference and planning purposes.

Snow Leopard Adventures

History

Snow Leopard Adventures Private Limited was established in 1990 in India and offers adventure tours throughout the year that emphasize safety, eco-friendly practices and education. Snow Leopard's tours are based out of Uttarakhand, India, and outbound trips are offered to other Himalayan destinations in Bhutan, Tibet and Nepal.

Activities

Activities include rafting, trekking, cycling, rappelling, skiing, fishing and various types of safaris. A wide portfolio of tours is offered that cater to corporate programs from retreats to outbound training, family adventures and school trips.

Beliefs/Practices

Snow Leopard Adventures practices include sound planning, emphasis on safety and environmental responsibility. "Eco-logical" efforts toward environmental conservation and sustainability are given emphasis. Snow Leopard established SAVE (Study and Value Environment) in order to bring school children together to make them more sensitive to the environment. Additionally, Snow Leopard has written its own Ecological and Safety Codes of Conduct and incorporate environmental-awareness and eco-friendly education into all tours.

Interview Responses

Summer Temperatures Occurring Earlier

- Snowfall in areas around 7,500 feet occurring in late January instead of late December, since 2001 (excluding 2007).
- Temperatures at 7,500 feet increase up to as much as 35°C in place of 32°C. Evergreen species in the Doon Valley region declined from 70% in 1958 to around 25% today. Increased temperatures are partly responsible for changing vegetation patterns, changing soil patterns and a drop in the water table.

- Increased temperatures result in less guest occupancy on trips two weeks earlier than usual. In Shivpuri (at 1,000 feet) temperatures rise up to as much as 42°C in the summer. Guest numbers on tours decline the third week of May instead of the first week of June.

Water Scarcity at Higher Camps

In a region near Chamba, the combination of existing lower water tables due to geography, alpine canopy and stream use by local village and resort development near Kanatal has exacerbated water scarcity, resulting in vehicles having to travel as much as twenty miles to fetch water for tours since 2002. Change in vegetation patterns has reduced the soil retention capacity and led to a drop in the water table.

Extreme Precipitation Events

Stormy conditions near beach sites are being felt two months prior to the close of rafting season that officially closes for the monsoons (end of June). This has resulted in a loss in revenue from dismantling campsites earlier than usual. However, extreme precipitation events rarely negatively affect water-based activities.

Changes in Flora

Observations by locals indicate that Ban Oak (*Quercus leucotricophora*) is being replaced by Chir (*Pinus roxburghii*) from a 90:10 to nearly 50:50 ratio. Pines have larger root systems, which lead to a drop in the water table, while oaks keep in moisture and humidity. This escalates the warming process in the region.

Adaptation Techniques

To Cope with Increased Temperatures

- Water-based activities are scheduled for the hottest period of the day.
- Beach campsites are preferred due to milder temperatures from their proximity to rivers. However, extreme storm conditions can offset the gain.

- Conferences for corporate training modules are held inside concrete structures.
- Games and other activities take place in the evening during cooler temperatures.

To Cope with Water Scarcity

Tourists are limited to one bucket of hot water a day.

Extreme Precipitation

Camp staff is trained to handle emergency situations and extra tents are kept at fixed campsites away from the beach campsites in case of evacuation. Since weather patterns are highly unpredictable, substitute products to replace lost tour revenue are being created.

Reduced Flora and Wildlife

A healthy amount of vegetation is maintained in and around campsites to protect the environment and offer an alternative habitat to birds and other animals. As an example, bird watching became an unintended additional activity as a result of attempts to reduce dust from nearby highway development at the Shivpuri campsite by maintaining approximately fifty different species of trees and plants. In turn, this allowed migrating birds to inhabit the site in addition to reducing dust.

Business Model Trends and Strategy

Operations

Operations include adaptive strategies with mitigation as described below.

Adaptation

Itinerary timeline is adjusted to cope with temperature increase. Structures and sites are used to alleviate discomfort of increased temperatures. Staff are trained to react to extreme weather events. Create and promote products that are in less-affected regions or which are tied to cultural or historical sites.

Mitigation

Driving reduced when possible in order to reduce emissions (e.g. trekking up the last mile to temporary campsites where roads are barely passable). Use solar lanterns in place of kerosene at beach campsites. Instead of investing in infrastructure for electricity at high-altitude campsites, use a generator for two or three hours a day.

Philosophy

Remain prepared and vigilant to unpredictable weather. Educate and keep staff updated on latest climate trends and research in the Himalayas. Reduce consumption wherever possible and maintain the environment. Educate travelers and community members on climate-change and individual best practices. Influence stakeholder policy to promote environmental conservation.

Advice and Future Thoughts

- Continue proactive and sustainable practices while educating staff and guests on climate change and the environment.
- Frequent weather audits need to be conducted, once every quarter, and on these lines safety and feasibility measures at sensitive campsite areas (like beaches) need to be evaluated.
- Visitor-number control procedures are a must in fragile higher-altitude regions. In low-lying regions with less traveler environmental impact, regulation numbers can be formulated on a daily basis in response to the season and the number of tour operators.
- Adventure tour products will inevitably experience change based on their dependence on the preservation of natural resources. "The only concern is that product components are being curtailed due to unpredictable weather patterns, but from a broader perspective, it can be considered a positive development that is actually pushing reform measures."

- Continued marketing and sales promotions combined with product-mix diversification.

Recommendations for other Companies Based on the Snow Leopard Adventures Experience with Climate Change in India

Companies must focus on key questions for how to cope and develop their business, rather than become complacent or resort to a reactionary strategy. Failure to adapt now can become more costly as a result of not having the necessary resources in place ahead of time. A commonly referenced example of failing to adapt in business comes from Kodak and digital-photography technology. Kodak was first to possess the digital camera, but failed to understand its customer base and offer it as a product for the masses. Kodak's strategy resided in complacency around its proven products. Adventure tour companies need to adapt in terms of climate change as necessary and not miss the opportunity as Kodak did in terms of realizing the value of change.

Adventure travel companies rely on very specific core competencies based on the environments in which they operate – such as whitewater river rafting or high-altitude climbing. Competitors should find it difficult to enter the market and compete against such companies because of the rare, non-substitutable or costly-to-imitate aspects of their core competencies (Sloman and Hinde, 2007: 284). Instead of solely focusing on competition, climate change poses threats and opportunities that can either enhance or diminish niche skills and product offerings. Adventure travel companies need to treat climate-change impacts as one of their greatest competitors.

Creating climate-change expertise through a committee or designated individual staff can provide competitive advantage against climate change. Outsourcing expertise may prove costly for a small tour company and it is rare for a consultant to possess such specific knowledge about a niche company and climate change. Building internal expertise by dedicating time and effort to educating leadership and staff is vital. Analysis of internal and external factors should drive strategy.

As part of the education process and the need to manage difficult-to-predict weather and climate scenarios, research is key. The appointed climate-change person(s) should work throughout the organization to track and maintain climate, weather, physical changes encountered and other methods/examples of climate-change impacts. Additionally, the climate-change specialist(s) should be well informed about the latest research and encouraged to interact with other tour operators, academic institutions, NGOs and local communities to better understand the climate setting. Annual reports and educational events can become deliverables as well. Accurate information will provide more weight for effectively engaging in policy discussion and in dealing with the media and customers.

Strategic simulation exercises for coping with various degrees of climate impact are effective tools for planning and decision-making. A team can discuss or work out rough plans for minimal to extremely high climate-change impacts. Addressing the natural environment and resource impacts on tours and what can be done ahead of time will reduce organizational myopia toward the need to adapt. Resource and fiscal planning will also be important. Resources and capabilities may be assessed in terms of strategic importance and relative strength in order to determine the potential for sustainable competitive advantage against climate change (Grant, 2008: 157).

Education on climate change and measures to adapt as a community is important for relationships between tour operators, other value-chain members and stakeholders. Encouraging combined efforts can be achieved when all parties are knowledgeable about the stakes at hand. Policy development between stakeholders and tour operators is necessary to advance sustainable adaptation beneficial to all parties. Additionally, strategic alliances between tour operators in the same regions and/or members in their value chain may reduce costs and improve adaptation results.

While the product offers for such niche adventure tour companies typically remain their core business, strategy must be focused on adapting to maintain the business through product resilience combined with new product creation for some companies. Snow Leopard Adventure provides an example of a company that has begun to diversify its product mix in order to maintain and attract customers. Adventure tour companies compete more through product differentiation than cost leadership as a result of the unique products that they feature. Providing products and services to a niche client base will require change to maintain market share and profitability while competing with the tourism industry.

Defining strategic direction based on informed approaches and shared effort is vital for sustaining any adventure travel business. As climate change inevitably influences and reshapes business models, companies need to incorporate climate-change adaptation into their strategic planning.

Moutinho provides a useful summary in "Strategic Management in Tourism," where he writes about managing in an uncertain environment:

> An industry faced with turbulent environments cannot cope simply by relying on the accumulated experience of those within it, or on conventional formal strategic planning processes to develop strategies which can cope. In the future, strategies based on prediction and predetermined planning will have to give way to strategies which emphasize preparation, detection of environmental change, flexibility and responsiveness. (Moutinho, 2000: 36)

Operation

Operational changes are very specific to location and types of activities. Regardless, common themes and flexibility to alter transportation, accommodation, resources, programs, schedules and other factors based on climate impacts need to be understood.

Adapting activities and product offerings will be vital to maintaining a customer base, as original activities may no longer exist over time or during certain seasons. A key to reducing the impact of lost revenue from trip limitations is to create new activities or combine activities to complement gaps in original tour models from climate change. Diversification of product offerings can spread risk and provide protection from seasonal and extreme weather impacts. Additionally, decentralization of operations and employee education can promote successful flexibility to adjust tours on the spot and ensure customer satisfaction.

Sound environmental practice is an essential part of normal operations, including water conservation, waste storage and removal, bio-security and other practices. Snow Leopard planted trees initially to reduce dust from nearby highway development and unintentionally ended up creating a habitat to sustain wildlife. Building on this lesson, companies can offer solutions to their environment and create attractions. Snow Leopard Adventures ended up with the added activity of bird watching as a result. Despite the challenge to sustain core business activities, opportunities exist to promote further adventure and environment-based products.

Policy involvement and stakeholder management are important business activities. Interaction with other tour operators and the local community can strengthen relationships toward confronting climate change. Sustainable-supply and value-chain management should become a focus to not only reduce emissions, but to encourage good environmental practices. Through codes of conduct and assessment of supplier compliance, companies may collaborate with their suppliers to satisfy customers and gain competitive advantage as well (Font et al., 2007: 262).

Continuous improvement is an inherent business goal to maintain competitive advantage and provide positive customer experience while reducing costs. Climate change is bound to further impact the supply-and-demand elements of adventure travel. The supply of natural resources, communities and sites to visit needs to be balanced with demand changes as a result of climate and customer psychographics. The ability for tour operators to shape both dimensions will be a result of modifications to maintain and/or create new activities in a fragile environment while providing truly positive experiences. Managing resources and traveler demand while overcoming environmental change may be achieved based on forecasts, shared information and flexibility. Monitoring the effectiveness of adaptation and management techniques is essential for achieving optimal levels of service and diffusing vulnerabilities.

Overall business strategy should promote operational adaptation and response to reduce threats and increase opportunities.

Marketing

As travelers begin to travel to different regions as a result of temperature and weather changes, marketing plays an important role in maintaining business success. Market trends and customer needs will be increasingly influenced by climate change and companies will have to adapt their marketing plan. Product mix, promotion and price are variables that can be altered to retain and increase customers, while product place and distribution tend to be fixed.

An active campaign to highlight a diversified and reliable portfolio of adventure activities is important. Repositioning and redirection of activities based on sites and season should be considered in terms of local climate impacts, but also in terms of what travelers demand on a global level and when. Product promotion should provide clear information and generate excitement for activities that may not make the most sense in the minds of traditional travelers. Contingency planning will provide preparation for extreme-weather incidents and communication about them. Companies should continually promote the environment and adventure as reasons for travelers to challenge themselves and seek enjoyment in the face of climate-change conditions. For example, Snow Leopard explained the following approach as "instead of just focusing upon the adrenaline rush of rafting, greater emphasis is given to activities like treks, highlighting that the experience proves to be more enlightening and rewarding from a broader perspective" (Joshi, 2009).

Travelers are interested in climate change and what is happening to the environment; therefore, companies should continue to educate and promote environmental programs. Adventure travel itineraries that include conservation activities provide an opportunity for travelers to help maintain the environment as well as the tour operator's activity base. Other programs via the Internet and media can educate and assist "wannabe" travelers to contribute to environmental causes from home. Local, community engagement is another important media relations and marketing event to attract attention for the business and promote the environment and climate adaptation. Interacting with schools and other organizations can increase network effects to attract customers as well.

In some cases, targeting local customers versus global customers may provide more sustainable business. As niche companies, adventure tour operators will most likely not consider changing their target audience. Most of Snow Leopard's customers are from India, but promotion may be pushed more toward international travelers unaccustomed to the weather of India. It may be easier at times to encourage people in the surrounding area to engage in activities during off-peak seasons with discount offers. Redefining the target customers and product mix will become important in maintaining customers for tour operators in regions facing adverse climate-change impacts.

Price is difficult to change or compete on for traditionally niche companies; however, packages and other deals during off-peak seasons or incentives to travel despite higher temperatures or weather variability should be considered. An important aspect of marketing in response to climate change is providing a sense of security. Accurate and honest information is essential to maintain company image. Travel insurance against adverse weather and other climate-change mitigation policies for customers should be offered if possible. Such campaign incentives can reduce the number of customers avoiding such trips and encourage others to join trips that otherwise would not have.

Customer service is also an integral part of maintaining and encouraging future customers via word-of-mouth promotion or other sales campaigns. Adventure tour company staff will be better prepared to deal with climate change through effective change management and leadership emphasis on climate education and discussion. Providing employees with an understanding of the "why" behind strategic, operational and marketing decisions can ensure increased satisfaction, job focus and improved customer service.

Conclusions

Strategy, operations and marketing are areas that tourism companies should consider now and use to assist in determining and preparing for impacts of climate change. Adventure companies are beginning to adapt to climate-change impacts in order to sustain their business and their experiences provide useful lessons for other tourism businesses.

As tour operators make small changes now, they should remain proactive and dynamically manage their future-business strategy. While confronting climate change, leadership, communication and education will assist in reducing costs, easing operational change and sustaining a potentially fragile business over the long run.

References

Font, X., R. Tapper, K. Schwartz & M. Kornilaki (2006). Sustainable Supply Chain Management in Tourism. *Business Strategy and the Environment* 17: 260–271.

Grant, R. M. (2008). *Contemporary Strategy Analysis*, 6th ed. Oxford: Blackwell Publishing.

Joshi, Parth (2009). Interview with Parth Joshi, Executive member from Snow Leopard Adventures Private Limited, by Ryan Piotrowski for MBA 2008 Individual Project, University of Cambridge, Judge Business School, 24 July to 24 August 2009.

Moutinho, L. (2000). *Strategic Management in Tourism.* Oxford: CABI Publishing.

Sloman, J. and K. Hinde (2007). *Economics for Business*, 4th ed. Harlow, England: FT Prentice Hall.

WMO (2009). WMO Statement on the Status of the Global Climate in 2008, http://www.wmo.int/wcc3/documents/1039_en.pdf.

WTO and UNEP (2008). *Climate Change and Tourism: Responding to Global Challenges*. Madrid, Spain: World Tourism Organization and United Nations Environment Programme.

PART V

ACCOMMODATION

CHAPTER TWENTY-NINE

THE ITC GREEN CENTER: INSPIRED BY NATURE

NIRANJAN KHATRI[1]

Introduction

> The road we have long been traveling is deceptively easy, a smooth superhighway on which we progress with great speed, but at its end lies disaster. The other fork of the road – the one 'less traveled by' – offers our last, our only chance to reach a destination that assures the preservation of our earth. (Rachel Carlson, *Silent Spring*)

ITC Hotels commenced their journey on the less-traveled road some twenty-two years ago; in the bargain, they discovered unknown opportunities leading to new ways of doing old things, and learning the art of eco-designing, which is nothing but the art of seeing opportunities where none exist.

Conceptualization, modeling and scaling up have been the forte of the hotel chain. The ecological challenges thrown up by modern industrial lifestyles have brought in their wake innumerable challenges that need to be addressed creatively and innovatively by process, policy and design innovation, which will hopefully result in reducing our collective environmental footprint.

The company, as an ecological pioneer, has throughout the years integrated environmental imperatives into its entire management system right from roots to shoots. In a very short span of time, the company has established more than 3,000 check dams in rural parts of India to enhance water security for the marginalized farmers and helped more than four million farmers to enhance their earning capacity through knowledge-empowerment techniques. Its paper division has reduced water

[1] GM Welcomenviron Initiatives, ITC Ltd. Hotels Division HQ and ITC Green Center.

consumption from 250 tons of water to 55 tons of water per ton of paper produced.

ITC Hotels and Green Center

Business Standard has identified the ITC Hotels chain as the greenest chain in India. With the already established "green lining" that exists in the company, it was easy for the management to plunge into the green-building concept that emerged in India in 2002.

The work on our new ITC office project had already commenced and management had to make some course corrections to adopt the criteria laid by USGBC. This bold green initiative was taken and implemented despite time and cost overruns.

Construction of the ITC Green Center commenced in October 2002. We were in the basement in 2003 when we heard about the green-building concept from the Confederation of Indian Industry. The most important learning that we derived from the green building was as follows:

We did an *energy-sensitivity exercise* for the building (which we had never done before), even before the building was up, which gave us a deep insight of what will be heat-loaded at different times of the day and what measures to take in order to decrease the heat load in the height of summer. This led to the use of energy-efficient glass and other green material that helped us to enhance our energy efficiency.

The second most important learning was about *project design* in oriental countries. Drawings are not frozen from the start, leading to inefficiencies by design intent during the operational phase, due to drawings from different consultants trickling in as the project work progresses. LEED (Leadership in Energy, Environment Design) criteria states that all drawings must be frozen on paper, and all corrections must be carried out at the drawing board.

The areas that we have addressed to make the green building are as follows:

Energy Efficiency and Atmosphere: In order to reduce energy consumption by design intent over existing ASHRAE standards, we introduced the concept of day lighting in the building in conjunction with the use of high-performance glass that keeps the heat out and brings in light. In the process, we consume approximately 135,000 units of energy per annum, as opposed to 635,000 units of energy, if the building was made with a standard business approach.

The collateral benefit of using the daylight is not only to reduce the operating cost, but also to help the occupants to enhance productivity. Studies in the United States indicate that people who work with daylight are healthier vis-à-vis people who work during the day with artificial lights.

Energy-efficient bulbs installed in the building are T-5 and CFL, with motion-sensor lighting.

Lamp-posts in the landscape area are capped on top so that the light falls on the ground and does not pollute the atmosphere. This is to ensure that birds get darkness to sleep at night; sensitivity to nature has to become a hallmark of sustainable development.

Reflective paint, also called, high albedo paint, has been applied on the rooftop, which reduces heat gain by 47%, thereby reducing the size of the air-conditioning plant.

High performance chillers with a COP of 6.1 have been installed in the building, which again addresses lower lifecycle cost of operation.

Green Materials: The paints, adhesives and solvents used in the building are low in volatile-organic compounds, which help avoid "sick-building" syndrome. Conventional paints, adhesives and solvents are embedded with harmful chemicals that exude gases throughout their lifetime.

The building has used forest-stewardship council-certified wood, which implies the use of wood from forests that are harvested in a sustainable manner as opposed to clear felling, and has extensively redeployed old furniture from its old building, thereby reducing the pressure on forests.

Water: ITC Green Center by design intent has reduced water consumption by 40%. It is a zero-discharge building and rainwater goes into the aquifer. The water used in the building is treated through the Sewage Treatment Plant and re-used for horticulture, AC cooling tower and for flushing.

We have installed waterless urinals that help us to save 300 kiloliters of water per annum. To further work toward water frugality by design intent we have reduced the flow rate in the taps from six liters per minute to two liters per minute.

We have plants that consume less water. The parking areas have interlocking tiles so that grass grows in between and gives the opportunity for rainwater to seep in.

In a manner of speaking, we have introduced the concept of conspicuous conservation of water!

Education: Since the green-building movement was a new concept in 2003, the architects and consultants were not familiar with the concepts as per the criteria of LEED. We trained more than 3,000 people to understand the finer nuances of green building so that the concept takes roots in the country and gets scaled up rapidly so that climate change and other ecological challenges are addressed collectively by all stakeholders.

Innovation: Small innovative ideas were implemented at the project stage, keeping in mind our triple bottom line – economic, environment and social. We demonstrated to the project workers the use of a solar-concentrator drip-irrigation device and sprinkler system not requiring energy. The intention behind this demonstration was that in India, the construction laborers are also farm laborers and if they take this idea back home it will help them to enhance their productivity and reduce the chores for women who normally walk four kilometers to collect wood for cooking.

Once the building was ready, we installed touch-screen computers in the lobby so that any person visiting the building can learn about the green building by touching the screen and getting answers in English and Hindi.

We were aware that many green-building concepts are expensive, and therefore we have loaded the cost-effective ideas on the touch-screen computer for the average homeowner. The contents of the green-building concepts are also available on our website www.itchotels.in. It is also the first and largest platinum-rated building.

Visitors from far and wide have come to see our building, including U.S. Secretary of State Hillary Clinton who called the building a "monument for tomorrow."

With exploding populations, increased earning capacity and higher aspirations, the pressure on natural resources will go up exponentially. As rightly pointed out by many enlightened people, such growth is not sustainable. Therefore, creative measures for regulating the use of finite natural resources are required.

In a country where the construction industry is growing at the rate of about 10% per annum, a few green buildings will not suffice. Keeping the above perspective in place, ITC Green Center has been instrumental in engaging all stakeholders in green-building practices and introducing green-building concepts to lawmakers so that changed policies and economic instruments of GOI results in green building becoming part of building bylaws, as opposed to simply a few organizations taking up the green path in construction.

During the construction stage, 3,000 stakeholders of ITC Green Center were sensitized to principles of eco-design. In addition, ITC Hotels conducts regular eco-design programs for schools, colleges, SMEs (Small and Medium Enterprises) and chambers of commerce as part of its eco-responsible best practices, so that a cascading effect takes place in the shortest span of time.

> During the next quarter century, the most significant net contribution to a greener world will be made by industry. Not every company is there yet, but most are trying. Those that aren't trying won't be a problem simply because they won't be around long term. (Ed Woolard, former Chairman of DuPont)

CHAPTER THIRTY

A HARMONIZED ACCOMMODATION-GRADING SYSTEM THAT REDUCES CO_2 EMISSIONS ACROSS SOUTHERN AFRICA

JAMES MACGREGOR[1]

Abstract

The development of a comprehensive, harmonized accommodation-grading system will be implemented by the fifteen nations of RETOSA (Regional Tourism Organization of Southern Africa) by 2015. In 2011, a Framework Document including a four-year action plan from 2011 to 2015 was prepared. This was followed by preparation of approximately 1,850 standards for six basic categories of accommodation. The program is possibly the most comprehensive set of standards ever developed because it also integrates sixty responsible tourism and forty universal-accessibility standards with the basic quality-assurance standards. This will provide southern African member states with a grading system that reflects the highest level of environmental and social responsibility while also providing facilities that are accessible to the disabled. Furthermore, the standards reflect the most rigorous international-level standards and will be mandatory when the program is launched. The responsible tourism standards are designed to produce a 20% reduction in CO_2, which will reduce emissions by approximately 1 billion kg/year by 2016.

Funding

- Preparation of Framework Document: Commonwealth Secretariat: Ms. Chanda Chellah, Advisor (Trade)
- Preparation of Harmonized Accommodation-Grading Standards: RETOSA

---

[1] President, ecoplan:net sarl, Morocco.

- Overall Project Management: RETOSA, Mr. Simba Mandinyenya, Planning and Development Director
- Project Consultant: James MacGregor, President, ecoplan:net sarl

1. Overview

Despite continued market demand for recognized and credible international tourist accommodation-grading systems, there continues to be substantial confusion about the merits of the various available systems. Furthermore, in many cases there is an obvious inconsistency from one national system to another. A three-star rating in one country may not reflect the same classification in another. As well as this, with an annual increase in online bookings, travel consumers more than ever require a recognized accommodation-grading system that is reliable. Despite the increased use of user-generated content on the Internet (TripAdvisor, Priceline.com, etc.), the time-tested star-rating system still has substantial value and, for many travelers, is more important than ever in their decision-making and choice of accommodation.

In 2010, RETOSA (Regional Tourism Organization of Southern Africa) recognized the importance of accommodation-grading systems and decided to create a completely harmonized system across the fifteen countries in southern Africa, and including Mauritius and Seychelles.

The overall goal of this ambitious project was to establish rigorous and comprehensive international-level grading standards in concert with a credible and coordinated grading scheme within each member nation.

2. The Processes of Building the Harmonized Scheme

The preparation of the harmonized accommodation-grading scheme was developed in four phases comprising: *Phase 1*: the preparation of a framework for the scheme, incorporating the identification of major issues and the resulting recommendations; *Phase 2*: the preparation of the individual standards; *Phase 3*: the integration of the various member-nation grading systems into one harmonized grading scheme; and *Phase 4*: the launch of the overall program (2015). This article will address Phases 1 and 2.

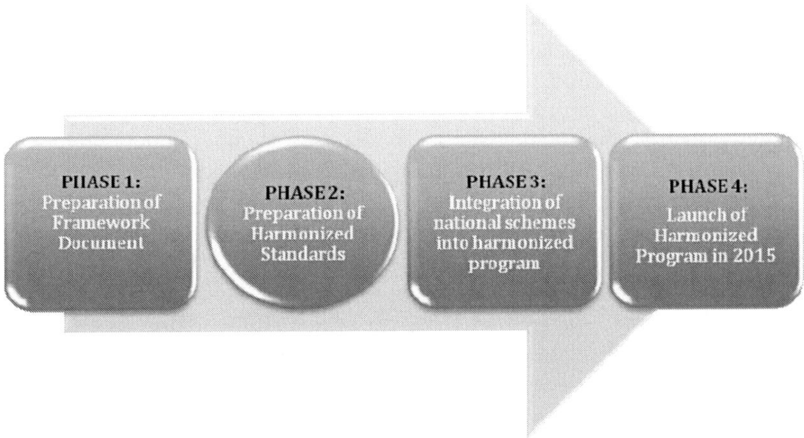

Figure 30-1. Stages of implementation.

2.1. Phase 1: The Preparation of a Framework Document

The framework document was launched with a workshop in October 2010 where representatives from the fifteen member states gathered in Johannesburg to present the status of their individual national schemes as well as identify the major issues associated with establishing a harmonized program across the region.

They included the need for recommendations to:

1. Select either a voluntary or mandatory harmonized grading scheme.
2. Identify the needs for auditors and supporting auditor/assessor training.
3. Create an overall management structure that would enable RETOSA to develop and administer the system.
4. Identify marketing channels and mechanisms to promote the system to consumers and travel trade.
5. Identify a feedback mechanism structure to ensure the continual improvement of the scheme.
6. Integrate responsible tourism and universal-accessibility standards with quality-assurance standards.
7. Identify funding sources to develop the program.

The Commonwealth Secretariat funded the preparation of the framework document. Following an international competition, James MacGregor from the firm ecoplan:net was hired to undertake the research and prepare the required framework. As a first step in the preparation of the framework, the consultant visited a number of member states, including Tanzania, Zambia, South Africa, Swaziland, Mozambique and Mauritius. This enabled a better understanding of the status of national grading schemes as well as the obstacles to developing both the national and regional standards scheme. The status of the national schemes varied widely, from South Africa, Botswana and Namibia, which have mature and well-developed systems, to Tanzania, which is integrating its proposed scheme into that of the East African Community, to Swaziland, which is in the early stages of development.

The country visits as well as reviews of other international systems provided the basis for a "Recommendations Report" that offered a preliminary response and a direction to the above-mentioned issues. These recommendations were subsequently submitted at a two-day RETOSA Technical Committee Workshop in June 2011. A brief summary of the results of the discussions included:

(A) There was overwhelming support for the development of a mandatory, rather than voluntary, harmonized systems across the member nations. There is currently a 50–50 split, and a three-phase process was proposed to have all systems mandatory by 2015.

(B) Lack of qualified auditors is one of the more serious issues facing individual-member-nation grading systems. This will be addressed during the implementation of the program with the design and delivery of auditor training programs throughout the region.

(C) RETOSA will be the organization tasked with coordinating the integration of various national grading systems into one harmonized grading program. However, it will require the complete support from all other stakeholders, including: ministries of tourism and national tourism authorities; national and regional tourism and accommodation associations; NGOs and individual accommodation providers. Collectively, these various stakeholders can offer a substantial contribution to the preparation of the harmonized accommodation-grading scheme.

(D) The overall marketing and promotion of the harmonized program will be the responsibility of RETOSA; however, it will require substantial support from individual member countries as well as their associated accommodation associations and providers.

(E) A feedback mechanism will be set up to enable the consumer and the travel trade (foreign and receptive tour operators and wholesalers) to evaluate the property within the context of the prescribed rating. Feedback will be provided at the ministry/tourism authority, tourism accommodation association and accommodation-provider level. The results will be consolidated by RETOSA and contribute to the on-going upgrading of the accommodation standards.

(F) Social and environmentally responsible standards, including fair-trade practices, as well as universal-accessibility standards, will be integrated into the national and regional harmonized accommodation-grading program. Furthermore they will be given substantial influence in determining the overall rating of each property. With upwards of 20% to 25% of source markets having residents with disabilities, it is now considered mandatory to have at least basic universal-accessibility standards in place. The standards would be integrated into the overall system making it one of the first in the world to be comprehensive and inclusive.

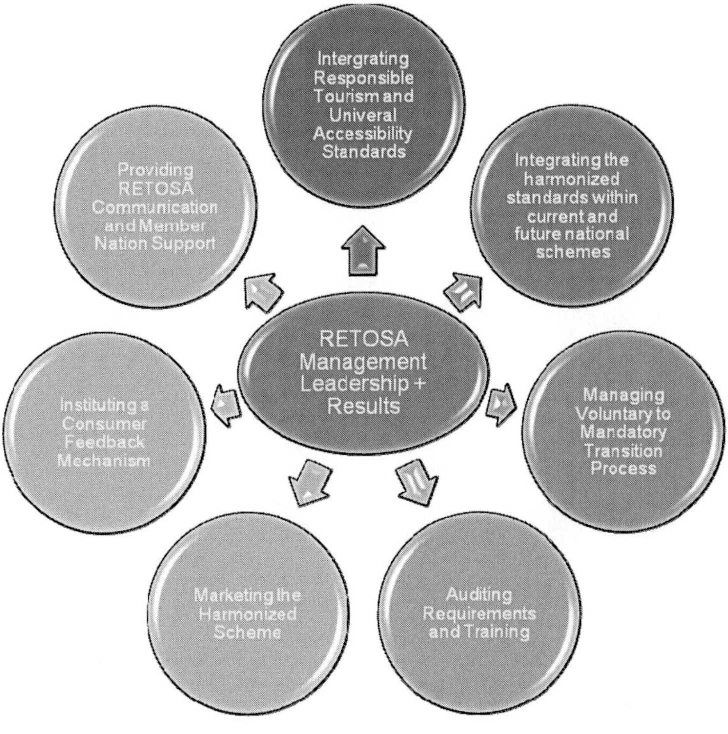

Figure 30-2. The RETOSA accommodation-rating harmonizing-framework diagram.

(G) Funding will come from several different sources. For instance, project support would be from donors for activities, such as training workshops and the preparation of technical manuals, development of marketing strategy and communication plans, etc. Project-based funding could also be used to both promote the scheme to the regional and international markets as well enhance communication between member nations.

Following a complete review by the technical committee the consultant prepared the final framework document, which was presented to the representatives of the member nations in early September 2011. The presentation also included an outline of the action plan and cost estimates required to launch the program by early 2015. Activities were outlined on a quarter-by-quarter basis from late 2011 through the middle of 2015.

2.2. Phase 2: The Preparation of the Grading Standards

Ecoplan:net was also hired, by RETOSA, to prepare the individual standards for six selected categories. These categories included:

1. Urban hotels: Those located in urban or near urban locations with particular services for the business traveler. Typically full service.

2. Resorts (vacation hotel): Typically located within a tourism destination and adjacent to a feature of interest such as lakes, beaches, rivers or tourism attractions. Usually full service and include a variety of recreation activities.

3. Self-catering apartments: Including the required amenities for meal preparation but also including housekeeping services.

4. Nature lodges: Full-service accommodation facility located within or adjacent to natural areas. Sub categories could include game or safari lodges, eco-lodges and game reserves.

5. Guesthouses: A facility that offers more limited lodging services but includes a common area shared by guests and host and may or may not include housekeeping services. Breakfast provided.

6. Campgrounds: An area with campsites or pads designed for mobile accommodation (motor home or caravan) and/or tents, providing essential facilities and services and select recreational activities.

A number of well-known and respected accommodation-grading systems at the international level were also analyzed and consequently provided some direction for the harmonization scheme. They included, in part, the Qualmark system from New Zealand, the AA rating system in the United Kingdom, the new French hotel-rating scheme, as well as select standards from the Brazil, Irish and Scottish programs. At the regional level, South Africa, Botswana, Zimbabwe and Namibia provided an excellent source of standards. Finally the East Africa Community (ECA) standards were considered as an important source, particularly with the possibility of integrating East Africa standards with the southern African in, say, 2016.

2.3. Phase 3: The Integration of Responsible Tourism and Universal-Accessibility Standards with the Quality-Assurance Standards

The complete incorporation of responsible tourism and universal-accessibility standards with quality assurance represents one of the first efforts, worldwide, to combine a comprehensive set of standards that embraces the entire accommodation product. Several countries have an independent ecotourism, sustainability or green-certification program; however, these environmental programs are typically separate from the basic accommodation-rating system.

Of significance are the integration of approximately sixty responsible tourism and the forty universal-accessibility standards for each category of accommodation. These standards represent approximately 30% of the total number of standards so their impact is significant. Several newer grading systems, developed within the past few years, have added a few environmental standards but they have relatively little influence on the final rating of the property. The responsible-tourism standards, within this RETOSA harmonized system, will however be specific and comprehensive. The universal-accessibility standards refer to actual dimensions and spatial requirements that need to be respected to achieve a grade.

This program identifies more than 1,850 standards across the six accommodation categories.

Collectively they will ensure a substantially level protection for the local environment, greater integration of the facility into the community and assured access to many more individuals with disabilities.

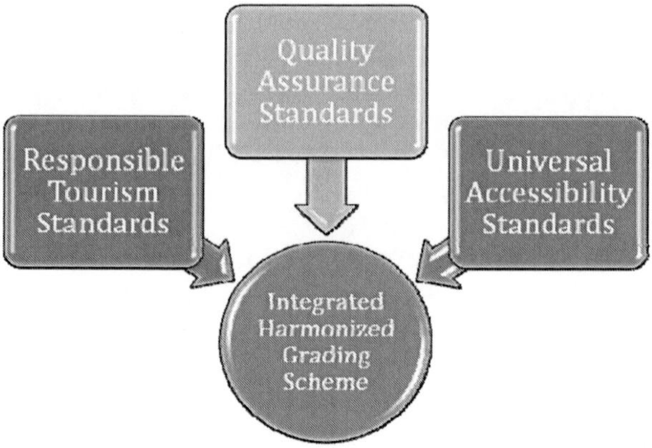

Figure 30-3. Integrated harmonized grading scheme.

2.4. Reducing Carbon Emissions Across Southern Africa

The southern African tourist accommodation sector does not itself generate comparatively significant levels of CO_2. Nevertheless, a concentrated effort across the approximately 20,000 properties (by 2015) can still make a meaningful contribution to reducing CO_2. Few accommodation-grading systems take this issue into account; however, when designing the system within the context of growing concern for climate change it is possible to integrate particular standards that will contribute directly to reducing accommodation-related CO_2. The responsible tourism standards, in particular, have been specifically designed to reduce carbon emissions. In fact the scheme should result in an overall reduction of CO_2 by approximately 20% by 2015. This would seem to be in keeping with UNWTO proposals to reduce accommodation-related greenhouse gases by 30% in 2020. Furthermore, the World Tourism and Travel Council (WTTC) proposes a 50% reduction in accommodation greenhouse gases by 2035. The RETOSA projections, therefore, seem to be realistic and in keeping with other international agencies.

By using the average daily CO_2 production figures from the World Resource Institute it is possible to calculate the decrease in emissions. It appears that these standards, particularly because they are mandatory, would have the significant effect of reducing emissions from the accommodation sector. It is estimated that CO_2 emissions will be reduced by approximately 1 billion kg of CO_2 per year across southern Africa by 2016.

Furthermore, implementing the standards would not necessarily be onerous on the operations of even the smaller facilities such as guesthouses and self-catering apartments.

In most cases the standards will actually result in reduction of the overall operational costs. Specific standards that have been identified to reduce CO_2 include:

- Generating at least 50% of energy from alternative sources in four- and five-star properties.
- Extensive use of energy-saving appliances, compact-fluorescence or LED lighting and the use of equipment such as solar ovens where applicable, such as in the lodge operations.
- Substantial reduction in transportation-related fuel consumption through local purchases; this is a significant consideration in southern Africa where so many products used by the accommodation sector, throughout the region, come from Johannesburg, usually by transport truck.
- Use of carbon credits for administrative and marketing-related travel.
- Energy and water conservation-education programs in local communities.
- Concerted visitor-awareness programs to encourage reducing water and energy requirements.
- On-going climate change-related discussions with the local communities to ensure the residents are doing their best to reduce CO_2 emissions.
- The introduction of short-duration local tours using bicycles and low-impact water transport. These trips also encourage an extended length of stay thus requiring less travel.
- Preparation of manuals that demonstrate to employees and the adjacent communities the various techniques for reducing greenhouse gases (GHG).
- Annual audits to identify how targets can be met with respect to local purchasing, reduced travel, improved energy conservation, etc.
- The regular use of reusable rather than disposable materials and equipment.
- Working with distributors and suppliers to ensure as much recycling as possible.

- Energy reduction through decreasing demand for hot water and preference for natural ventilation and ceiling fans over air conditioning.

Many of the currently accepted standards, such as a towel and linen re-use programs, room sensors and the use of compact-fluorescent light bulbs have been integrated into the overall program.

3. Benefits to the Southern Africa Tourism Industry and the Accommodation Sector

Since this harmonization scheme is the first of its kind in the world, when fully developed, it should be able to solicit substantial media attention and exposure for southern Africa.

However, the greater benefits will come from those long-haul travel consumers who appreciate that the grading system is equivalent or more rigorous than those systems found in Europe and North America. This is particularly important to both the individual FIT traveler as well as the tour operator since much of the travel to the region is multi-destination and typically includes visits to several countries. The travel consumer wants to know that a three-star property in one southern African country is equivalent to a three-star property in another location.

The notion of a harmonized grading system that blankets all of these nations reinforces the concept of southern Africa as a unified destination. This is particularly important for the long-haul traveler who prefers an extended stay (ten days or more) and a visit to several different countries.

Other benefits that will contribute substantially to the improvement of the southern Africa accommodation product and its competitiveness include:

- The new standards will stimulate significant upgrades across all accommodation categories. Facilities that wish to emulate international standards must make the effort to ensure that, say, their four-star property is equivalent to the same grade in the source markets.
- The development of the new harmonized system will include the delivery of training programs for auditors and grading officers. This information will then be distributed to the accommodation suppliers and will further contribute to product upgrades and improvements.

- With a harmonized system in place, southern Africa will become competitive with other multi-nation regional destinations where there may be a lack of consistency in product quality across each country.
- The pricing structure across the various nations will be more credible. Despite slight area conditions, the rates for a three-star guesthouse in one country should not be noticeably different from those in another.

4. Conclusion

The harmonization of accommodation-grading standards across this multi-nation destination region is an essential to position southern Africa in the international marketplace; however, the initiative requires a major effort on behalf of numerous stakeholders at both the regional and national level, but there is commitment at all levels. Under the direction of RETOSA, the various nations are moving quickly to adopt or integrate the new standards.

This effort demonstrates that southern Africa believes unequivocally that tourism will play a central role in the economic development of the region. As with any product, quality assurance and environmental responsibility are critical to the commercial and social success of the business. As tourism expands exponentially throughout the next two decades, particularly with travel from BRICS nations, southern Africa is clearly positioned to be a major destination for these emerging markets and maintain or strengthen their market share in their traditional markets.

Harmonization of the accommodation standards and the resulting product upgrades and improvements are a significant part of that process.

CHAPTER THIRTY-ONE

RESPONSIBLE TOURISM PRACTICES IN THE NON-HOTEL ACCOMMODATION (NHA) SECTOR IN PORT ELIZABETH, SOUTH AFRICA

HUGH BARTIS AND CAROLE BALDIE[1]

Abstract

This paper focuses on responsible-tourism practices in the non-hotel accommodation (NHA) sector in Port Elizabeth, South Africa. The NHA sector includes all the guesthouses, bed-and-breakfast establishments and self-catering accommodation establishments which collectively constitute this sector. The primary purpose of the paper was to examine the factors that inhibit the NHA sector from embracing the adoption of responsible-tourism (RT) guidelines as set out by the South African government. The awareness and implementation of RT principles is critical for the tourism industry as it serves as an indicator as to whether this sector can contribute toward sustainability in reducing its carbon footprint. The reduction in the carbon footprint could then lead to less strain on the environment, including aspects such as climate change.

In order to address the primary purpose (main problem) of the paper, a questionnaire was designed and emailed to respondents. A response rate of 24% was obtained and the results indicated some understanding of the term "Responsible Tourism," but lacked knowledge of any governmental guideline, policy documents or organizations advocating responsible tourism. In addition, efforts were made by the NHA sector in terms of being energy efficient, conserving water and reducing waste management; however, social responsibility within the NHA sector was a low priority.

From this study, it can be deduced that for government, especially at the local level, significant efforts will have to be made to ensure that the NHA sector fully comprehends the principles of RT and the significant role RT plays in terms of sustainability. This would ultimately benefit the

[1] Nelson Mandela Metropolitan University.

tourism industry, as it will lead to more RT practices and enhanced sustainability.

1. Introduction

The tourism industry is one of the fastest-growing industries; however. the impacts that result are wide-ranging and controversial (Evans, Campbell & Stonehouse, 2003). The Earth Summit of 1992 in Rio de Janeiro, Brazil, included sustainable tourism on the agenda. It was highlighted at the summit that special attention to the relationship between environmental conservation and protection was necessary, in order to have a sustainable tourism industry. The conference concluded that the tourism industry needed to be aware of the negative impacts, and that it was necessary to encourage the adoption of more responsible-tourism behavior in the future in order for tourism to be a sustainable industry.

In response to this and the democratization of South Africa, the 1996 White Paper on "Development and Promotion of Tourism in South Africa" was drafted. The then South African Department of Environmental Affairs and Tourism (DEAT) published a book outlining Responsible Tourism Guidelines (DEAT, 2002). This provided information on the ethics and codes of practice that would be expected from the tourism industry in the future. The first conference on responsible tourism was hosted in Cape Town in 2002, prior to the 2002 World Summit on Sustainable Development (WSSD). This conference led to the "Cape Town Declaration" where responsible tourism was defined as a three-tiered approach, namely social, economical and environmental.

Globally, South Africa is seen as being at the forefront of responsible-tourism initiatives (Frey, 2008). Despite this, there seems to be limited awareness of or participation in responsible-tourism practices. This paper focuses on responsible-tourism practices in the non-hotel accommodation (NHA) sub-sector (guesthouse and bed-and-breakfast establishments and backpacker lodges) in Port Elizabeth, South Africa. The objectives of the study are:

- To ascertain the degree of responsible-tourism awareness among non-hotel accommodation establishments.
- To determine how members within the non-hotel accommodation sector interpret and understand the term "Responsible Tourism."
- To investigate the level or degree of responsible-tourism implementation at identified establishments.

- To understand which factors encouraged and motivated non-hotel accommodation establishment to practice responsible tourism.
- To explore whether the non-hotel accommodation establishments supported local communities.

2. Context

The implementation of responsible tourism is considered a major challenge. Tourism practitioners and tourism product owners are challenged with understanding the concept, as well as the implementation of responsible-tourism practices. In a study conducted by Van der Merwe and Wocke (2006) about responsible-tourism practices within the hotel industry in South Africa, serious reservations were raised about what the concept means and the initiatives related to responsible tourism.

According to Goodwin (2009), there is a major gap between the definition of this concept and the way it is used and abused by practitioners. Thus, the issue of responsible tourism, specifically the confusion around the concept and implementation, is not necessarily restricted to South Africa. On the positive side, however, it would seem that consumers, including tourists, may have a better grasp of the concept of responsible tourism. However, this study focuses on tourism-product owners, specifically the owners and managers of non-hotel accommodation (NHA) establishments.

Responsible tourism strives towards minimizing the negative impacts on the environment, society and economy, and facilitating respect and benefits to the tourists, hosts and the destination. Therefore, responsible tourism, in essence, is based on the "triple bottom line," which would include three pillars, namely the economy, society and the environment. It is the interplay between economic responsibility, social responsibility and environmental responsibility on which tourism must be hinged in order to create a more sustainable tourism industry.

2.1. Three Pillars of Responsibility

2.1.1. Economic Responsibility

The tourism industry is seen as a good job-creating and labor-intensive industry that can create direct employment in restaurants, at attractions and accommodation establishments. It is also able to create indirect jobs in the agricultural sector. Tourism in South Africa is already considered to be an important contributor to employment and wealth creation, with a huge flow-through effect, impacting on many sectors of the economy. According to Business Day, tourism contributed three percentage points to the GDP in 2005 (Business Day South Africa, 2009). According to the same report, the accommodation sector was a key contributor to this figure, contributing 22%.

The White Paper of 1996 proposes responsible tourism as the guiding principle for tourism development in this country. It also implies that a proactive approach by the principals of the tourism industry, which includes the non-hotel accommodation sector, be developed to market and manage this industry in a responsible manner (White Paper, 1996). This means that government and businesses involve local communities, provide employment, revise investment policies and relations with investors, decrease economic leakages and develop meaningful economic linkages by supplying local agricultural produce to non-hotel accommodation establishments. South African Tourism is the official destination-marketing organization for South Africa. In its mission, SA Tourism highlights sustainable economic and social empowerment for all South Africans and strives to make tourism the leading sector in the economy. As part of their mission, they also want to increase dispersal of visitors across South Africa, thereby spreading economic benefits and encouraging development. To successfully achieve these goals, Harris, Griffin and Williams (2002) stress that local communities usually only benefit from tourism if it is planned together with investor communities. By adopting a responsible economic approach, positive contributions to the economic development of a region will outweigh the negative impacts, thereby leading to a more sustainable future.

2.1.2. Social Responsibility

Social responsibility emphasizes greater recognition of the interests of the host community and stresses the negative impacts that tourists and their activities can have on the destination (Bennett, 2000). An observation can be made that the South African tourism industry has been slow to adopt the principles of social responsibility. Van der Merwe and Wocke (2007) researched the level of social tourism with a focus on the hotel industry. Their findings indicated that although the percentage of awareness and implementation were initially encouraging, further results from subsequent research revealed that socially responsible tourism actually has a low level of practice. For example, HIV/AIDS policies are not in place in many establishments (Frey, 2008), and safety and security issues remain problematic. Besides the importance of reducing inequality within the tourism industry, there should be an obvious interest in following socially responsible principles, and adopting more open and transparent business practices. The rationale behind this approach is that if the business should increase the involvement of communities in tourism developments, and subsequent benefits have a positive impact, support for the tourism enterprise will increase, crime levels will decrease and employee morale will be improved. The aims of corporate social responsibility should be to deliver sustainable value to the industry for the long-term benefit of both hosts and tourists. This can be achieved by building mutually beneficial partnerships where opportunities can be provided and fair-price policies can be implemented. As stated by Tearfund (2002), investing in socially responsible tourism ethics has benefits for the business and the community; it improves business performance, reduces poverty, unemployment, and returns on investments are usually positive.

2.1.3. Environmental Responsibility

Central to environmental responsibility is the lifecycle of an enterprise or product. This applies to managing the business, and integrating the establishment with environmental management. All tourism enterprises can make significant contributions to environmental sustainability, for example by supporting producers of local products and advising guests how to make responsible purchase decisions (DEAT, 2002). Establishments must seek to minimize energy requirements, limit noise, light and air pollution, and consider the business neighborhood by minimizing any negative transformation of the environment, for instance, by planting indigenous vegetation and non-invasive species. All resources should be used in a sustainable fashion, meaning waste and over-consumption should be avoided. Establishments should encourage visitors to behave in a manner that respects natural heritage and has a low impact on the environment by discouraging certain activities (such as quad biking).

2.2. Who is Adopting Responsible-tourism Ethics?

Businesses are displaying a growing trend not only to look after their financial success, but also after the environmental and social dimensions of their business activities and performance (UNISA, 2008: 1). However, Frey (2008) mentions that the tourism industry has been strongly criticized for its lack of transformation. Government policies have been developed, and social marketing campaigns have been aimed at trying to influence management policies, but non-hotel accommodation establishments have not embraced the responsible-tourism ethic in a way that will contribute to making the industry sustainable. This does not appear to only be a South African problem. According to Van der Merwe and Wocke (2006), there seem to be very few tourism companies participating in globally responsible tourism initiatives. At a recent seminar held by UNISA, it was stated that the trend to adopt responsible business ethics, both locally and internationally, has grown significantly, and corporate citizenship has been introduced to describe a more balanced way of doing business; however, compared with other industries, tourism has been slow in adopting these ethics (UNISA, 2008).

2.3. Consumer Demand: The Only Hope for Sustainable Practices

Consumers might not be realizing that they are beginning to practice responsible tourism. There is a growing culture to travel and enjoy more authentic experiences. Tourists are beginning to request local restaurants where they can enjoy traditional food and meet the local people. Local guides are being requested, particularly for certain areas such as township tours (McGrath, 2004).

3. Study Area: Design and Process

The study area for this project was Port Elizabeth, which is the fifth largest city in South Africa and located along the east coast of South Africa. It boasts a population of about 1.2 million, and is seen as the gateway to the Garden Route, with a thriving safari sector, boasting the Big 5 (lion, leopard, buffalo, rhino and elephant) and many game reserves. In addition to a growing tourism industry, the city is also known for its automobile industry.

To access the NHA sector, a number of sources were used, including the Port Elizabeth Metro Bed-and-Breakfast Association (PEMMBA), Nelson Mandela Bay Tourism (the local tourism authority), a listing from the Tourism Grading Council from 2008 and a listing from the Portfolio Bed-and-Breakfast Collection.

In order to address the primary purpose (main problem) of the paper, a questionnaire was designed and emailed to respondents in the Port

Elizabeth area. This was followed by personal interviews with NHA establishments, where they had not completed the emailed questionnaires. The survey was conducted over a two-month period. A sample of 120 NHA establishments was targeted and a response rate of 24% was obtained.

4. Results and Discussion

4.1. Level of Awareness of Responsible Tourism

The intention was to establish the actual level of awareness within the non-hotel accommodation industry. Fifty-two percent had some sort of policy guidelines in place regarding responsible tourism, although only 7% were aware that there was an official handbook available. Ninety-three percent had therefore never heard of the handbook before. Probing questions led the researcher to understand that "policy guidelines" were practiced according to knowledge gained through the media; advertising, such as "save-water" campaigns; and various levels of one's own conscience. All respondents seemed to recognize the need to employ some sort of responsible-tourism practice, but the motivation is largely due to strategic business reasons rather than simply the "feel-good factor" that was rated at 28%.

Only 3% of respondents indicated that they were members of a sustainable organization. There was almost no knowledge of what the FEDHASA Imvelo Awards and the Heritage Ratings program stood for, and there was no evidence to suggest any non-hotel accommodation owner or manager had made any effort to investigate the criteria necessary to belong to any sustainable organization.

An interesting note is that the majority of establishments surveyed were members of the local tourism authority, namely Nelson Mandela Bay Tourism (NMBT) and the Port Elizabeth Bed-and-Breakfast Association (PEMBBA). The researcher was advised that belonging to associations did not bring any benefits and there was limited support from local authorities.

4.2. Understanding Responsible Tourism

There appears to be confusion amongst the NHA establishments regarding the understanding of responsible tourism. Forty-nine percent strongly believed that responsible tourism is about creating sustainability for the future. This is in line with DEAT's Responsible Tourism Guidelines, as mentioned above. A significant proportion, 32%, understood it to signify service delivery. The minority (19%) of respondents had no thoughts or ideas on the subject, and this is indeed cause for concern.

4.3. Implementing Responsible Tourism

According to Goodwin (2009: 1), there appears to be a major gap between the definition of the responsible-tourism concept and the way it is used and abused by practitioners. Van der Merwe and Wocke (2006) confirmed this perception in their research conducted on responsible-tourism practices within the hotel industry, when they explored reasons "for the limited uptake of responsible tourism despite South Africa being regarded as a leader in the field of responsible-tourism policy."

In the survey, 69% viewed the level of importance regarding the implementation of responsible tourism as personally important, while 62% deemed it the right thing to do. Fifty-nine percent felt that implementing responsible tourism is necessary for future sustainability of the industry, linked strongly with the fact that it is also important for business. The sustainability concept appears to have different interpretations, and it is with this in mind that 48% of respondents felt that the environmental factor was the most important. It is significant that most of the respondents felt that implementing responsible tourism for environmental reasons is necessary.

The same level of importance was, however, not attached to improving relationships with the local community. This could be because the managers and owners have not yet considered the vital part that the local community can play in the future success of their non-hotel accommodation establishment.

Thirty-eight percent were motivated to implement responsible tourism due to the cost-saving factor.

It would appear from the collected data that the guest-influence factor plays a minor role in implementation practices. This could indicate that guests are not adopting a culture of responsibility by demanding good-practice ethics.

4.4. Motivation for Responsible Tourism

Tourism is a high-impact industry. It is also considered to be an industry that can make significant contributions to solving environmental and social problems; however, survey results indicate that the most important factors that contribute to the practice of responsible tourism were not widely known by non-hotel accommodation owners and managers.

Environmental needs, specifically related to waste and water management (ranked the highest), and cultural needs (ranked second) were deemed to be the most important factors that influence the practice of responsible tourism. Environmental needs pertaining to the flora and fauna of Port Elizabeth, humanitarian and community needs had a very low ranking by comparison.

This shows that there is a fairly large awareness level regarding waste disposal and water management. The cost of these environmental factors is carried by each particular non-hotel accommodation establishment and therefore has a direct bearing on their financial resources. These factors are deemed to be more important than those that do not directly affect the bottom line.

Although 59% of respondents encouraged positive guest behavior, almost all had never thought about influencing supplier's behavior. In other words, they were of the view that they had limited influence on the existing relationship between suppliers and themselves.

4.5. Supporting Local Communities

The researcher was attempting to investigate the level of responsible tourism on a social level that might be practiced within the non-hotel accommodation industry. They could choose any of the descriptions provided, and there was an option to provide any additional information in writing.

Sixty-six percent do not support any type of community initiative, which equates to 19 of the 29 respondents. Seventeen percent of respondents supported a local school, and one respondent supported a clinic in a nearby disadvantaged community. This was because their chef's family lived in the area, and he had brought this clinic's plight to their attention. Seventeen percent responded "other."

The findings suggest that the lack of enthusiasm and commitment in terms of implementing responsible tourism is primarily due to limited awareness, and additionally, a lack of understanding regarding the responsible-tourism concept. It is disconcerting that although there are certain awareness campaigns and initiatives in place, so many within the non-hotel accommodation sector of the tourism industry appear confused about the concepts and demonstrated this lack of understanding. This is likely to impede any adoption of responsible-tourism guidelines, and could be the reason for the non-hotel accommodation industry's slow response in the adoption of these principles

None of the non-hotel accommodation establishments had set any targets or objectives for implementing any responsible-tourism guidelines; however, gauging from the responses, all the accommodation establishments practiced some form of responsible tourism, albeit at various levels. It was noted that these particular activities were providing definite cost-saving benefits to the establishment.

The primary factor that appeared to motivate a non-hotel accommodation establishment to practice any form of responsible tourism was the cost-reduction benefits. Unwittingly, most non-hotel accommodation establishments were adopting some form of responsible-tourism practice in the form of energy-saving light bulbs, switching off appliances when not used and practicing water-conservation methods.

In previous literature, it was suggested that the degree of responsible tourism that is practiced could be consumer-driven. One accommodation establishment, in particular, was very pro-active in socially responsible tourism. Guests contacted the owner prior to their arrival requesting information regarding historically disadvantaged communities and they became involved making positive contributions. Although this was not the initial intention of this particular accommodation establishment, it remains a fact that their social-responsibility program has brought more business and therefore more financial benefits to all those involved. It was noted that this particular accommodation establishment was a larger one, and had been operating in the industry for a number of years. The owner of this establishment indicated that they were thriving during this time of economic recession, and were increasing the size of their business.

A viewpoint often expressed by the respondents was that the associations that they belong to show a lack of any supportive policy or planning framework. The non-hotel accommodation owners and managers thought that associations such as Nelson Mandela Bay Tourism (NMBT) should be advising its members on topical issues, and communicating with them on such topics as per the guidelines found in the DEAT Handbook on Responsible Tourism.

5. Recommendations

- Continuous assessments relating to the understanding of the various causes that prevent the non-hotel accommodation industry from embracing the principles and practices of responsible tourism need to be conducted. Conducting evaluations could result in a continual improvement in awareness levels and stimulate implementation.

- Due to the negativity regarding government and the role that it plays in tourism, it would not be a good strategy to implement legislation regarding the responsible-tourism guidelines. It could produce a more positive outcome if the non-hotel accommodation owners and managers were encouraged to buy into the concept, and a provincial Responsible Tourism Strategy, to address specific issues was set up as a benchmark for tourism enterprises.
- Membership organizations need to pay close attention as to how they communicate with their members, especially about initiatives, such as responsible tourism, which could benefit the tourism industry.

6. Conclusion

This study has offered a snapshot view of the present status, constraints and success factors regarding responsible-tourism ethics and implementation within the non-hotel accommodation industry in Port Elizabeth. As this industry continues to grow, there is a need to develop a better understanding regarding responsible tourism and what it entails, before policy can be implemented. This fact will remain the largest inhibiting factor that will affect progress in the adoption of the responsible-tourism principles. Further study with a larger sample to gain a better understanding for the reasons regarding lack of awareness is necessary.

Responsible tourism has not entirely been embraced by the non-hotel accommodation industry, and there are serious shortfalls in the training of those who are able to implement change. The reason appears to be a lack of awareness and knowledge regarding how to move forward and embrace the guidelines as set out by the 2002 Responsible Tourism Handbook and the government policy document, namely the 1996 White Paper on the Development and Promotion of Tourism in South Africa. Once tourism is managed responsibly, positive impacts will be felt in economic, social and environmental areas. Education is one of the key factors that limit the positive impacts. There is scope for further research in this particular field and it is highly recommended. The research could immediately identify areas where differences can cause maximum change affecting all sectors in the tourism industry in a positive manner.

References

Bennett, J. (2000). *Managing Tourism Services*, 2nd ed. Pretoria: Van Schaik.

Business Day South Africa (2009). Tourism Contribution to GDP in Focus, http://allafrica.com/stories/200905110033.html.

DEAT [Department of Environmental Affairs and Tourism] (2003). *Responsible Tourism Handbook*. Pretoria: Government Printer.

—. (2002). Responsible Tourism Guidelines, http://www.info.gov.za.

Evans, N., D. Campbell & G. Stonehouse (2003). *Strategic Management for Travel and Tourism*. England: Butterworth-Heinemann.

Frey, N. (2008). Responsible tourism management: The missing link between attitude and behavior in and emerging market. Paper presented as the Second International Responsible Tourism Conference, March 2008, School of Management Studies, UCT.

Goodwin, H. (2009). Taking Responsibility for Tourism, http://www .haroldgoodwin.info.

Harris, R., T. Griffin & P. Williams (2002). *Sustainable Tourism: A Global Perspective*. United Kingdom: Butterworth Heinemann.

McGrath, G. (2004). Is consumer demand the only hope for sustainable tourism? http://www.travelmole.com/news/html.

Republic of South Africa. Department of Environmental Affairs and Tourism (1996). The Development and Promotion of Tourism (White Paper).

—. (2002). National Responsible Tourism Development Guidelines for South Africa.

Tearfund (2002). Worlds Apart: A Call to Responsible Global Tourism. London: Tearfund.

UNISA (2008). Understanding Corporate Citizenship. UNISA Center for corporate citizenship, seminar conducted 2008.

Van der Merwe, M., & A. Wocke (2006). An investigation into responsible tourism practices in the South African hotel industry. *S. Afr. J. Bus. Manage* 38, no. 2.

PART VI

CREATING AWARENESS

CHAPTER THIRTY-TWO

"TREES FOR ZAMBIA": MEETING CLIMATE-CHANGE CHALLENGES THROUGH "VOLUNTOURISM," KNOWLEDGE SHARING AND COLLABORATION

LAUREN O'DONNELL[1]

Abstract

"Trees for Zambia" is a "voluntourism" project taking place in Livingstone, Zambia, for three weeks in July 2012. More than just a volunteer-abroad opportunity, this project will be a conference of action and learning. Days will be spent planting trees at the Victoria Falls World Heritage Site, in an area of the Dambwa Forest, in a subsistence farming village and at local schools as a method of practical education. In the evenings, speakers will present on issues such as climate change, deforestation and sustainability; workshops will be conducted on propagation, seed harvesting and various other areas within forest ecology; local authorities and farmers as well as volunteers will share experiences and environmental debate will be encouraged. "Trees for Zambia" is the start of an ongoing project in Zambia to promote sustainability and a culture of planting trees.

This is a pilot project for similar campaigns in Southern Africa in the future. Its style and roll-out aims to create a platform through which various sectors of the tourism industry (and other industries) can be part of solutions to climate change. Addressing climate-change mitigation and the need for adaptation in vulnerable areas, Trees for Zambia aims to bring diverse parties together to collaborate, share knowledge, act, inspire, spread awareness and work together for a healthier planet.

[1] Director, Greenpop.

Figure 33-1. Greenpop Reforestation Festival, South Africa, 2011. This saw volunteers getting involved in tree planting. Photo credit: Ruth De Freitas.

Tourism as Part of the Solution to Climate Change

Tourism's Footprint

The tourism industry faces challenges with regards to climate change and, by its nature, has a hefty carbon footprint. The industry's rapid growth provides both a challenge and an opportunity. The tourist community has taken up this challenge and has visibly stepped up its response to climate change. Rather than only seeing a problem, there is now a clear understanding that the industry can be part of the solution to climate change, by reducing its greenhouse gas emissions as well as by helping the communities where tourism represents a major economic source to prepare for and adapt to the changing climate (Simpson et al., 2008).

"Voluntourism"

As one of the fastest-growing forms of travel in the world, voluntourism has the potential to help vulnerable communities prepare for and adapt to the changing climate. "Voluntourists" even have the potential to be the drivers of important projects that might otherwise not have been possible, through manpower and financial support.

Highlighting the Problem

With tourism being one of the world's largest industries, there is great potential for highlighting climate change and making people aware of the risks, consequences and ways in which they can be part of the solution. The media also has a role to play in highlighting climate change in tourism hotspots.

Involving Tourists and Tourist Institutions

Creating concrete ways for tourists and local tourism bodies to get involved in mitigation and adaptation to climate change has the potential to attract much support, action and collaboration.

"Trees for Zambia": A Voluntourism Project by Greenpop

Introduction to "Trees for Zambia"

Greenpop, a southern African tree-planting social enterprise, is coordinating a project called "Trees for Zambia" in Livingstone, in July 2012. This exciting venture will see 5,000 or more indigenous and fruit trees planted in three weeks within the Victoria Falls World Heritage site, a forest concession area, the Mukuni subsistence farming village and surrounding farms and in schools around Livingstone. Volunteers from around the world will be invited to join the local community, alongside key environmental speakers and teachers. Four days of each week will be spent planting and educating and one day will see volunteers and community members attending forest-based teaching, outdoor lectures and educational workshops around climate issues and adaptation. During these workshops, prominent speakers will talk about pertinent environmental issues, volunteers will share their knowledge, community members will talk about their experiences with climate adaptation, and environmental films will be screened. Essentially it will be a conference of action, sharing and learning.

Why Zambia? Background and Problem

Zambia has seen rampant deforestation in recent decades. According to the UNDP, the country has the second-highest deforestation rate in Africa, and the fifth highest in the world. More than 4,000 km^2 of Zambia's forests were lost every year from 1990 to 2005 – with signs that the rate has recently increased (UNDP, 2008).

The Livingstone area is one of the worst-affected areas. The Forestry Department puts the rate of deforestation in the Livingstone area at 5% annually and believes it has a bearing on local temperature and rainfall. According to a paper by the Zambian National Heritage Conservation Commission, this "is also likely to affect the Mosi-oa-Tunya World Heritage Site. The Falls are in an area experiencing notable impacts of climate change and variability. The 2005/2006 rainfall season recorded a notable flood disaster over the Kazungula district, a district upstream of the Zambezi, where thousands of families had to be evacuated to higher grounds."

Implications of Deforestation

Deforestation and climate change result in a change in the local ecosystem, a loss of biodiversity and increased incidence of both drought and flooding. It also means a decline in food security due to a reduction of agricultural yields and the loss of forest resources such as wild fruits. Additionally, women have to walk further to find wood, resulting in lower productivity, while children are also being left alone at home.

The Cause of Deforestation

There are two main causes of deforestation – tree cutting for timber and charcoal and poor soil management through traditional farming methods.

Tree cutting: Since electricity is expensive in Zambia, it is not available to many people. Charcoal is therefore the primary power source for all domestic heating and cooking. Wood and charcoal make up approximately 70% of Zambia's national energy demand, and the country is losing approximately 8,000 hectares of forest a year. The current and estimated shortages in electricity production forecasted for the next ten years could cause a further increase in forest degradation (Ministry of Finance and National Planning, 2008).

Poor soil management: Unsustainable farming methods, particularly maize mono-cropping, employed in Zambia for many years, have led to extensive land degradation – soil erosion, loss of soil structure and soil micro-organisms, acidification, oxidation of organic matter and compaction (GART, Brief 2). The Food and Agriculture Organization (FAO) notes that the 65% increase in maize production between 1981 and 1991 came from the expansion of cultivated land rather than increased yields (Conservation Farming Unit, Brief 2), and with this expansion comes deforestation.

According to the brief on reversing environmental degradation through conservation farming and conservation agriculture (GART, Brief 2), when land degradation occurs, families:

- abandon farmlands and migrate, encroaching on forests (trees are cut down to make way for farming and families temporarily restore their crop yields by mining out the nutrient reserve accumulated by nature);
- encroach into areas of ecological significance, including designated forest reserves and wildlife sanctuaries to farm; and
- engage in poaching, off-season fishing or charcoal-burning to supplement incomes.

Effects on Tourism

Zambia, and the area of Livingstone in particular, is a tourist hotspot. People flock to this little town to see the mighty Victoria Falls, get involved in various adventure activities and experience the beauty of the Zambezi River and surrounds. While there are varied, indirect and far-reaching ways in which the issues outlined above can affect tourism, the most glaring is perhaps the fact that Zambia's draw card, the natural beauty, biodiversity and wildlife, are under threat.

With the aim to improve the situation through planting, knowledge-sharing and collaboration through "Trees for Zambia," Greenpop will start a substantial and tangible drive to encourage education for improved adaptation to climate change. Furthermore, the project aims to contribute towards climate-change mitigation through the planting of indigenous trees.

Planting Trees

Planting Indigenous Trees in Forest Concession Areas

In Zambia, protected forest areas are under threat. People are cutting down trees for charcoal-making, wood fires and agriculture. Permits to fell trees are available, but are expensive and often not an option for people battling to make ends meet. Charcoal-makers also have to travel a lot further to find suitable trees for burning as these are running out.

The Trees for Zambia project will see 1,000 trees planted in Dambwa forest near a lion-release location and 1,000 trees planted in the Peace Park at the Victoria Falls. The aim is to highlight the importance of replanting in forest areas to replace the pioneer trees and allow for new growth, regeneration of soil content and biodiversity. The natural forests of Zambia are part of the country's heritage, and education surrounding replacing trees in these protected areas is vital for conserving the environment, protecting Zambia's forest assets as well as sustaining tourism in these beautiful areas.

Planting in Subsistence Farming Villages

The Trees for Zambia project will see 1,000 or more Faidherbia Albida trees planted within the crops of subsistence farmers around Livingstone.

The practice of planting these particular trees in fields forms part of an important sustainable-farming technique that is being adopted by the Department of Agriculture in Zambia and among small-scale farmers to try and reverse the situation.

Through leaf and pod fall, these trees fix nitrates and micro-organisms into the soil and improve fertility, eventually resulting in better, more sustainable yields.

Figure 33-2. The Greenpop team visited subsistence farming communities in June 2011 to discuss their experiences with mono-culture farming techniques, the challenges they now face with soil fertility and the potential of planting Faidherbia Albida trees in their fields to act as a natural fertilizers. Community members were very aware of the technique but requested help with seed harvesting, planting and education. Photo credit: James Cumming, Resource Economist at OneWorld Sustainable Investments.

Promotion of Conservation Farming Methods

The Trees for Zambia project will promote the use of this and other conservation farming methods in an effort to encourage subsistence farmers to move away from traditional methods that cause soil degradation and deforestation. We will host three half-day workshops for small-scale farmers in the Livingstone area. Through these workshops, we will teach farmers about the following farming methods:

- residue retention
- reduced tillage
- dry-season land preparation and early planting
- rainwater harvesting
- precise application of seeds and nutrients
- crop diversification and rotation
- use of Faidherbia Albida, a deep-rooting indigenous tree, as a natural fertilizer (farmers will be shown how to collect seeds for the Faidherbia Albida, and how to plant and grow the trees)

The use of these methods will help to reverse land degradation, bring soil back to life, enhance household food security, increase the resilience of crops to droughts, enable sedentary farming and eliminate the need for families to migrate and cut down forests.

Planting Fruit Trees in Schools

Education on climate-change issues is of vital importance for successful future adaptation and mitigation. Practical education, through the act of planting trees, is tangible and memorable. During the Trees for Zambia project, indigenous and fruit trees will be planted at schools.

Figure 33-3. Students learn through the practice of planting trees. Photo credit: Carmen Visser.

Planting indigenous trees with learners:

- Teaches valuable environmental lessons about the importance of trees.
- Reconnects learners with nature.

- Beautifies schools and communities instilling pride of place.
- Sequesters carbon and provides an activity through which to teach this complex concept.

Planting fruit trees at schools:

- Provides fruit and nutrition for school children.
- Aids concentration, uplifts learning experiences and improves school attendance.
- Provides children with the valuable experience of planting, growing and caring for something as well as seeing the positive results.
- Provides practical environmental education on the benefits of trees, growing produce and caring for our environment.

Knowledge-sharing

In the run-up to the three-week campaign (starting four to six months before the planting), educational material will be provided to the Livingstone Department of Education who will be facilitating the delivery of these worksheets to schools to help youth to understand what is currently happening in their direct environment and how this has come about. An experienced nursery manager in Livingstone is providing the information on indigenous trees and their uses to be added to this material. The Department of Education has advised that material will be adapted by teachers to suit their class levels and curriculum.

During the three-week campaign, while planting and in the evenings, workshops will be facilitated for the volunteers in order to share global knowledge about climate issues as well local concerns about what is happening in this space. A representative from the UNDP, local NGOs and experts on environmental issues will be speaking.

Topics include:

- What climate change is, adaptation insights and ways to mitigate pending risk.
- What the effects of climate change are on communities, especially family farming communities.
- How deforestation and climate change affect wildlife.
- How to adapt to these effects, minimize risk and enhance resilience.

- How to start solving the climate-change crisis through carbon sequestration projects and the importance of community tree planting to replace trees that have been cut down for future generations.
- Seed collection, propagation and rearing of saplings.
- The uses of trees (increase fertility in agriculture, carbon sequestering, medicinal, animal fodder, etc.).

Following the three-week campaign, the Ministry of Agriculture, the Forestry Department and the Zambia Environmental Management Agency will be assisting us with sustaining long-term relationships with parties involved. During the three weeks, eco-heroes within communities will be singled out and will become the point of contact between the ministries and the communities. They will also head up forum discussions and monitor the success, problems and questions surrounding the trees. The Ministry of Agriculture will work with these contacts to evaluate the farming techniques.

Media will also play a role in knowledge-sharing during and after the event. The Zambian Broadcasting Commission showed an interest in Trees for Zambia when the Greenpop team was in Livingstone for a reconnaissance visit in June 2011. They produced a story for the evening news, which was instrumental in making local schoolteachers and learners aware of and excited about the campaign. Local media companies will be contacted in the run up to July 2012 to be part of Trees for Zambia and share the events with as many people as possible. Makhulu Productions, a South African–based, socially minded production company, will also be documenting Trees for Zambia to create a full-length documentary as well as shorter pieces for distribution globally. We hope to highlight the situation in Zambia as well as promote meaningful tourism options that address climate issues.

Collaboration

Trees for Zambia will be a collaborative solution – involving tourism bodies, government, business and individuals. Through bringing interested parties together to do something positive for climate-change mitigation, we aim to facilitate knowledge-sharing that will reach a wide local and international audience.

By creating a project that offers value to all stakeholders and is implemented in a strategically beneficial and economically streamlined way, we hope to capture a process which can be effectively transposed to fit many similar applications. July 2012 will be a pilot and we will replicate this annually and increase the number of trees and awareness sharing possibilities exponentially.

Tourism Bodies

Voluntourism Agencies

The project will be advertised all over the world through voluntourism agencies. As well as providing a climate-change project for them to put to the market, Greenpop is proposing that these companies also offer voluntourists traveling to other projects the opportunity to offset their carbon emissions – by giving trees to a project like Trees for Zambia.

Hotels and Lodges in Zambia

The trees for the project will be sold in various ways. One method will be through local hotels and lodges in and around Livingstone. They have been invited to purchase trees themselves to compensate for their carbon footprints. On top of that, a large number of lodges and hotels are offering their clients the chance to give a tree by adding it to their bill. The involvement of the local tourist industry is vital in spreading awareness of the project in the local community, as well as internationally. Their involvement will also create positive PR. Again, this will act as a catalyst for spreading awareness about climate change, highlighting the problem to the industry and getting more parties involved in the solution.

Local Government

It is of utmost importance that Zambian authorities are involved in the project. Greenpop has already met with the Ministry of Agriculture, the Forestry Department, the National Heritage Conservation Commission, the Zambia Environmental Management Agency and a Regional Tourism Development and Research Officer in Livingstone, all of whom have expressed support for the project. These partnerships are valuable for the sustainability of the project and successful knowledge-sharing between authorities and communities.

Figure 33-4. Greenpop is using creative ways to sell the trees to tourists visiting Livingstone so that they can be involved in improving the area and be part of Trees for Zambia.

Business

We are offering local and international companies the chance to become a sponsor of Trees for Zambia by buying trees. In return for their support, we offer companies positive exposure through various marketing tools. Sponsorship of the project is particularly relevant and beneficial for timber companies and companies operating internationally looking to invest in a carbon-offsetting or corporate social-responsibility project.

Individuals

We expect this project to attract individuals from different backgrounds and nationalities with an interest in climate change. This will complement the knowledge-sharing and bring varying perspectives together. Voluntourism, nowadays, is not limited to students and graduates but also attracts professionals looking to do something meaningful while traveling. Local community members will be very much involved in the planting and education as well, and a platform for cross-cultural learning will be created.

Figure 33-5. Greenpop's Reforestation Festival, June 2010. This attracted a wide range of voluntourists. Photo credit: Ruth De Freitas.

Making it Happen: Parties Involved in Producing the Event

Greenpop: Greenpop is a tree-planting social enterprise that celebrates greening and environmental education. The Greenpop team is passionate about ground-level environmental awareness, community upliftment, social integration, urban rejuvenation and contributing positively towards adapting to climate change. Part of the mandate at Greenpop is to home in on areas within the SADC region which are facing environmental catastrophe through deforestation. One of Greenpop's missions is to inspire a greener and more conscious movement through creating platforms, projects and activities that people want to be involved in – focusing on solutions rather than problems. Spreading the message through experiences that are enjoyable, memorable and inclusive is the organization's aim. Greenpop will be coordinating the Trees for Zambia project and plans to roll out similar projects in sub-Saharan Africa.

Zambezi Nkuku: Zambezi Nkuku is a chicken farm and tree nursery in Livingstone, and Greenpop's logistical partner on the ground for Trees for Zambia. The trees are already being reared in their nursery and the beneficiaries have chosen the tree varieties. In addition, Zambezi Nkuku has assisted in partnering Trees for Zambia with various other key players, like Zambian ministries, which will ensure the smooth running of this project.

Derek Pollard: Derek Pollard is an environmental program manager. His core task is to develop relations between Greenpop and parties on the ground in Livingstone and to establish sustainable planting techniques specific to the context. Derek will be creating much of the educational material aimed at subsistence farmers.

Zambian Authorities: It is of utmost importance to us that we partner with Zambian authorities. Please see the section on "Local Government" for more information.

Discussion

The Second International Conference on Climate Change and Tourism (Davos, Switzerland, October 2007) outlined the need for the tourism sector to rapidly respond to climate change if it is to develop in a sustainable manner, which will require actions to mitigate greenhouse gas emissions from the tourism sector; adapt tourism businesses and destinations to changing climate conditions; apply existing and new technologies to improve energy efficiency; and secure financial resources to assist regions and countries in need (UNWTO, 2008).

Trees for Zambia will be an annual voluntourism project involved with mitigation of carbon emissions through tree planting and conservation farming; and spreading awareness and knowledge about adaptation to climate change through face-to-face cross-cultural sharing. This project will be made possible through voluntourism as well as other support from the tourism industry and is an example of a platform through which the industry can move towards the goals set out at in Davos in 2007.

Greenpop and the coordinators of Trees for Zambia invite partnerships, support and collaboration on this project and others like it in the future.

References

GART, Brief 2. Reversing Environmental Degradation through Conservation Farming and Conservation Agriculture. GART, Conservation Farming Unit, Zambia.

Ministry of Finance and National Planning (2008). The Zambia Millennium Development Goals Progress Report. Lusaka, Zambia.

Simpson, M. C., S. Gössling, D. Scott, C. M. Hall & E. Gladin (2008). *Climate Change Adaptation and Mitigation in the Tourism Sector:*

Frameworks, Tools and Practices. UNEP, University of Oxford; UNWTO, WMO: Paris, France.

The National Heritage Conservation Commission & the Provincial Meteorological Office, Livingstone, Zambia. (2007). Report on the Impact of Climate Change on the Mosi-oa-Tunya World Heritage Site. Livingstone, Zambia.

UNDP (2008). *Human Development Report 2007/8*. New York: UNDP.

UNWTO [United Nations World Tourism Organization] (2008). Climate Change and Tourism – Responding to Global Challenges. Madrid, Spain.

CHAPTER THIRTY-THREE

THE QUALITYCOAST AWARD PROGRAM

FOPPE J. SEEKLES[1]

Introduction

QualityCoast aims to establish a worldwide network of coastal communities that share the same values and practices on sustainable development, holistically integrating their natural, cultural and social values and at the same time maintaining high standards in the quality of their tourism. It has been developed through the CoPraNet project, a European Union project with twenty-one partners from eleven countries under the lead of the Coastal and Marine Union (EUCC) – one of the largest coastal expert networks in the world. QualityCoast provides new ways of cooperation between coastal communities whereby they come together in a network of like-minded communities to share their experiences and exchange best practices in order to improve sustainability in the fields of nature, environment and socio-economics. As well as this, their performance in these fields can be connected to international tourism marketing through the QualityCoast Award.

The QualityCoast Awards Program

QualityCoast utilizes 110 indicators to assist the tourist in choosing the best possible destination for their holiday based upon the tourist's individual desires and interests. These indicators are the building blocks of the several award levels (bronze, silver, gold and platinum). The awards are mentioned in magazines of tour operators (such as TUI) and are actually working as a marketing tool.

Twenty QualityCoast criteria have been adopted to provide a measure of the sustainability of coastal communities and the quality of information

[1] General Ambassador of the Coastal & Marine Union, EUCC.

available. Each criterion provides information on a distinctive aspect of sustainability.

The criteria are divided into three categories:

- Nature and Biodiversity
- Environment
- Socio-Economics

The criteria themselves will be the subject of periodic reviews, a process that will involve the QualityCoast destinations.

Nature

1. Natural Values: Presence of internationally and nationally important habitats and species that occur within the territory.
2. Information and Education: Information sources and facilities that promote interaction with the natural values in and around the territory.
3. Contact with Nature: Appropriate level of accessibility of natural areas for low-pressure recreational access and nature experience.
4. Green Policies: The extent to which natural values are enhanced through cross-sector approaches.
5. Open Landscapes: Presence of open landscapes and existing trends in land-use and land-cover patterns.
6. Quiet Places: Presence of areas where people can escape from traffic, crowds and noise.

The identity of the destination is an increasingly important issue for the tourist. Reviewing this, starting 2011, we are giving identity special attention in the categories of *cultural heritage*, *territory and tradition*, *local identity*, *quiet places*, and *hospitality and satisfaction*.

Environment

7. Tourism Pressure: The pressure and the impacts of tourism on the territory and the way the applicant is controlling tourism pressure.
8. *Business Involvement*: The extent to which the local business community is involved in efforts for environment, sustainable development and marine conservation.
9. Blue Flags and Beaches: Environmental beach management, meeting bathing-water quality standards of Directive 2006/7/EC and Blue Flag.

10. Water Management: Sources of water, environmental and efficient use of water, supply, wastewater treatment and reuse of depurated water.
11. Sustainable Transportation: Availability and promotion of sustainable transport modes.
12. Waste and Recycling: Policies and structures for waste collection, recycling and re-use.
13. Climate and Energy: Policy aimed at reducing greenhouse gases and promoting renewable energy.

Socio-economics

14. Cultural Heritage: Presence and level of protection of elements that are connected to the history of the territory.
15. Territory and Tradition: Presence of valuable landscapes that are man-made or man-formed and traditional use of the territory.
16. Local Identity: Presence and attention for a distinct physical and/or social aspect that is considered typical for the territory.
17. Leisure and Recreation: The way sustainability aspects are taken into account in leisure and recreation activities and facilities in the territory.
18. Community Participation: Local initiatives to promote community participation and stakeholder involvement in management and decision-making.
19. Satisfaction and Hospitality: The appreciation level of residents and tourists regarding the territory.
20. Health and Safety: Presence of a healthy, safe and secure environment and preventive measures.

Since 2007, twenty-nine coastal communities from ten European countries have received the QualityCoast Award. The only region where all communities have met the QC Award requirements (2010) is the Azores. The QualityCoast BasiQ scan is now performed for international (coastal) tourism destinations in the following countries: Cyprus, Greece, Slovenia, Malta, France, Spain, Portugal, Italy, The Netherlands and the Netherlands Antilles.

Developing the QualityCoast Partnership

In January 2011, EUCC started to develop a partnership of NGOs to implement a first series of pilots applying the QualityCoast BasiQ indicator set (1.0) in a number of countries. The NGOs are independent of the coastal communities to be assessed.

So far the QualityCoast partnership consists of:

- AKTI Project and Research Center, Nicosia, Cyprus
- Black Sea Coastal Center, Constanta, Romania
- Centro Mediterraneo EUCC, Barcelona, Spain
- Dutch Caribbean Nature Alliance, Bonaire, Netherlands Antilles (to be confirmed)
- EUCC-France, Lille, France
- EUCC-Germany, Warnemünde, Germany
- EUCC-Malta, Sliema, Malta
- Vereniging Kust & Zee, Leiden, the Netherlands

2011 QualityCoast Ceremony and Workshop

The 3rd International QualityCoast Award Ceremony and Workshop was recently held in the medieval castle of Kouklia (Cyprus), near the Rock of Aphrodite. Representatives from local and regional authorities exchanged experiences in the field or sustainability in tourism. EU Commissioner Dr. J. Potocnik and the Minister of Environment of Cyprus addressed the audience and presented the QualityCoast Awards and the QC BasiQ Certificates on behalf of EUCC's President, Minister of State Johan Vande Lanotte (Belgium) and the international QC Jury. Award and certificate winners represented more than ten countries.

More Information

For more information please see the QualityCoast website for partners and applicants, www.qualitycoast.net; the QualityCoast website for the public, www.qualitycoast.info; Coastal & Marine Union – EUCC, www.eucc.net; or contact the QualityCoast Secretariat at EUCC in Leiden (NL).

Affiliated Projects

The INTERREG IV B co-funded IMCORE project: This has just completed its first year of activities in 2010. It aims to promote an innovative and sustainable approach to reduce the ecological, social and economic impacts of climate change on the coastal resources of northwest Europe.

This approach is based on the association between a research group and a local authority, working together and bringing in multidisciplinary views and experiences to achieve more integrated solutions to local management. These partnerships were named Expert Couplet Nodes (ECN) during the previous project COREPOINT and are being further formalized and developed during IMCORE.

Each of the nine ECNs established in coastal areas of the United Kingdom, Ireland, Belgium and France are addressing particular local issues and developing strategies to adapt to climate change at a local scale but at the same time are designing and testing various tools and techniques. These include stakeholder mapping and future scenarios of potential climate-change impacts, whose value can then be shown and made available to other local and regional authorities.

As a partner in the project, EUCC is assessing the engagement and process within the ECNs, defining guidelines and recommendations on this approach and collaborating in the development of a set of indicators to monitor their process. These will be one of the project products integrating the Multimedia Distance Learning Tool – the final output of IMCORE – a comprehensive resource of information and tools designed to help coastal managers planning for climate-change impacts in their coastal areas. For more information, go online at www.imcore.eu or contact info@imcore .eu.

ACCMA – Climate Change Adaptation in Morocco: As elsewhere in the Mediterranean, Morocco's coastal province of Nador is vulnerable to the impacts of climate change. ACCMA is a research project to develop capacity for, and contribute to, policy- and decision-making for strategic coastal land-use planning and management, with the purpose of reducing the vulnerability of coastal communities to the impacts of sea level rise, coastal flooding, and related extreme-weather events.

The project aims to advance the science and technology that underpin preparations for, and responses to climate-related events, and contribute to the information systems that guide policies of public protection. Its results will inform the Nador ICZM Action Plan developed by the EUCC-led project CAP Nador, and support the assessment and implementation of adaptation measures in a regional context.

These goals are being pursued through development of vulnerability assessments, participatory deliberation of land-use and climate-change-adaptation strategies, and the formation of capacity through technology transfer, awareness rising, and institutional strengthening. EUCC is playing an important role throughout the project by supporting information activities, stressing the policy perspective and advising the lead partner ENFI, a Moroccan research institute. For more information, go online at www.pik-potsdam.de/~kropp/accma/.

CHAPTER THIRTY-FOUR

DEVELOPING WEB-BASED PLATFORMS TO PROMOTE RESEARCH AND COMMUNICATION ABOUT TOURISM AND CLIMATE CHANGE IN SOUTHERN AFRICA[1]

HELEN PURKITT[2]

Abstract

The chapter describes the organization of a web-based project designed to promote collaborative research, learning, communication and action about Botswana tourism. Examples from this project are used in the final section of the paper to illustrate how data collection, research results, policy initiatives and communications about tourism and climate change in southern Africa can be organized on a dedicated website. The Botswana project used freeware, i.e., Google Sites and Drupal, an html content-management system, to archive specialized research and knowledge and for online communication and collaborations. This website was developed

---

[1] This chapter is based on two presentations made at the 5th IIPT African Conference: "Meeting the Challenges of Climate Change to Tourism in Africa and the Developing World," held in Lusaka, Zambia, May 2011 and hosted by the Zambia government.

Disclaimer: The views expressed in this article are personal ones of the author and do not reflect the views of the U.S. government, the U.S. Department of Defense, the U.S. Naval Academy or the University of Botswana or its affiliated research institutes, the International Tourism Research Center (ITRC) or the Okavango Research Institute (ORI).

Acknowledgements: This project builds upon earlier research supported by a grant from the Rapid Reaction Technology Office/Advanced Technology, Office of the U.S. Secretary of Defense (OSD). The Botswana Tourism Research project was supported by a U.S. Fulbright Africa Research grant. Mr. Lenny Chavhunduka, Digital Treasure (Lenny@digitaltreasure.co.bw) developed the Drupal site.

[2] Professor, Dept. of Political Science, U.S. Naval Academy, Annapolis, MD 21402.

as a Fulbright Research Project for the International Tourism Research Center (ITRC) at the University of Botswana in Gaborone in 2010 and is available online at www.BotswanaTourismResearch.org.

1. Introduction

The incredible growth of social-networking sites, instant text messaging, collaborative chat rooms, blogs, Facebook, MySpace, Twitter and increasingly powerful cell phones that serve as software and Internet platforms all underscore the fact that the way we communicate, store our knowledge and organize for action is undergoing remarkable change. The extensive use of the Internet by President Obama's 2008 campaign for fundraising, communicating and getting-out-the-vote, as well as the central role played by web-based tools during the revolutions in several Arab countries during 2011, are timely reminders of the fact that the Internet is now a central tool in making, changing and implementing public policies. The irreversible and highly disruptive changes associated with climate change suggest that it is important to also harness the Internet to promote research, communication, policy changes and manage needed adaptations due to future disruptive effects of climate change and its impact on tourism as well. Nowhere are such systems needed more than in Africa where the disruptive effects of climate change are already being experienced.

This paper describes an ongoing project designed to harness some of the latest web-based freeware tools to facilitate collaborative learning, research and communication among geographically dispersed participants with common interests. The project offers examples that may be useful in future efforts to construct a web-based database to promote research, communication and actions to promote tourism and manage climate change in southern Africa. The next section describes a web-based site designed to facilitate tourism research for the ITRC at the University of Botswana in Gaborone. This content-management system, designed for use by students, researchers and serious eco-tourists, was developed as part of a Fulbright Research Project for the ITRC during 2010. This project is available online at www.BotswanaTourismResearch.org.[1] The final section discusses how the basic design of this database could be used to develop a web-based site geared towards the interests of tourism and climate-change researchers to promote research, communication and actions geared toward understanding, monitoring and mitigating the adverse effects of climate change on tourism in one or more regions of Africa.

2. Developing a Specialized Content-management System for Botswana Tourism

The database described in this section is designed for tourism researchers, students and serious eco-tourists interested in contributing to knowledge about Botswana's environmental and tourism sector. The initial prototype website was developed using Google Sites freeware because it required no programming expertise. This prototype website is available at: http://www.purkitt.org/. Figure 34-1 presents the front page of this initial website.

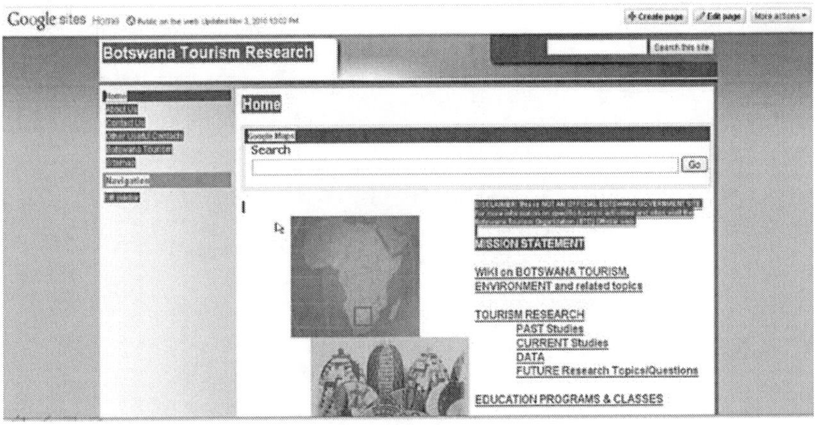

Figure 34-1. Google Sites prototype for Botswana Tourism Research.

The use of Google Sites was helpful for designing the preliminary front page because this freeware permits the importation of numerous apps such as Google Map, weather updates, etc. It is also possible to upload pictures and other non-textual materials without having to write specialized programs. Google Sites is very easy to use and one can quickly design and implement a specialized wiki. While easy-to-develop, Google Sites did not have all of the functionality needed to develop any embedded wiki data within the website; therefore, another content-management system was written using Drupal freeware modules. A screenshot of the front page of the final website, available at http://www.Botswanatourismreserach.org, is shown in Figure 34-2.

This specialized website consists of a series of linked web pages including: (1) a specialized wiki on Botswana tourism and environmental research open to anyone interested in accessing or contributing to it online;

(2) an archive of relevant research, news articles, reports, websites, etc. (3) a "where am I" picture quiz of tourist sites in Botswana; and (4) communication tools, including a blog and forum. The picture quiz was based on a former column in Africa National Geographic. This feature was incorporated into the front page of this website home page in order to capture the interest of potential users.[2] Another reason for its inclusion is because too many citizens and long-time residents in Botswana have never visited many of the most famous tourist locations in their own country. Preliminary feedback from University of Botswana students at the end of 2010 during an all-university fair suggests that it will remain one of the most popular features of the website.

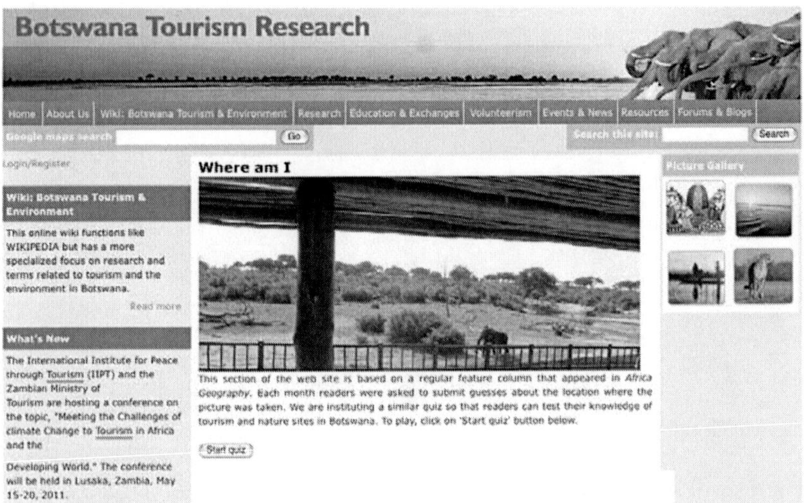

Figure 34-2. Final template for Botswana Tourism Research.

Additional information is accessible from the front page through drop-down menus. Current information contained in the drop-down menus include: (1) a list of relevant actors and websites involved in tourism in Botswana; (2) archival space for past, relevant research, summaries of ongoing research and suggestions for future research; (3) a series of informative drop-down menus with information about in-country and international education programs related to tourism and the environmental programs and information about exchanges; (4) drop-down menus for relevant news stories; and (5) pages to support both forums on specific topics and a blog. Links to other social-networking sites (i.e., Facebook, Twitter, etc.) will be added as demand for these links arise.

The most unique feature of the content-management system is a specialized wiki embedded within the system. On the left-hand side of the front page (See Figure 34-2 or go to www.Botswanatourismresearch.org), a user can easily enter and view articles in the wiki or register and contribute to topics related to Botswana tourism and environmental topics. Figure 34-3 displays the current state of the wiki that one accesses from the front page. Midshipmen at the U.S. Naval Academy wrote additional articles on general topics displayed in the index. Students majoring in Environmental Science at the University of Botswana in Gaborone are going to be adding more specific information related to Botswana to the general articles. The goal is to spur cross-continental electronic collaborations among students as they collaborate to write additional articles in the future.[3]

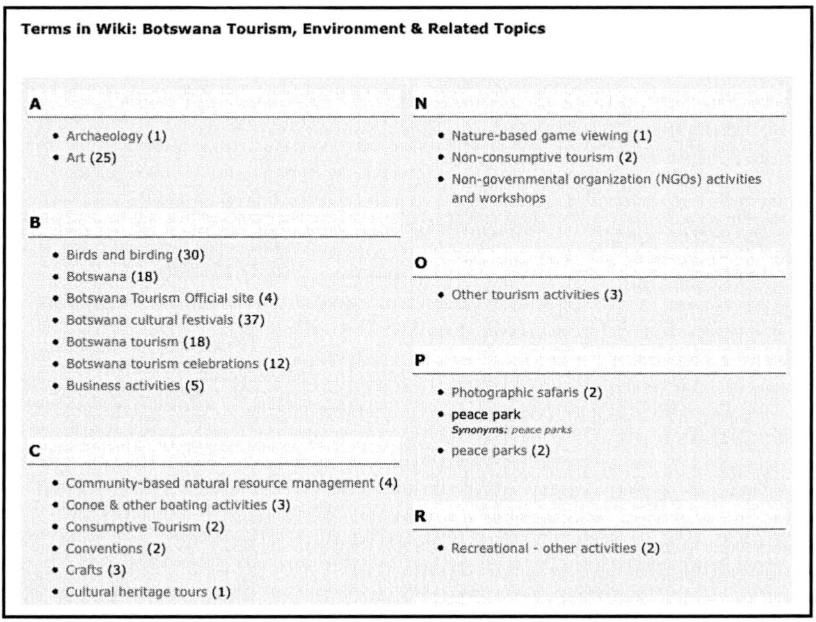

Terms in Wiki: Botswana Tourism, Environment & Related Topics

A

- Archaeology (1)
- Art (25)

B

- Birds and birding (30)
- Botswana (18)
- Botswana Tourism Official site (4)
- Botswana cultural festivals (37)
- Botswana tourism (18)
- Botswana tourism celebrations (12)
- Business activities (5)

C

- Community-based natural resource management (4)
- Conoe & other boating activities (3)
- Consumptive Tourism (2)
- Conventions (2)
- Crafts (3)
- Cultural heritage tours (1)

N

- Nature-based game viewing (1)
- Non-consumptive tourism (2)
- Non-governmental organization (NGOs) activities and workshops

O

- Other tourism activities (3)

P

- Photographic safaris (2)
- **peace park**
  Synonyms: peace parks
- peace parks (2)

R

- Recreational - other activities (2)

Figure 34-3. Keywords currently listed in the Botswana Tourism Wiki.

Unlike Wikipedia, the basic unit of this content management is the keyword rather then the article. One purpose in using this organizational logic was to see which keywords would be contributed to a bottom-up community-developed thesaurus. Time will tell if this was the best logic to use to link related content information in the wiki. Preliminary work suggests that it may have been best to stick with the article-based approach of Wikipedia. The outline template for articles currently being uploaded into the wiki is displayed in Figure 34-4. The exemplar article is about climate change. The organization of this specialized wiki outline is based on the categories similar to those used in Wikipedia articles. We felt this was best since many contributors would already be familiar with the organization of Wikipedia articles. Like Wikipedia, articles can be revised over time as additional users contribute to this specialized wiki.

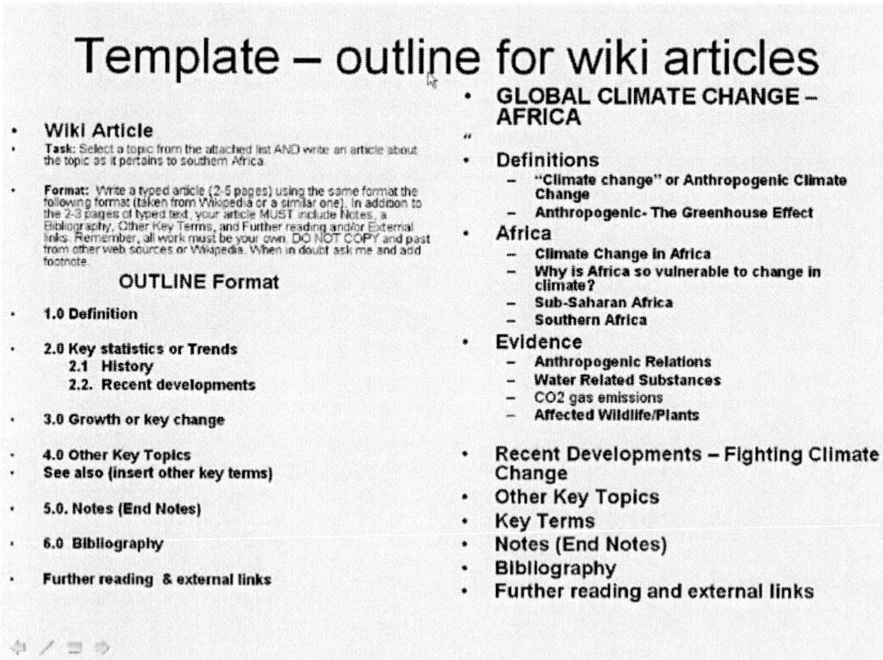

Figure 34-4. Botswana Tourism Research Wiki article outline.

3. A Wiki on Tourism and Climate Change in Southern Africa

Although the Botswana Tourism Research website is geared toward the interests of tourism and environmental researchers in Botswana, the basic design could easily be modified to serve the needs of researchers, students, policy-makers and serious eco-tourism visitors concerned about climate change and its impact on tourism in southern Africa or the entire African continent. The Google Sites approach described in the prior section may be the most attractive option for individuals who want to launch their own websites at no cost and have no prior programming experience.[4] However, for those with some html programming knowledge or resources to pay a programmer, the Botswana Tourism Research site written in Drupal (www.Botswanatourismresearch.org) would be a better alternative since the module approach of this content-management system permits more applications and additional features to be added as new needs arise. This section describes some ways that a web-based system could facilitate the synthesis of research, knowledge-accumulation and communication among those interested in the impact of climate change on different areas of Africa and worldwide.[5]

A specialized wiki would serve as a useful resource for both the novice and expert user. All that is required is a knowledgeable "Wiki Master" who has the time to manage and check the accuracy of wiki entries from volunteer contributors. Figure 34-5 contains excerpts from a wiki article written by an undergraduate student at the U.S. Naval Academy who was enrolled in an African Politics course during the spring of 2011. This student, and other class members, wrote wiki articles on assigned topics related to tourism and environmental issues in Africa. In coming months,

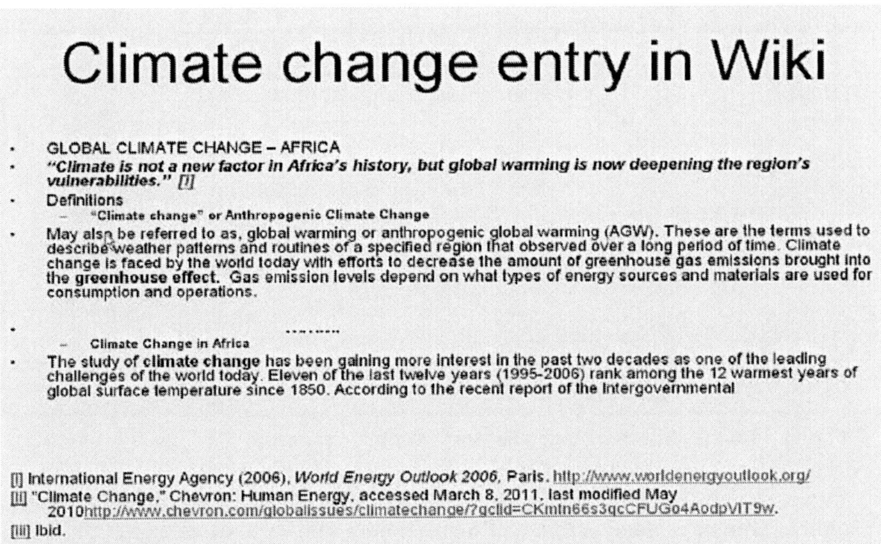

Figure 34-5. Excerpt from general article on climate change.

student volunteers from the Department of Environmental Science at the University of Botswana will be adding more specialized information to these general articles about how climate change is impacting tourism in Botswana.[6] These excerpts illustrate how useful such articles can be for providing both general and very specific information in a single article. Figure 34-5 provides some preliminary definitions extracted from the article while Figure 34-6 contains charts from the same article. These charts illustrate how easily new information and summaries, in nearly any format (pictures, films, etc.), of recent trends can easily be added to update an existing article.

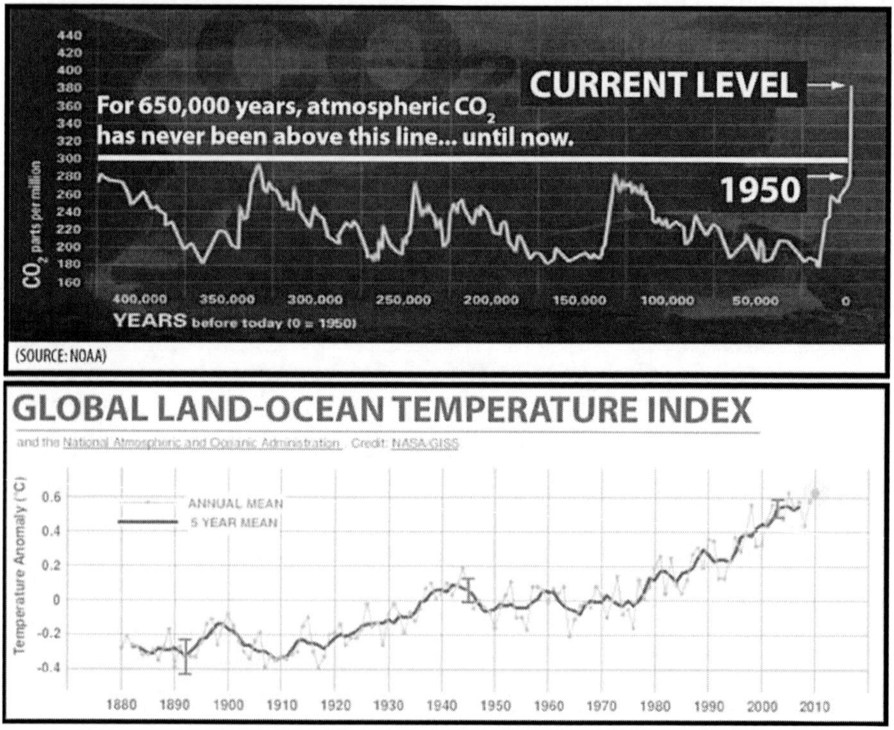

Figure 34-6. Excerpt from general article on climate change.

There are several advantages for shifting knowledge and research studies online. One advantage is that it allows for new studies to be integrated into an existing wiki database quickly and made available to the widest possible audience. For example, a new study on Climate Change and Tourism Policy in OECD Countries (OECD 2011) that was published while this chapter was being revised is now integrated and linked to the above wiki article that will soon be uploaded to the Botswana Tourism Research site.[7]

The ability of such sites to serve as free-access repositories for research that might otherwise not be available to users should also not be underestimated. There are hundreds and possibly even thousands of research studies that are relevant to policy issues related to tourism and climate change in Africa that are likely not to be read because they are not in electronic form or are too expensive for African libraries or governments to purchase. These studies include not only scientific studies but also efforts to develop systematic methodologies that may help national policy-makers and others to better manage and prepare for the dislocating effects of climate change. It is important to find ways for decision-makers and others working in Africa to have quick access to these works given the fact that the continent is the most vulnerable to the adverse effects of climate change. An estimated 75 to 250 million Africans will be exposed to increased water stress due to climate change by 2020 and estimated yields in African states that are dependent on rain-fed agriculture could be reduced by up to 50%.[8] These are only two of numerous statistics that can be cited to illustrate how urgent it is for African governments and their advisers to gain access to policy-relevant research about the key factors, risks and likely future scenarios that will be necessary to manage the dislocating effects from climate change.

Table 34-1 illustrates the results from a recent effort to develop a comprehensive, policy-relevant system using six policy categories and ten performance indicators to measure the environmental burden of disease, water and air pollution; biodiversity and habitat; forestry; fisheries; agriculture; and climate change and country's policies to maintain human and eco-system viability based on both quantitative and qualitative analytical techniques. While one may disagree with this researcher's methodology, data sources, etc., few would quibble with the idea that this research should be more widely available. Currently, it is in a book chapter in a very expensive edited volume that is accessible to only a few; however, if there was a non-profit site dedicated to compiling relevant research, the holders of the copyright would very likely allow this and many other studies to be uploaded and available to all for free.[9]

Table 34-1. Ackerman's Environmental Policy Index scores for African states (Ackerman 2009).

Rank	Country	Score	Rank	Country	Score
1/58	Mauritius	78.1	23/123	Ethiopia	58.8
2/59	Tunisia	78.1	24/126	Nigeria	56.2
3/64	Gabon	77.3	25/127	Benin	56.1
4/66	Algeria	77.0	26/128	C.A.P.	56.0
5/71	Egypt	76.3	27/129	Sudan	55.5
6/82	Morocco	72.1	28/130	Zambia	55.1
7/86	Ghana	70.2	29/131	Rwanda	54.9
8/88	Namibia	70.6	30/132	Burundi	54.7
9/93	Congo	69.7	31/133	Madagascar	54.6
10/95	Zimbabwe	69.3	32/134	Mozambiq.	53.9
11/96	Kenya	69.0	33/138	Guinea	51.3
12/97	S. Africa	69.0	34/139	Djibouti	50.5
13/98	Botswana	68.7	35/140	G. Bissau	49.7
14/103	Cote d'Iv.	65.2	36/142	D.R.C.	47.3
15/113	Tanzania	63.9	37/143	Chad	45.9
16/114	Cameroon	63.8	38/144	B. Faso	44.3
17/115	Senegal	62.8	39/145	Mali	44.3

Another advantage of an online website dedicated to synthesizing and capturing up-to-date knowledge about climate change and tourism is that is can also serve as a repository for several different types of data. The Okavango Research Institute, formerly the Harry Oppenheimer Okavango Research Center (HOORC), is a world-class research institute that has been studying various aspects of the Okavango and collecting and using different types of data for decades. In the area of flood research, Dr. Piotr Wolski collects near-real-time data on a regular basis that allows him to make daily, monthly and annual forecasts based on a variety of data sets, including discharge and water levels, Delta flood in MODIS images, current meteorological conditions and daily rainfall in the Okavango region. Much of the daily data comes from volunteers, particularly those in Okavango safari camps, who send data to Dr. Wolski on a daily or near daily basis. Figure 34-7 displays an Island Safari Hydrograph using data from these sources. Such "bottom-up" strategies offer low-cost methods for collecting data needed for several research and forecasting tasks.[10]

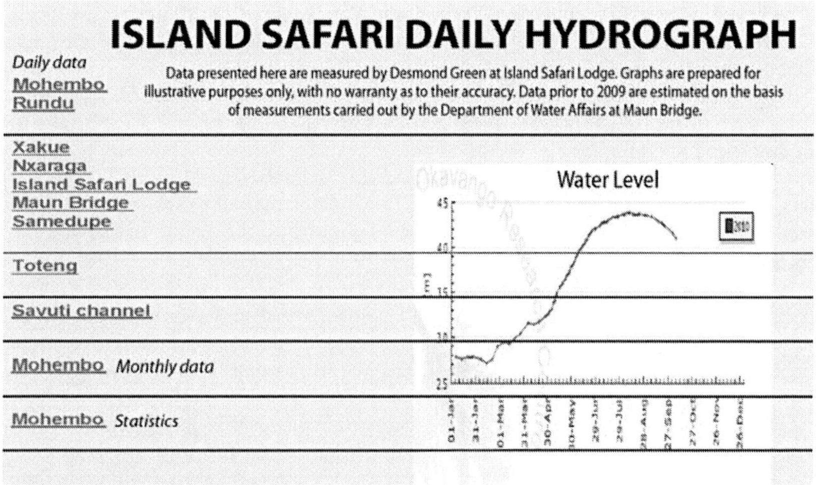

Figure 34-7. Daily hydrograph.

One can easily imagine creative ways to collect relevant "bottom-up" data from volunteers, along with inputs from professionals, to start documenting and monitoring changes in precipitation, climate and movements of people that may be relevant to climate-related changes and locations where tourists are being affected. Such networks of data collection might also be enlisted to operate as "human sensors" in an effort to establish early-warning systems to reduce the vulnerability of certain groups, including tourists, in different areas of Africa. Such systems could also use the communication capabilities of a specialized online website to help disseminate subsequent announcements and information during periods of extreme dislocation.

Notes

[1] For further details about this project, contact Prof. Helen Purkitt, Department of Political Science, U.S. Naval Academy, Annapolis, MD 21402; purkitt@usna.edu; or Letlhogonolo Ralafala, Coordinator, International Tourism Research Center (ITRC), University of Botswana at letlhogonolo.ralefala@mopipi.ub.bw.

[2] Mr. Lenny Chavhunduka, Digital Treasure, programmed the Drupal content-management systems using the initial prototype and some of his own design concepts. Lenny is a Zimbabwean who lives in Gaborone, Botswana, and frequently travels throughout the region. Lenny Chavhunduka's contact information is Lenny@digitaltreasure.co.bw; Plot. 5266, Village, Gaborone; Tel/Fax: +267 3191451.

[3] For more information about the work of University of Botswana students, contact Dr. Naomi Moswete, Environmental Science at the University of Botswana, Gaborone, moatshen@mopipi.ub.bw.
[4] The initial Botswana Tourism Research website developed using Google Sites offers one example of how to organize a customized database on African tourism or any specialized topic:
see http://sites.google.com/site/botswanaecotourismdatabasebed/home.
One large negative of using this freeware, however, is that all of the information is stored on a Google-owned server. Thus, Google has the right to use all the information stored on the site.
[5] An experienced computer scientist working at USNA on a related project developed an access page for the project using Drupal and concluded that it is probably one of the best freeware programs available to the public. As noted previously, the programmer who designed the Botswana Tourism Research in Drupal was Lenny Chavhunduka.
[6] This initial wiki article was written by 1/C A. Lagunzad. With her permission, we are uploading her article entitled "Global Climate Change" into the Botswana Tourism Research wiki. Students working with Dr. Naomi Moswete will be adding more specific information related to Botswana tourism to this and other articles currently being uploaded into the Botswana Tourism Research Wiki.
[7] Climate Change and Tourism Policy in OECD Countries. Organization for Economic Co-operation and Development/United Nations Environment Program. 2011. Available at no cost at http://www.oecd.org/dataoecd/8/28/48681944.pdf.
[8] Statistics taken from Ackerman 2009.
[9] Ackerman 2009 is published by Cambria Press. This publisher, like many in the West, is likely to look favorably on such requests so long as the purpose is a non-profit one.
[10] Another ORI project that is likely to be relevant to those interested in climate change and tourism throughout southern Africa is the Climate, Land-use, Institutions and People, or CLIP, project. The CLIP project is a four-year study effort started in 2009 to understand and predict the impact of climate variability and climate change on land-use and land-cover change associated with socio-economic institutions as they seek to adapt to climate change in the areas surrounding the Okavango, Kwando-Linyanti and Zambezi river basins. Professor Jane Southworth, University of Florida, and Dr. Lin Cassidy, University of Botswana, are the co-principal investigators for this project.

For more information on research programs and publications sponsored by the Okavango Research Institute, go to www.orc.ub.bw/research.html. For details about the bottom-up flood-data project contact Dr. Wolski at pwolski@orc.ub.bw.
[11] For recommendations regarding a remarkably similar type of system for West Africa, see "Participants Want Early Warning Systems on Climate Established in West Africa," Economic Community of West African States, http://www.comm.ecowas.int/.

References

Ackerman, John (2009). The most vulnerable continent: Africa and global climate change. In *African Environmental and Human Security in the 21st Century*, ed. H. Purkitt, ch. 6. Amherst, NY: Cambria Press.

OECD (2011). Climate Change and Tourism Policy in OECD Countries. Organization for Economic Co-operation and Development/United Nations Environment Program., http://www.oecd.org/dataoecd/8/28/48681944.pdf.

Lagunzad, A. (n.d.). Global climate change. Unpublished wiki article prepared for African International Relations, U.S. Naval Academy, Annapolis MD, Spring, 2011.

Mendelsohn, J. M., C. VanderPost, L. Ramberg, M. Murray-Hudson, P. Wolski, K. Mosepele, (2010). *Okavango Delta: Floods of Life*. Windhoek, Namibia: RAISON.

CHAPTER THIRTY-FIVE

HEROES AND ROLES IN CLIMATE CHANGE

GREGORY B. GALLAGHER[1],
WITH JEAN-MICHEL COUSTEAU
AND OCEAN FUTURES SOCIETY

Heroes

Most of us grow up with heroes of various sorts impacting our lives at a very early age. These champions may take the form of a special relative or public figure, but the impact of these mythical characters may last a lifetime. This was the case with me, in the person of Jacques-Yves Cousteau (1910–1997),[1] legendary scientist, environmentalist, inventor and filmmaker.

Figure 35-1. Jacques Cousteau. Photo credit: Associated Press.

---

[1] Writer, photojournalist and new-media producer; Executive Producer of Ambassadors of the Environment Transmedia Series.

Captain Cousteau's influence on me as a young child paved the way for my own explorations, as well as my enjoyment of and involvement in the natural world around me. The lessons I learned while watching his televised travels around the world aboard his ship Calypso continue to teach me half a century later. His principles have an equally important relevance for the planet's current climate-change challenges.

One of the most compelling elements of Jacques-Yves Cousteau's work was his development of the tools allowing him to study the oceans, and archive his findings for future generations. He is credited with co-inventing the aqua-lung[2] scuba diving apparatus (1943), which quickly became commonplace for amateurs and professionals throughout the world. Scientists and consumers alike owe a huge debt to him for making all waterways approachable, fun and quantifiable.

The lesson here is the miracle of invention to reach our goals, to engage our senses and bring together resources required, in a one-mindedness to succeed. Having heroes is no longer enough, we must be our own hero, and embrace the climate-change emergency on an individual level. Each of us shares the planet, and together we likewise share the outcome of this challenge as individuals, families, and nations.

Roles

In my own case, the personal impact of watching Captain Cousteau's adventurous escapades broadcast during the 1950s and 60s, ignited in my personality a sense of eagerly wanting to know what was high up in the treetops, under the surface of the water or around the next corner along an unfamiliar road. Curiosity about our world was indelibly tattooed on my very being.

Irony of ironies, as my own life story evolves into the current decade, it finds me working as a professional photojournalist for clients such as National Geographic Society, Smithsonian Institute, Royal Canadian Geographical Society, Jane Goodall Institute and many others. While on a magazine assignment on the island of Grand Cayman in the Caribbean, I came upon a program for children called Ambassadors of the Environment,[3] which would carve a surprising role for me at the center of the climate-change drama.

Ambassadors of the Environment

Figure 35-2. AOTE Iguana Boy. Photo credit: Ocean Futures Society.

Designed by Jacques Cousteau's son Jean-Michel, together with his Ocean Futures Society in California, and in particular his Director of Science and Education, Dr. Richard Murphy,[4] this comprehensive set of experiential games and activities was nothing short of life-changing for me. Before knowing about this curriculum, climate change and environment troubles in general, were matters being dealt with somewhere "out there" by the scientific community, not by me as an individual.

I did not feel that direct involvement in creating change was front-and-center in my life. My wife disagrees, and argues that my writing informs and advances the cause in a personal and professional manner. When I witnessed the Ambassadors of the Environment process, it sparked something in me, and made me realize I could use my life's experience, and the resources at my disposal (just like Jacques Cousteau did when inventing the scuba gear), and forge ahead on a truly distinctive level, to contribute to correcting environmental problems like climate change.

Back-story

As one of the original Canadian writers of the Sesame Street TV Show,[5] as I toured the Cayman Island facility, my imagination ignited immediately. It did not take long for me to come up with the idea to "digitize" this children's curriculum. I saw it in my mind's eye being made available interactively to kids everywhere using the latest technologies as the delivery systems. I believe if real change is to occur across global barriers, children are the key. My goal is to help kids grow into conscious citizens for a peaceful world.

For my dream to become real, adults must provide the necessary information in "simplespeak," as well as easy-to-use techniques, weaving it together in a fun process for children to connect with the environment. Jean-Michel Cousteau and Ocean Futures Society have accomplished part of this, and the recognition of this fact is happening on numerous global fronts. ABC Travel News placed the Ambassadors of the Environment in first position on their Top 10 Kids Resort Programs List for 2010. Currently, this experiential program is available in ten locations worldwide: Brazil, Hawaii, French Polynesia (cruise ships), Greece, California (x3), France, Grand Cayman Island and Turks and Caicos.

My first baby step to success came when I decided to write a letter to Jean-Michel Cousteau, asking him for the rights to digitize his program, and deliver it to the world's children. By the time I finished my letter, I had a simple game plan to present to him. I would wait until I got back home to finesse the letter, talk to my wife and get her collaboration, and send off the proposal by email care of the Ocean Futures Society website.

Waiting Game

I am not a patient person, and possess limited ability to focus on a project for longer than perhaps three or four months at most. After that, unless there is a profound evolution in the work, I want to move on to something else. But with the Ambassadors of the Environment I found a new level of focus somewhere inside me.

It seemed to me this marvelous curriculum, if served to children in the best context, could spawn a worldwide environmental turnaround. In my mind's eye I was watching a hybrid media game-changer coming together, what one friend calls "Planet Earth Meets Sesame Street for the Twenty-first Century."

It took one whole year of pursuing Jean-Michel Cousteau with my idea, aided by Dr. Richard Murphy, who responded immediately. Once Jean-Michel had read my proposal to digitize the Ambassadors of the Environment into a transmedia series of interactive content, he opened his heart and his organization to me.

Figure 35-3. AOTE Pontoon. Photo credit: Ocean Futures Society.

We signed an exclusive agreement giving me the rights to develop the Ambassadors of the Environment into a transmedia children's series of content, using live-action, digital puppetry, original animation, and global music, in a rollicking series engaging kids in environmental tasks, games and activities. One of the biggest elements of my dream is the interactivity of the series content. I want kids everywhere to be involved using their digital devices, telling us their stories, sharing local music, games and traditions from the culture and environment they call home.

Ultimate Partnership

Suddenly, I found myself as an ambassador of the environment, building a project game plan to present to the world. But first, I had to secure my wife's blessing, enthusiasm and involvement. Any marriage is an ultimate partnership, and I needed Linda's magical creative touch if the project was going to move forward. We talked long and hard about how to make our effort effective. I could see the wheels turning inside her, and feel her support of the idea.

Ultimately, we decided to be the change we wanted to see in our world, beginning at home. Our commitment was designed to invest two years of our time and our personal savings to allow work on the project full-time. There was no other way by which a project of this size and complexity could properly be presented and developed. I agreed to take several months to clear my freelancer's desk of commitments, and then I would dedicate every workday to the Ambassadors of the Environment Transmedia Series.

Linda, an accomplished painter, would sacrifice her art time for the planet on this project. She invested her time and energy by preparing and maintaining the household, personal relationships and acting as my editor and consultant, allowing this process to unfold. She came up with new ideas for environmental games the kids could play, fundraising concepts, methods by which her art would be auctioned to raise funding, and other innovative input. Without my wife's dedication, spirit, and kindness, I would not have been able to focus on presenting the project to three levels of government, top-tier media companies plus private and institutional investors around the world.

Action Stations

Once my desktop was cleared of obligations, I began to study the best way to unveil our unique concept. Obviously, our world is enthralled by the visual medium, so the decision was made to produce a short video presenting the series. But there was no money to do this, so how could we achieve such a modest goal?

I met with a young Montreal filmmaker, Sharif Mirshak, to present my concept verbally, and asked if he would work on my video production for free. He listened intently and agreed that he and his partner, Noé Sardat, would shoot, edit and produce my video as part of their tiny company ParaFilms.[6] This would create the ultimate marketing tool, bringing together the creative elements of my "vision" into a brief seven-minute video. This document would hopefully ignite people's interest in the climate change issue and other environmental problems, and how to begin educating our children in ways to repair our collective relationship with nature.

I wrote the script, acted as the narrator on the video – receiving enormous help from Jean-Michel Cousteau's film editor, Jim Knowlton[7] – and helped oversee appropriate other visuals to include. We had to borrow screen shots from websites, and since my eventual presentations would be a private process, we justified this decision for the greater good the series would create long-term. Our labor resulted in a seven-minute video everyone is proud of, and our eternal thanks continue to be directed to Sharif, Noé and Jim for their generosity.

Simple Wisdom

Paramount to the core principles of the Ambassadors of the Environment is a set of goals ideally suited for grassroots climate changes and beyond:

- To give students a positive view of themselves.
- To develop knowledge and respect for the outdoors world.
- To give students life skills in communication, teamwork, responsibility and friendship building that will help them throughout life.
- To inspire youth to live more sustainably and take responsibility for their future.

Achieving a positive view inside a young person is a daunting task in any country, but critical if that child is to mature into a healthy, peaceful adult. By involvement in the experiential tasks of Ambassadors of the Environment, children see firsthand what respect is all about. The process for self-respect becomes less theoretical in the act of appreciating nature's wonders, and performing specific tasks as a team with others to achieve simple goals. Over time, the kids learn they too are an integral part of this magical environment, and by practicing inter-dependence with their fellow students, reflect the cooperation visible throughout the natural kingdom. Simple wisdom begins to germinate inside these younger citizens.

Figure 35-4. AOTE Sea Conk Girl. Photo credit: Ocean Futures Society.

Climate of Sustainability

It is my belief that we must teach children how to connect with the notion of sustainability to properly realize significant climate-change progress. To accomplish this, the best place to begin may be inside the child's first environment – his or her own physical body. Ambassadors of the Environment makes this task easier by providing concrete examples of the systems and processes we see at work in nature and inside our bodies everyday. Here are just a few examples:

- *Everything runs on energy*: stars, cities, machines and people (skin, bones, muscles and organs)
- *There is no waste in nature*: nature recycles everything (so do our bodies)
- *Biodiversity is essential*: each species has an important job (each part of the body too)
- *Everything is connected*: all species depend on others (interdependency of body/mind/spirit)

Imagine children around the world absorbing the lessons of nature at a very young age, and holding their parents, families and communities accountable. This is not such a far-fetched notion. Never before has the possibility of communicating with large numbers of young people been so real. Given the peaceful, constructive input from programs like Ambassadors of the Environment will surely reap untold benefits long into the future for our troubled planet.

Solutions

The IIPT's 2011 Lusaka Declaration on Sustainable Tourism Development, Climate Change and Peace articulates numerous goals for the travel and tourism industry worldwide to achieve without delay. From supporting the Davos Declaration – Responding to Global Challenges, to developing "Green Growth" transformations, waste management systems, embracing the wisdom, knowledge and values of aboriginal peoples, appreciating we are all custodians of one common home – planet Earth, and to calling on all travelers to be ambassadors of peace.

The remarkable synergy of the Lusaka Declaration and the Ambassadors of the Environment curriculum is a celebration of solutions. Former Vice President of the United States Al Gore proved like-minded environmentalists are able to connect using the latest technologies, when he attracted more than 8.6 million website "views" from people around the world in only twenty-four hours of his recent Climate Reality Project.[8]

Finale

From my point of view, all the necessary components are in place to successfully address the climate-change problem, as well as many other environmental challenges:

- The list of modern environmental heroes who are ready, willing and able to lead the climate-change turnaround grows daily.
- The common-sense wisdom of the elders can be immediately shared via state-of-the-art technology tools accessible on all continents.
- Our roles are precisely articulated in the Ambassadors of the Environment programs at work now with community and tourism partners worldwide.

Our fragile Mother Earth needs a simple decision from each of us, on a very personal level, for her future to be a peaceful, healthy one. I invite you to take some time in a quiet, private place, like I did two years ago in the aftermath of witnessing the Ambassadors of the Environment program firsthand, to examine your own relationship with the planet. It is not a complicated issue, rather a straightforward personal decision-making act. But this sacred reflection will be heard and felt for generations on all continents, and especially on the path you walk from here forwards.

In the foreword to the 2011 book Canada's National Parks, CEO of Parks Canada,[9] Mr. Alan Latourelle writes:

> Since the first humans evolved on this planet we have managed to inhabit practically every corner of the globe. In so doing we have, through experimentation and best efforts, developed an incredible diversity of cultures adapted to various climates, landscapes, and other ecological challenges that the natural world has presented to us. To understand a culture one must examine the land that shaped it. To understand the Aboriginal peoples of Canada and the many who subsequently settled and adapted here, one needs to understand Canada's landscape – its climate, forests, plants, wildlife, and spiritual places.

If we cut out the word "Canada" this text could apply to all nations, continents and peoples. Our recipe for success in dealing with climate change or any of our environmental challenges is set. We must only decide to activate it as soon as possible. I am confident we will be able to accomplish this together in peace.

Notes

[1] http://en.wikipedia.org/wiki/Jacques_Cousteau.
[2] www.cousteau.org/technology/aqua-lung.
[3] www.oceanfutures.org/learning/ambassadors-environment.
[4] http://www.rcmurph.com.
[5] http://en.wikipedia.org/wiki/Sesame_Street.
[6] http://vimeo.com/parafilms/albums.
[7] www.jimknowlton.com.
[8] http://climaterealityproject.org.
[9] http://www.pc.gc.ca.

CHAPTER THIRTY-SIX

EDUCATION THROUGH "TOURISM AS A CULTURE OF PEACE"

JULIA MORTON-MARR[1]

Abstract

The scientists from the Global Issues Project of Canadian Pugwash and Science for Peace in Toronto Canada have, for twenty years, followed climate-change statistics. Climate change is now a fact, as discussed at the Roundtable on Climate Change and Energy in the Wasan Action Framework (see Appendix I to this chapter). The Intergovernmental Panel on Climate Change also supported this. Tourism around the globe must pay attention, as the many resources that are used for visitors may be needed for local cities. Educators have been part of the problem. Educators must now be part of the solution – at every level. Curricula must teach how to protect basic resources, not how to abuse them for economic gain. It is a major responsibility to teach students how to reverse the negative global concerns into positives. This paper presents possible solutions for local actions.

Introduction

If global societies are to survive climate change, they must support the web of life because it serves human purposes (Burkhardt, Dyson & Morton-Marr, 2006). Education is seen as part of the paradigm shift for climate change and energy (see Appendix II to this chapter). It would be helpful to implement climate-change curricula that can be delivered to all schools around the world at the same time, stimulating awareness and action. The International Holistic Tourism Education Center's (IHTEC) suggestion is to introduce the International School Peace Gardens

---

[1] Founding President, International Holistic Tourism Education Center – IHTEC and the International School Peace Gardens.

curriculum globally. This is a place of annual dedication for solving climate-change issues. See Figure 36-1.

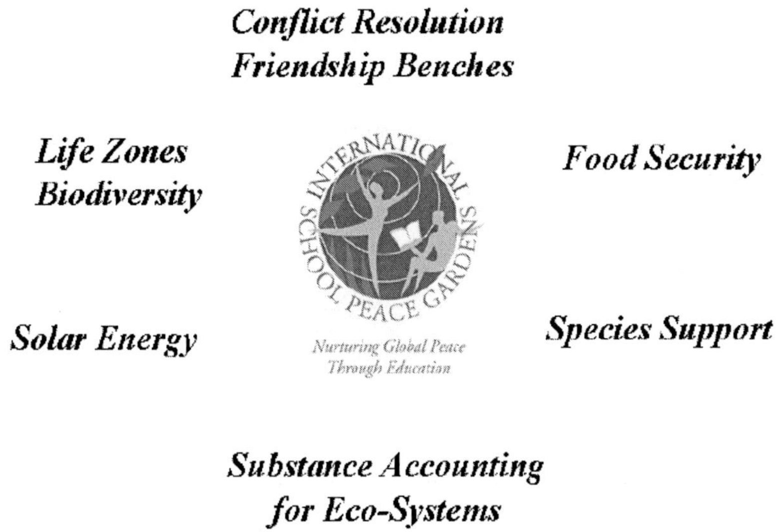

Figure 36-1. Peace-garden concept.

Global Education is Dimming

The global educational IQ may be dimming as a result of our carbon-based energy use and the current climatic changes. Earth's citizens are currently in a "war-based economy." It is hoped that a "peace-based economy," using alternative energies, such as solar, wind and geothermal, will change the paradigm of a war-industrial society to a peaceful-sustainable society, especially through the tourism industry (as a culture of peace).

IHTEC has been working on solutions that have been developed to give existing curricula some support. IHTEC supports the Ontario Ministry of Education in their initiatives for positive action in communities, such as eco-schools and the Education Alliance for a Sustainable Ontario.

IHTEC's modules in "Global Sustainability Education" teach how to clean up and try to never create the same crisis on earth again. Young people need a message of hope and something positive to help them make the transition from the "entitlement generation" to a "peaceful generation."

Technology

Everything we focus on for the future needs to cool the planet. Technology is one of those positive things schools can do. Schools need to teach how to use web-based communication technology internationally to reduce CO_2 emissions. Most schools have computers and there is a push in poorer countries to use cell technology with access to the web. Transnational online education is now possible (Hogan, 2012). Web conferencing technologies, such as www.TalkingCommunities.com and www.iVisit.com, will enhance solving environmental concerns and sharing societal achievements. IHTEC is using these technologies globally.

Curriculum

There is an urgent need for curricula change to support humans and the planet's web of life. Current species loss is between 30,000 and 50,000 annually. IHTEC's suggestion is to plant food for species along migrating flyways, mindful of local biodiversity. Figure 36-2 shows the issues.

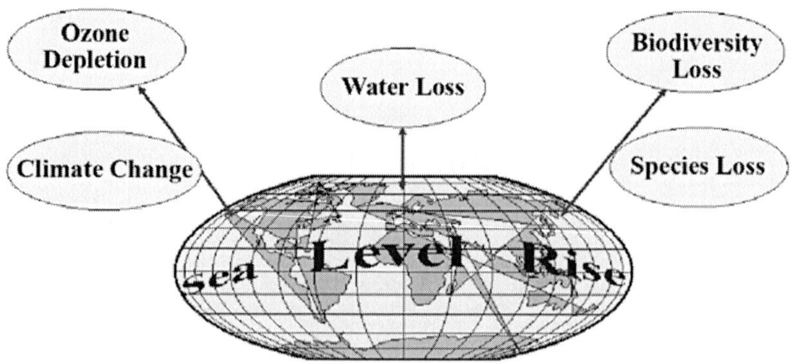

Figure 36-2. Hot Planet Pod for climate-change curriculum.

Curriculum changes and implementation is vital to society's health. UNESCO has traveled the world during the "UN Decade of Education for Sustainable Development 2005–2014," stimulating change within faculties of education. This means that the existing curriculum has begun to be modified to include environmental education. More attention is needed on peace and sustainability education. Additionally, all teachers will need to be re-trained in peace and sustainability education. Public education is vital. University and college students will require at least one subject within their degree. Young women, who are not included in decision making in some places around the world, are to be fully included and know their role within sustainability.

Forests

Forests are the lungs of the earth and they are still being burnt for agriculture. Creation of permanent forest estates may help to conserve them. It is felt that there will be more unintended consequences if forests are not replanted. At the Science for Peace Roundtable on Forests held at the University of Toronto, in September 2006, John McRuer, Canadian Association of the Club of Rome, identified that there is a major loss of quality and quantity in global forests. When the last forest is gone, humans will be gone. This means that biomass is not an option for energy farming. At the present usage, forests will not be able to meet the demands of our great grandchildren as the population increases. Roads into forests should be prevented. We must check traditional knowledge and management before continuing any tourism development. The roundtable advises that we need to strengthen national laws, as 90% of all logging is illegal. Soils globally need humus to restore them. This includes bio-carbon sequestration to make soils fertile again.

Water

Humans are already over the planet's water footprint (see Figure 36-3). The planet has only *x* amount of total water. Populations need *y* amount. Nature, including all other species, needs *z*. Humans are already using more than their share, which will prevent local water from keeping the hydrological cycle working, causing droughts. Greater care of all rivers and streams is needed.

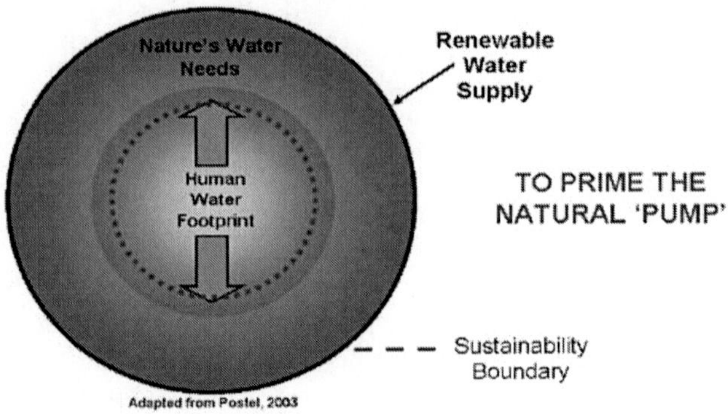

Figure 36-3. Nature needs H_2O (Adapted from Postel 2003 in Maas, T., 2008).

The importance of preventing pollution in underground rivers and aquifers has often been ignored. Freshwater is lost when all rivers run into oceans. IHTEC has created a water dedication for schools.

Oceans are 75% damaged with many dead spots with no oxygen. The ocean absorbs carbon, but there is a limit to the oxygen-exchange process, which is very serious. The carbon sinks into the benthos and produces methane (greenhouse gasses) that may burp back into the atmosphere (Mitchell, 2009). Oceans around South Australia are already increasing in acidification, which is affecting all marine life, especially shellfish. This is apart from reducing carbon emissions and sea-level rise. Additionally, plankton and marine grasses die from excessive UV light. Recent sea-level rise has been experienced in the Hawaiian Islands, with fish on low-lying inland roads. IHTEC suggests Marine Peace Parks.

Food

The world has approximately forty to fifty-nine days of food available for the population at a time (according to the World Food Program). This will decrease due to climate change. Hawaii has only two weeks' food at any one time. Peace here is essential for sustainability. Food holds virtual water and is transported globally. It is better for the land if we eat food grown nearby. All planting should solve a global problem. Everyone needs to know how to live in a desert as the planet heats up. IHTEC presented a written statement on Food and Population at the Climate Change Conference in Copenhagen (Benking, 2009). IHTEC suggests Food Security Gardens.

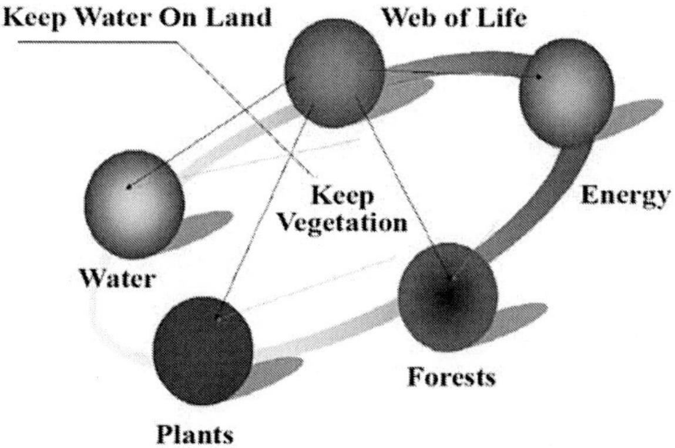

Figure 36-4. Cool Planet Pod.

Conclusion

Will a new energy source stop climate change, resource depletion and economic poverty and war? There is always hope, but it will take a major crisis to get everyone working together. How will we change the outcome from a "Culture of War" to a "Culture of Peace"? IHTEC hopes that we can all work together and make the changes that are needed. As in the Cool Planet Pod (see Figure 36-4), we must protect the web of life, keep all biodiversity and local vegetation and pay careful attention to water.

References

Bukhardt, H., R. Dyson & J. Morton-Marr (2006). Handbook on Critical Global Issues and Viable Solutions. Council on Global Issues, http://www.ihtec.org/index.php?id=208.

Benking, H. (2009). Council on Global Issues. Climate Change Conference in Denmark.

Canadian Pugwash & Science for Peace (2008). Freshwater Declaration, http://www.ihtec.org/index.php?id=214.

Council on Global Issues (2009). IHTEC Food and Population Statement, http://c-g-i.info/IHTECFood-Population-RoundTable.pdf.

Hogan, R. (2012). *Transnational Distance Learning and Building New Markets for Universities*, edited by Robert Hogan. IGI Publishers.

Hopkins, C. (n.d.). UNESCO: UN Decade of Education for Sustainable Development.

IHTEC Water Dedication: http://www.ihtec-ispg.org/PrimaryPages/WatershedHome.html.

Maas, T. (2008). Science for Peace and Canadian Pugwash, Round Table on Fresh Water, http://www.ihtec.org/index.php?id=214.

McRuer, J. (2006). Models, Dogma and Claptrap (p. 5), http://www.ihtec.org/fileadmin/archives/IHTEC/documents/ForestRoundtableNarative_060910.3.pdf.

Mitchell, A. (2009). *Sea Sick: The Hidden Crisis in the Global Ocean.* McClelland & Stewart.

Appendix I

13–16 September, 2007, Wasan Island, Muskoka Lakes, Ontario, Canada
Sponsored by Science for Peace, David Suzuki Foundation and Breuninger Foundation

CLIMATE CHANGE AND ENERGY: THE WASAN ACTION FRAMEWORK

Declaration and Recommendations of the Interdisciplinary Round Table on Climate Change and Energy Strategies

THE FIRST CLEAR WARNINGS OF DANGER FROM GREENHOUSE GAS EMISSIONS DUE TO HUMAN ACTIVITY EMERGED 25 YEARS AGO. Prudence would have called for precautionary action at that time to slow down the growth of these emissions. Since then, the scientific understanding of the impact of human activity on global warming has been overwhelmingly confirmed; key predictions based on that understanding have started to occur. Evidence has emerged that the potential impacts of global warming will be much more severe than was predicted even five years ago.

Individuals, corporations and all levels of government around the world have a duty to act as global citizens in the face of the danger posed to life on Earth and to the well-being of the human race as whole.

1 WE DECLARE that human-induced climate change and energy security, in particular peaking of the world oil supply, are crucial issues requiring immediate action.

2 WE DECLARE agreement with the statement by the Intergovernmental Panel on Climate Change (IPCC) Working Group 1 regarding the physical basis of climate change: "Most of the observed increase in global average temperatures since the mid-20th century is very likely due to the observed increase in anthropogenic greenhouse gas concentrations."[1]

3 WE IDENTIFY AS THE ROOT CAUSES OF THIS CRISIS:

a) the large per-capita over-consumption and waste of natural resources in the industrialized countries
b) the growth paradigm (economic growth for its own sake)
c) the large and growing human population
d) the very large dependence on fossil-fuel-based energy

e) the resistance by vested interests to necessary change in energy technology
f) the lack of appropriate political leadership
g) the lack of global governance to protect the global commons

4 WE PROPOSE A GLOBAL SOLUTION FRAMEWORK: We must begin immediately to:

a) curb over-consumption and give priority to efficiency, conservation and the avoidance of waste
b) promote lower birthrates by empowerment of women through educational, economic and social measures, including access to birth-control information and services
c) focus globally and locally on developing low-impact renewable-energy infrastructure and technologies (e.g., small-scale biomass, geothermal, hydro, ocean energy, solar, wind, etc.) to their full potential, so as to avoid large-scale bio-fuel usage and nuclear energy
d) reduce carbon emissions by creating a just and universal framework through the implementation of appropriate incentives, government regulation, legislation and taxation
e) preserve forests, especially tropical rainforests

5 WE URGE IMPLEMENTATION OF THE FOLLOWING SOLUTIONS:

a) All levels of government as well as the UN and international organizations should embrace the Wasan Action Framework.
b) Media, corporations, the educational system from kindergarten to university and all civil society should collaborate in implementing this Wasan Action Framework.

[1] IPCC, 2007: Summary for Policymakers. In: Climate Change 2007: The Physical Science Basis. Contribution of Working Group I to the Fourth Assessment Report of the Intergovernmental Panel on Climate Change [Solomon, S., D. Qin, M. Manning, Z. Chen, M. Marquis, K.B. Averyt, M.Tignor and H.L. Miller (eds.)]. Cambridge University Press, Cambridge, United Kingdom and New York, NY, USA. Page 10.

Appendix II

Global Sustainability Education

Definitions of Global Sustainability Education

Global Sustainability Education (GSE) has two main parts – first, ecological integrity and social integrity. Both of these are well defined in the Earth Charter. In 2003, UNESCO adopted this soft law. Included in these parts are the two primary values, life and procreation, that foster a bio-centric worldview. GSE also includes the two pillars of sustainability – peace and resources.

Second, we include Ehrlich's Formula:

Human Impact = Population x Consumption x Technology

"The Ehrlich Formula covers the human impact on the environment. The second pillar of sustainability is social integrity or social peace. Environmental degradation leads to resource scarcity, which in turn leads to loss of social integrity and violent conflicts, according to T. Homer-Dixon, of the University of Toronto. Injustice is another way of losing social peace; I see injustice as the main cause of terrorism," says Prof. Helmut Burkhardt.

As a GSE Curriculum is developed, and before any action is implemented, each current issue and its impact requires curriculum discussions on:

1. The effect of a current population increase on the environment and how it impacts to the "eco-system" and all species in the "web of life."
2. The impact on the local economy of human's affluence and the consumerism involved with the earth's resources.
3. The local impact of current and future technology on the local environment and society.

The GSE Curriculum for all students would require these elements and the use of mathematical and statistical knowledge of the current problem, to enable conflicts to be solved peacefully.

Holistic Tourism Education

Holistic Education examines the inter-disciplinary relationships between various branches of knowledge.

All of IHTEC's programs contain systemic thinking through a Global Sustainability Curriculum. These include the two primary values, life and procreation, that foster a bio-centric worldview, and include the two pillars of sustainability, peace and resources. These are linked with the values of the Credo of the Peaceful Traveler to complete the GSE concept.

The core curriculum consists of concepts around a "Culture of Peace through Tourism" as a central or "pivot curriculum." The curriculum is linked through the International School Peace Gardens, which also links to GSE curriculum in the following ways:

1. Environmental Integrity = environmental studies
2. Social Integrity = conflict resolution, inter-cultural understanding
3. Earth Charter = Environmental Bill of Rights and local laws, national laws, UN conventions, agreements and international laws
4. School Curriculum = language arts, performing arts, science, math, geography, history, technology, etc.

Global Sustainability Education in the ISPG Program

International School Peace Gardens (ISPG) Curriculum began in 1993. On February 14, 1995, Eric Foster and members of the school planted the first School Peace Garden that was linked to the launch of the Environmental Bill of Rights (EBR), in Ontario. The EBR tree became the "Peace Tree" in their peace grove or Bosco Sacro, a place for conflict resolution.

The practical application of Global Sustainability Education, includes the Earth Charter and can be implemented through the ISPG Curriculum as follows:

Environmental Integrity

- Link to your local, national and world heritage parks as the knowledge base
- Conflict resolution using "Friendship Benches"
- Knowledge of what grows in each area ("Life Zone Biodiversity") is vital to the survival of all species. e.g., Carolinian Life Zone, Desert Life Zones
- Food security, i.e., planting foods that suit local soils (ISPG)

- Species support, i.e., planting food supplies for migrating species and creating water supplies (Creature Corridors)
- Water and soil protection (Watershed Peace Pathways)
- Oceans and coastal areas (Marine Peace Parks)
- Rainforests (Rainbow Rainforests Rock)

Social Integrity

- Conflict resolution, inter-cultural understanding through language arts/performing arts

Population, Affluence and Consumerism

- Substance accounting for eco-systems, i.e., accounting for the gains and losses in their ISPG and identifying what problems may arise

Technology

- Solar energy, i.e., use solar ovens, lights, rocks, on or near school buildings; build a solar car; use solar for science experiments

CHAPTER THIRTY-SEVEN

PRINT MEDIA AWARENESS CAMPAIGN ON IMPACTS OF CLIMATE CHANGE IN AFRICA

APOLINARY TAIRO[1]

Introduction

Tanzania is considered one of the premier tourism destinations in Africa, with its wildlife and coastal attractions providing the second-largest source of foreign exchange. More than 25% of Tanzania's surface area is devoted to conservation of some of the world's greatest concentrations of large mammals (including elephants), a variety of birds and indigenous flora. Wildlife is one of the most valuable resources of Tanzania, generating significant revenues and foreign exchange via tourism.

"Climate change" remains a new terminology to most journalists and media professionals in Africa. This new terminology reflects the real situation in most media houses in sub-Saharan Africa, where governments lack resources and proper mechanisms to engage journalists in campaigning on impacts of climate change in respective countries. The exception is South Africa, where considerable reporting has been done on the impacts of climate change.

Print media, which is still dominant and most influential compared to electronic media (internet, radio, television, blogs, etc.) in Africa, has so far played a leading role in educating and informing the public on effects of environmental deterioration and related human impacts. But little has been done on climate change.

Why African Media Fails to Campaign on Climate Change

There are several reasons why the print media fails to take a leading role in campaigning for education of climate change in most African countries.

[1] Journalist and eTurboNews Representative, Tanzania.

1. Lack of Climate-change Awareness: Most journalists and other media professionals are not well informed of climate-change impacts. Some fail to differentiate between climate change and environmental degradation. Several seminars, public campaigns, special trainings and in-house campaigns have been conducted on environmental degradation among media professionals, while less or little has been done to educate media professionals on climate change and its impacts. Most news covered in newspapers and magazines is sourced from European and American media outlets, not African sources. Between 2005 and 2010, less than 200 news articles and features were covered in East African print media (Kenya, Uganda and Tanzania) directly reflecting the effects of climate change in the region (East Africa).

2. Poor Coordination between African Governments and the Media: African governments do little or nothing to engage the media on educative campaigns for public awareness on impacts of climate change in Africa. In Tanzania, as an example, climate-change information and data are coordinated under the Vice President's Office through the minister of state responsible for environment. But, unfortunately, no information or news comes from the minister except briefings on conferences and other events. No scientific data or research information is released to the media for use as a public awareness campaign. Most research information on climate change remains as secret documents in government files.

3. Donor Participation: Compared to the environmental degradation campaigns since the 1990 Rio Summit, in which donor countries sponsored the media houses and journalists to campaign for environmental protection, little is being done to support reporters to write news on impacts of climate change in respective African countries. Key donors in Tanzania and Africa – The European Union, USAID, SIDA, DFID, GTZ and NORAD – have sponsored and organized a number of seminars, workshops and special training sessions on environmental protection for media houses, staff writers, individual journalists and correspondents. But nothing similar has been done for climate change.

4. Poor Resources in Media Houses: Journalists, correspondents and contributors are interested in covering areas that have been affected by climate change, but fail to carry out such assignments due to lack of resources, including working equipment – portable computers fitted with internet for quick communication, lack of funds to cover travel costs, etc. Compared with other countries in Europe, the Americas, South East Asia and the Pacific Islands, Africa is less developed with regards communication, while a major part of the continent is forested land or arid land where traveling from one corner of a country to another takes days or weeks. This situation adds more to the cost of travel.

Role Played by Print Media in Campaigning for Climate Change

Despite all the hurdles facing journalists in sub-Saharan Africa, there have been a remarkable number of articles and media releases on the impacts of climate change in Africa. In Tanzania there have been a number of articles on climate-change impacts, although most of them do not appear on a regular basis as compared with sports and political news.

The effects of El Niño rains in 1998 reflected the most coverage in Tanzanian and East African newspapers. A number of articles were published in several newspapers including the *Daily Nation* (Kenya), *The Standard* (Kenya), *Daily News* (Tanzania), *The Guardian* (Tanzania), *The East African* (East Africa) and the popular Kiswahili newspapers, all reflecting on the negative effects of the rains in most parts of the East African region.

The journalists did not relate the El Niño rains to climate change, but took the rains as part of common weather variations. People in East Africa did not have any idea that the El Niño rains could be the effects of climate change.

Roads leading to key wildlife parks in Tanzania, including the Serengeti National Park, Ngorongoro Crater, Selous Game Reserve, Ruaha National Parks and several others were badly damaged, causing delays or itinerary cancellations from visitors booked to visit the areas.

Permanent droughts, the effects of El Niño rains and the deaths of masses of wildlife in key tourist sites received little coverage, as journalists and other media professionals failed to tell their readers in detail about the effects of climate change in relation to the El Niño phenomena.

Natural disasters and malaria outbreaks in northern regions of Tanzania (Arusha and Kilimanjaro) were reported while tourist sites experienced higher temperatures as a result of climate change. In those areas, tourists were required to take anti-malaria pills, apply mosquito repellant and use mosquito nets on their beds. The print media had not adequately reported on these situations to sensitize the public and tourists to these conditions.

The impact of climate change in Africa is also seen with islands being submerged because of a rise in the sea level, as well as periodic flooding, while in other areas we see decreasing water levels in lakes and rivers.

In Tanzania, for example, a tourist hotel on the Indian Ocean beaches in the capital city of Dar es Salaam (Hotel Africana) was totally submerged due to the rising levels of the Indian Ocean waters. Kunduchi Beach Hotel, also on the Indian Ocean beaches of Dar es Salaam, is facing the threat of submergence as the Indian Ocean waters rise. The media has made little or no mention of either situation.

Climate change is likely to pose a significant threat to the tourism sector. Most hotels are located along the coastline and any increase in sea level will affect them severely. For example, the Kunduchi and Bahari Beach resort hotels in Dar es Salaam have been so substantially eroded that a huge investment is being made to restore them.

Mount Kilimanjaro, Africa's tallest mountain, remains the only widely publicized tourist hot spot in East Africa. The mountain is the leading tourist attraction in Tanzania, accounting for 25,000 to 40,000 foreign and domestic tourists a year and sustaining the livelihood of four million people in Kenya and Tanzania through agricultural and business activities.

Global-warming effects are being felt in most parts of Africa with great impacts on tourist sites, including Tanzanian wildlife parks and the Mount Kilimanjaro ecosystem. Standing freely and majestically, with its snow gleaming in the sun, Mount Kilimanjaro is in great danger of losing its eye-catching glaciers. The mountain is located some 330 kilometers and three degrees south of the equator. In recent decades, Kilimanjaro glaciers have progressively shrunk as climate-change effects have caused the mountain snow to melt.

Fortunately, Mount Kilimanjaro is slowly regaining its snow after several years of drought in East Africa and the effects of climate change in the African continent. The snow is slowly mounting, providing beautiful views of Kibo peak and new hope to the Mount Kilimanjaro environmental watchdogs that the mountain may not lose its beautiful ice cap as scientists predicted.

Climate-change impacts on agriculture, and on common-pool resources such as forests, pastures and human health, are the most threatening ones from the viewpoint of the majority of the people in Tanzania and which need frequent coverage in the media.

The media must examine that factors causing and the impacts of climate change much more closely, together with the potential vulnerabilities of affected human populations, wildlife and natural resources as discussed above, and with much greater scrutiny of government officials and policies.

Challenges Ahead

Africa's bountiful natural resources and tourism assets are facing imminent threats due to climate changes. The media must mount a concerted public awareness campaign to broaden awareness on these issues.

The east African mountains of Ruwenzori and Elgon in Uganda, along with a section of other mountain ranges in the region, are losing their ecological heritage at an alarming rate due to global warming, posing great dangers to the regional economies.

The media – especially the print media – has a major role to play in sensitizing vulnerable communities, donor agencies and African governments so as to help in pulling together resources that will help to reduce the effects of climate change in the African continent.

PART VII

SUPPORTIVE ACADEMIC RESEARCH

CHAPTER THIRTY-EIGHT

THE IMPACT OF CLIMATE CHANGE ON TOURISM IN AFRICA

SMART N. UCHEGBU AND EJIKEME J. KANU[1]

Abstract

Today, new tourist centers and cities are planned to make them more environmentally friendly and sustainable; however, the strategies inherent in these plans do not adequately address the severe and radical changes in local and African climates being witnessed today. Some African countries rely heavily on tourism as their main source of revenue. The effects of climate change on tourism cause changes in the ecosystems and natural resources needed to sustain the tourism economy. The study reveals that there are some climate-induced challenges facing African countries in general. Natural disasters, such as rising sea levels, flooding, desertification, erosion and other health-related problems, are now rampant in African countries. These serious issues, most of which have been linked to the emission of greenhouse gases (GHGs), could be addressed in part by change of habit and new ways of thinking to reduce the effect on tourism in Africa.

1. Introduction

Today, new tourist centers and cities are planned to make them more environmentally friendly and sustainable; however, the strategies inherent in these plans do not adequately address the severe and radical changes in local and African climate patterns now being witnessed.

The effects of climate change are causing changes in the ecosystems and the natural resources needed to sustain the tourism economy. Climate change impacts that affect tourism in African countries include: beach

[1] Department of Urban and Regional Planning, University of Nigeria, Nsukka, Enugu State, Nigeria.

erosion, saline intrusion, droughts, flash floods and landslides, coral-reef bleaching, less productive fisheries and agricultural systems, rising sea levels, flooding, desertification, erosion and other health-related problems, all of which are now rampant in African countries.

These issues are linked to the emission of greenhouse gases (GHG) (Stainforth, 2005). Climate change is now at the forefront of debate, with dire warnings that worldwide temperatures may continue to rise throughout the next 50 to 100 years (Dakaica, 2002). Some of the factors have human influences and some are natural (Uchegbu, 2002); climate change is likely to continue for many centuries, changing human and animal life patterns (Uchegbu, 2009).

Tourism is closely connected to the environment and the climate itself. At the same time, tourism is a contributor GHG emissions, including emissions from transport, accommodation and activities (Beatley, 1998). According to United Nations World Tourism Organization, tourism's contribution to GHG emissions was estimated to be approximately 5% (UN, 1990). Tourism is a primary source of foreign exchange earnings in forty-six out of fifty of the world's Least Developed Countries (LDCs) (Pickering and Owen, 1995). The sector needs both to mitigate its contribution to climate change by reducing its emissions of greenhouse gases as well as its overall environmental footprint, and also to adapt to the impacts of climate change. Both aspects require substantial changes in the tourism production system.

2. Methodology

This paper adopted the survey design approach as its method of study. It is qualitative in nature, highlighting problems, causes and adaptation and mitigation strategies relative to the tourism sector based on a survey of environmental practitioners.

3. The Importance of Tourism to African Countries

To Brown (2001), tourism has the potential to lift people out of poverty through the employment and entrepreneurial opportunities it provides. The recognition of tourism's role in poverty alleviation has made it a substantial component of the international development and trade agenda.

Tourism is a vital part of the African and global economy (Uchegbu and Ugwuanyi, 2009). International tourism ranked as the fourth-largest industry in the world, after fuels, chemicals and automotive products

(Uchegbu and Ugwuanyi, 2009). Tourism has been established as one of the main drivers of growth in many developing economies and is a good source of income, especially for African countries.

The breadth of international travel also has greatly expanded in recent years to encompass African countries. In 1950, just fifteen destinations, primarily in Europe, accounted for 98% of all international arrivals, but by 2007 that figure had fallen to 57% (IPCC, 1996).

African countries have now become major growth areas. Tourism is a key foreign exchange earner for 83% of developing countries and the leading export earner for one-third of the world's poorest countries (IPCC, 1997). For the world's forty poorest countries, tourism is the second most important source of foreign exchange after oil (UNCHS, 2006).

4. Advantages of Tourism over Other Industries

Tourism is consumed at the point of production, so it directly benefits the communities that provide the services. It enables communities that are poor in material wealth but rich in culture, history and heritage to use their unique characteristics as an income-generating comparative advantage (Calthorpe, 1993).

Tourism has helped to provide employment and alleviate poverty, the first of the United Nations' Millennium Development Goals (MDGs) (UNCHS, 2006). It creates networks of different operations, from hotels and restaurants to adventure sports providers and food suppliers. This enables tourist centers to form complex and various supply chains of goods and services, supporting a versatile labor market with a variety of jobs for tour guides, translators, cooks, cleaners, drivers, hotel managers and other service-sector workers.

Many tourism jobs are flexible or seasonal and can be taken on in parallel with existing occupations, such as farming and trading. It tends to encourage development of multiple-use infrastructure that benefits the host community, including roads, healthcare facilities and sports centers, in addition to the hotels and high-end restaurants that cater to foreign visitors. With its tendency to produce flexible labor markets and offer diverse working opportunities, tourism can also help realize the second MDG – promoting gender equality. In Mali, the World Tourism Organization's Sustainable Project for Eliminating Poverty (ST-EP) program has supported an effort to train female artisans of Djenné, one of Mali's oldest and most visited towns.

Tourism has potential not only for protecting the natural environment, but also preserving historical, archaeological and religious monuments and stimulating the practice of local traditions, arts and crafts and cuisine. As tourism by definition involves the transfer of people, culture and ideas, it is ideally placed to foster effective global partnerships as stipulated by the eighth MDG.

In addition to advancing development goals, some have credited tourism with helping to build and sustain peace. Tourism is an instrument that creates a global language of peace through its tendency to promote communication between nations and cultures. Tourism can also help promote peace and stability in African countries by providing jobs, generating income, diversifying the economy, protecting the environment and promoting cross-cultural awareness.

Kenya identified the potential for tourism quickly, opening its first national park in 1947 and building a flourishing industry based on safaris. By the 1990s it was providing the sort of mass tourism commonly associated with the resorts of the Spanish coast. Kenya's eco-tourism has been able to deliver concrete benefits to its people by involving local communities. Kenyans also have been involved in developing tourism through programs, such as Parks Beyond Parks, that encourage the management and conservation of wildlife areas outside of parks and protected wilderness areas; they play the leading role in managing and conserving wildlife areas in their own communities.

5. The Effects of Climate Change on Tourism in Africa

The challenge of climate change is particularly significant for the African countries because the sector is both a victim of (i.e., increased health and natural disaster risks) and contributor to greenhouse gas emissions. African countries are vulnerable to the impacts of climate change, such as rise in temperature, rainfall, sea-level rises, coral bleaching, increased storm intensity, saline intrusion, food shortage and unemployment.

5.1. Beach Erosion

Beach erosion can result from a number of factors, including the simple inundation of the land by rising sea levels resulting from melting of the polar ice caps and increased frequency of harsh storms. One of the hot spots of beach erosion is found in Barbeach, Lagos, Nigeria (Pickering and Owen 1995).

The importance of coastal zones to the tourism industry and the need to protect such resources is not only vital to the economy of nations but presents a growing dilemma for many localities and regions. Beaches have become synonymous with tourism and, with current predictions of climate change and sea-level rise, they are under significant threat of erosion worldwide, as shown in Figure 38-1.

Figure 38-1. A gully site along Nkpor/Onitsha Express Way (Field work, 2010).

5.2. Rises in Temperature

Global warming has resulted in disappearing glaciers. A trickledown effect of temperatures leads to the presence of mosquitoes where they never used to exist. Higher temperatures bring about a decline in vegetation and cause wildlife to migrate toward areas that still have water. Rivers that were known to be permanent in the 1960s have now turned into seasonal rivers and are drying up.

Figure 38-2. A Gully at Uturu-Kanu Road near Abia State University, Uturu.

5.3. Changes in Tourist Preferences

Changing climate and weather patterns at tourist destinations and African countries can significantly affect the tourists' comfort and their travel decisions. Changing demand patterns and tourist flows have impacts on the tourism businesses and host communities, as well as "knock-off" effects on related sectors, such as agriculture, handicrafts or construction.

5.4. Coral Bleaching

Coral bleaching, otherwise known as whitening of coral reefs, is a phenomenon affecting coral reefs by which they lose their natural color as a result of increased water temperature or other environmental stressors such as pollutants. Coral bleaching occurs when ocean waters becomes too warm, causing stress on the zooxanthellae algae that live inside coral animals and provide them with food in a symbiotic relationship (Calthorpe, 1993).

As a result of this stress response, the relationship between the corals and the zooxanthellae breaks down: the corals lose their color, and become white. These bleached corals are still alive, but they are weakened. If excessively warm-water conditions continue for too long, the bleached corals will die.

Acidifying seas also threaten coral reefs as well as increased sedimentation resulting from the chopping down of forests or changing farming methods inland, leading to soil erosion that causes sediments to enter the rivers and then the sea.

Pollution is also a problem. Nutrients coming from sewage or from fertilizers encourage the growth of algae and plankton, which can out-compete the corals, or smother them by blocking out sunlight (Beatley, 1998) – thereby denying people a wide variety of benefits from reefs. Many coastal communities draw large portions of their food from fisheries. Coral reefs are also important to economies across the tropics, attracting tourists who enjoy diving, snorkeling, fishing or simply lounging on beautiful beaches.

Coral reefs are a source of white sand that protects beaches from erosion by reducing wave energy. The loss of coral reefs also produces "knock-off" effects through the marine food chain and associated ecosystems. This destruction of coral reefs ultimately threatens food security, livelihoods and the economic integrity of local and national economies.

5.5. Saline Intrusion

Climate change leads to saltwater moving into freshwater bodies. It has been observed in surface water, groundwater bodies and freshwater aquifers. It is caused by groundwater pumping from coastal wells, navigation channels or oil-field canals. The channels and canals provide conduits for saltwater to be brought into freshwater marshes. Saltwater intrusion can also occur as a result of a natural process such as a storm surge from a hurricane. Saltwater intrusion occurs in virtually all coastal aquifers, where they are in hydraulic continuity with seawater.

When freshwater is withdrawn at a faster rate, water tables are drawn down and hydrostatic pressure is reduced. When this happens near an ocean coastal area, saltwater from the ocean is pulled into the freshwater aquifer. The aquifer becomes contaminated with saltwater and coastal communities suffer as a result.

5.6. Food Shortages

The ecological and political reasons for production problems vary widely. They range from natural disasters, such as drought, flood or fungus, to political disasters, such as civil conflict, to misguided economic policies such as price controls – all of which effect production of essential foods (Huntley, 1991). Tourists avoid regions that have food scarcity and those who do come are deprived of the normal diversity of foods and cuisine indigenous to the area.

5.7. Unemployment

People whose jobs have been lost or interrupted because of climate change or resultant disasters become insecure, and, if little or no help is rendered in time by the government, might in some cases resort in desperation to acts such as theft, armed robbery, prostitution, etc.

5.8. Ecological Imbalance

Ecological imbalances as a result of economic development has been observed, such as destabilization and destruction of fragile environments. A change in the food chain may also lead to ecological imbalance (Environmental Agency, 1998). The factors responsible for imbalance include: pollution due to industrialization, deforestation, fuel consumption, disproportionate cultivation of land, expansion of urban areas, change of locations of water reservoir, etc.

Many industries – such as textiles, paper, sugar, steel, petroleum, cement, food and chemical manufacturing – release wastes into nearby water bodies, thus creating pollution and ecological imbalances that affect the air, water and soil condition. Invasion of soil surface by fertilizers and pesticides has led to imbalances in several food chains between micro- and macro-organisms as well as chemical reactions. Ecological consequences such as increased soil erosion create loss of surface soil and exposure of un-weathered infertile soil. This reduces crop yields and atmospheric-nitrogen fixation. Improper use of non-renewable resources, cutting down forests and jungles, chemical wastes left in the sea, etc., affect not only humans but all the living species on the planet.

5.9. Biodiversity Loss

Climate change, with its effects on temperature, precipitation and other weather/climate elements, will surely impact biodiversity. Biodiversity can be said to include all life forms, such as fungi, protozoa, bacteria, plants, insects, fish and mammals (Uchegbu, 2002).

Climate-change-induced rises in temperature may lead to sea-level rise with its consequent flooding, which may tend to disturb the biodiversity-ecosystem balance (Uchegbu, 2009).

Figure 38-3. Tiffindell, South Africa.

A decrease in rainfall and persistence of dry seasons may lead to drought conditions, which in turn negatively affect biodiversity by reducing the lifespan of livestock and other species – resulting in lost food resources for human populations.

5.10. Decrease in Snow Cover and Shrinking of Glaciers

A decrease in snow available for winter sports/activities requires in turn the use of snow-making equipment, with consequent further impacts on the environment – increased costs for ski areas, shortened winter-sport seasons and landscape aesthetics. This is found at Tiffindell, the highest mountain peak in the Cape, and a ski resort in South Africa (shown in Figure 38-3).

5.11. Water Availability

It has been argued that persistent climate change may lead to a drastic reduction in the level of streams, flow of streams/rivers, volume of underground water and the like (Ukpong, 1994).

Climate change may lead to shrinkage in the surface-water supply as well as the quantity and regularity of water supplies in some parts of Africa. This is the conclusion of a study entitled "Effective Planning and Management as Critical Factors in Urban Water Supply and Management in Umuahia and Aba, Abia State, Nigeria" (Uchegbu, 2009).

As shown in Table 38-1, 23.8% of the respondents in Aba reported the influence of climatic factors (especially temperature) on water sources to be great while in Umuahia this was 23.4%. But more respondents in Aba, 34.3%, reported either moderate or little influence. Reduced precipitation combined with increased evaporation in some regions is leading to water shortages, competition over water (between tourism and other sectors), desertification and increased wildfires threatening infrastructure.

Table 38-1. Climatic factors' influence on water sources (Fieldwork, 2006).

Response	Aba # Respondents	Percentage	Umuahia # Respondents	Percentage	Total %
Great	34	23.8	26	23.4	24
Moderate	49	34.3	32	12.8	32.4
Little	49	34.3	8	7.2	22.8
None	2	1.4	9	8.1	4.4
No response	15	10.5	36	32.4	20.4
Total	**149**	**100**	**111**	**100**	**100**

6. Adaptation and Mitigation Strategies of Climate Change in the Tourism Industry

Adaptation is a process by which strategies that aim to moderate, cope with and take advantage of the consequences of climate changes are enhanced, developed and implemented (UNEP, 1995). It is equally referred to as a change in natural or human systems in response to actual or expected climatic conditions that "moderates" or "exploits" beneficial opportunities (Karl and Trenberth, 2003). The design and implementation of effective adaptation strategies requires consideration of the adaptive capacity of a given system to respond successfully to climate variability and change. It includes adjustments in behavior, resources and technologies and requires deliberate policy decisions, plans and implementation by both the public and private sectors in a timely manner. Various mitigation/adaptation strategies include the following:

Re-think their choice of destinations: Tour operators can restructure the choice of destinations they offer, by replacing more distant ones with closer ones of similar attraction value.

Support low-carbon holiday options and carbon labeling: Tour operators should seek to market low-carbon holiday options. Packages and journeys can also be carbon-labeled, i.e., indicating the emissions caused by the option.

Develop new low-carbon products: Tour operators should seek to strategically develop attractive low-carbon packages (e.g., a train-based holiday).

Traveling less often and staying longer: Tourists could consider traveling less frequently and staying longer at their destination.

Minimize air travel: Any air travel that can be avoided will be a major factor in reducing individual emissions.

Reward airlines with sound environmental management: Tourists should favor airlines that are committed to serious environmental management, using aircraft with new fuel-efficient engines and airframe designs.

Reward pro-environmental and pro-community development tour operators: Tourists should favor tour operators engaging in pro-environmental management (for instance providing carbon labeling or certified packages), and those that provide community benefits particularly when operating in Africa.

Capacity building: There is a need to build capacity for adaptation and mitigation in response to climate change across government bodies, tourism institutions and organizations at national, regional and destination levels. To this end, outreach and communication efforts should be viewed as the first step in building capacity in African countries that climate change is likely to impact, and be impacted by, tourism.

Figure 38-4. Using bicycles.

Other adaptive measures: Changing one's environment; encouraging irrigation in drought-prone areas; change in behavioral patterns; living within the working place; designing multi-purpose buildings; shifting from fossil fuels to renewable energy; cutting back on the use of energy; using energy efficiently; car pooling; using bicycles; and walking/trekking.

7. Mitigation Strategies

A discussion of mitigation measures must include technological, economic and social changes and substitutions that can be employed to attain reduction in GHG emissions from tourism. These include: reducing energy use/focusing on energy conservation by changing transport behavior (for example, using more public transport, shifting to rail and coach instead of car and aircraft, choosing less distant destinations); changing management practices, for example by greater use of videoconferencing for business meetings rather than traveling to meetings; and generally eliminating GHG emissions by circumventing activities that can be avoided without significant change to the tourist experience (Blakely, 2005).

Other measures include:

- Increasing the use of renewable or carbon-neutral energy;
- Substituting fossil fuels with energy sources that are not finite and that cause lower emissions, such as biomass, hydro, wind and solar energy;
- Substitution of practices that account for large GHG emissions with practices that have a low-carbon footprint;
- Use of fuel-efficient cars by tour operators, hotels and resorts;
- Employing measures, such as effective ventilation and appropriate roofing, to reduce temperature in buildings;
- New buildings designed with the conservation of energy in mind;
- The use of more efficient lighting and energy-efficient appliances;
- Development of alternative energy sources, e.g., solar, wind, bio-fuel, and hydro – except large dams, cessation of gas flaring, windbreaks and shelterbelts;
- Use of buses and trains for urban transport;
- Soft coastal protection to prevent erosion (e.g., reforestation of mangroves or reef protection);
- Enhanced design, setting standards and planning guidelines for tourism establishments;
- Integrate climate-change factors into regulatory frameworks for tourism development, such as Environmental Impact Assessment for tourism infrastructure and establishments;
- Implementation of tourism development plans within the framework of Integrated Coastal Zone Management (ICZM) processes and spatial planning;
- Shade provision and crop diversification; and
- Reduction in tourism pressures on coral reefs.

Further methods include: encouraging water-conservation techniques such as rainwater storage, the use of water-saving devices and wastewater recycling; diversification of the tourism product to less climate-dependent and seasonal activities; encouraging eco-tourism; education/awareness among tourism businesses and their staff, as well as tourists; the introduction of programs on awareness and preparedness to face extreme climatic changes and disasters at the national and local levels through improved coordination between disaster management offices, tourism administrations, businesses and host communities (Uchegbu, 2010); the provision of information on climate change to the tourism sector through cooperation with national meteorological services; providing insurance coverage (or alternative schemes) for the recovery of infrastructural and other damage; drainage and watershed management to reduce flood and erosion risks; protected-area management, and other means of conservation of coastal ecosystems in order to enhance their resilience; protection of wildlife to maintain ecological balance; proper land-use planning; restriction of industrial production within industrial zones; pollution control through environmental laws/polices as well as their enforcement; switch to use of raw materials that reduce the production of carbon; and enforcing solid-waste recycling policies that embody the three Rs (reduce, reuse and recycle) (Uchegbu, 2009a).

8. Conclusion

The tourism sector is affected profoundly by climate change. It is therefore very important that African countries adapt to and mitigate against impending climate change to promote and sustain growth. To undertake this task, governments must come together to formulate policies which would ensure that the sector remains sustainable. It is essential that governments and policymakers become involved in the process at a very early stage, since climate change is an environmental and ultimately a developmental problem. With the formulation of mitigation and adaptation strategies and the appropriate policies in place, the tourism sector can play a key role in dealing with climate change and encouraging sustainable growth in the sector.

References

Beatley, T. (1998). The vision of sustainable communities. In *Cooperating with Nature: Confronting Natural Hazards with Landuse Planning for Sustainable Communities*, ed. R. J. Burby, pp. 233–262. Washington, DC: Joseph Henry Press.

Blakely, E. D. (2005). Climate change: Exploring the literature and research opportunities. In Proceedings of the conference Planning

Education and Sustainability, Sustainability Education, 30 September to 2 October. University of South Australia.

Brown, V. (2001). Planner and the planet: Reforming the people/planet relationship; Do planners have a role? *Australia Planner: Journal of the Royal Australia Planning Institute* 38, no. 3: 67–73.

Calthorpe, P. (1993). *The Next American Metropolis: Ecology, Community and the American Dream*. Princeton, NJ: Princeton Architectural Press.

Chan, W. W. & J. C. Lam (2003). Energy-saving supporting tourism: A case study of hotel swimming pool heat pump. *Journal of Sustainable Tourism* 11, no. 1: 74–83.

Crowley, T. J. (2000). Causes of climate change over the past 1000 years. *Science* 289, no. 5477.

Dakaica, S. (2002). The challenges of education sustainability of small island developing countries of the south Pacific. In Proceedings of the conference Planning Education and Sustainability, Sustainability Education, 30 September to 2 October. University of South Australia.

EEA [European Environmental Agency] (1998). Europe's Environment: The Second Assessment. Elsevier Science.

Gunder, M. (2005). After the post-modern abyss: Is the discourse of sustainability really planning's saving grace? Proceedings of the conference Planning Education and Sustainability, Sustainability Education, 30 September to 2 October. University of South Australia.

Huntley, B. (1991). How plants respond to climate change: Migration rates, individualism and the consequences for plants communities. *Annals of Botany* 67.

IPCC (1990). IPCC Special Report: The Regional Impacts of Climate Change; An Assessment of Vulnerability.

—. (1996a). IPCC Special Report: The Regional Impacts of Climate Change; An Assessment of Vulnerability; Nairobi.

—. (1996b). *Climate Change 1995: The Science of Climate Change; Contributions of Working Group I to the Second Assessment Report of IPCC*. Cambridge: Cambridge University Press.

—. (1997). *The Regional Impacts of Climate Change: An Assessment of Vulnerability*. Cambridge: Cambridge University Press.

Karl, T. R., & K. E. Trenberth (2003). Modern global climate change. *Science* 302, nos. 56–51.

Moan, J., A. Dah/back, T. Henriksen & K. Magnus (1989). Biological amplification factor for sunlight-induced non-melanoma skin cancer at high latitude. *Cancer Res* 49: 165–182.

Pickering, K. T. & L. A. Owen (1995). *An Introduction to Global Environmental Issues*. London: Routledge.

Stainforth, D. A. (2005). Uncertainty in predicting of the climate response to rising levels of greenhouse gases. *Nature* 433.

Uchegbu, S. N. (2010). Planning education and practice in Nigeria: The case of the University of Nigeria. *The Journal of Building and Land Development* [Special issue, June].

—. (2009). Effective planning and management as critical factors in urban water supply and management in Umuahia and Aba, Abia State, Nigeria. *Journal of the Physics and Chemistry of the Earth* 34.

—. (2009a). The relationship between the quality of surface water available for domestic consumption and climatic factors in umuahia and Aba, Abia State, Nigeria. *Journal of Applied Sciences* 12, no. 1.

—. (2002). *Environmental Management and Protection*, 2nd ed. Enugu Spotlite Publishers.

Uchegbu, S. N., & B. Ugwuanyi (2009b). Climate change: Problems and solutions. *Journal of the Nigerian Environmental Society (JNES)* 5 (Nigeria).

Ukpong, S. J. (1994). Global and Nigerian environmental problems analysis. *SIRF* (3 November).

UN (1990). *Global Environment Outlook 2000*. Earth Scan Publications Ltd.

UNCHS [National Center for Human Settlement] (2006). *Urbanization and Sustainable development in the 3rd World: An Unrecognized Global Issue*. Nairobi: UNCHS.

UNEP (1995). Environmental effects of ozone depletion 1994 assessment. *Ambio* 3.

WTO-UNEP-WMO (2008). Climate Change and Tourism: Responding to Global Challenges.

CHAPTER THIRTY-NINE

GREENING THE TOURISM SECTOR: AN EFFECTIVE MITIGATION MEASURE AGAINST CLIMATE CHANGE

ROSE MUKOGO[1]

Reflective Considerations

In today's industrialized world it is important that community, business, government and political leaders are sensitive to environmental considerations. This is for the benefit of current and future generations. Everyone is aware of the damage caused by climate change, as witnessed by weather phenomena such as the ice storms in the United States that saw the cancellations of thousands of flights, and the delays of flights in Europe due to snow-covered runways.

Why Tourism Greening?

This discussion looks at the efforts at mitigation undertaken by the tourism sector that have been introduced in reaction to climate change. At this stage, we can only speculate as to whether these measures by the tourism sector will sufficiently reduce the impact of climate change. The following questions arise:

- Why tourism greening? Is it not just another marketing gimmick – a sales pitch?
- How does it help the climate given that tourism is not manufacturing any products, nor does it use major industrial equipment/plant?
- Should it be enforced by law or through self-regulation?
- What actions should be taken and by whom?
- Who is responsible for its implementation?

[1] Managing Director, Green Tourism Services (Pvt)Ltd (GTS).

The sustainable tourism industry is inter-connected in three ways: environmental, socio-cultural and economic. There is a misconception that greening issues are limited to the environment, yet any socio-cultural and financial initiatives that impact on the bottom line of an organization may indirectly impact on climate change.

Tourism greening should not be used for marketing purposes, but should be given important consideration. People should embrace and share the message.

Background

At a meeting held in Madrid in 2006, the Secretary-General of the UNWTO Francesco Frangelli announced, "This is a call to the leaders in the public and private Tourism Sector and ultimately to travelers themselves. Climate change is real; its effects are proven; and the Tourism Sector has to play its part in contributing to the solution of the challenges it poses." This discussion outlines the symbiotic relationship between the tourism industry and climatic conditions.

Many recent weather events seem to point to the fact that climate change is real; therefore, all sectors contributing to economic activities need to ensure that strategies to combat the negative impact of climate change are developed and adopted. It is important to note that the phenomenal growth of the tourism sector has a potential to deplete the natural environment, thus impacting negatively on the prospects of the growth industry itself. This discussion therefore proposes strategies to combat the negative impact of climate change on tourism development, so that they may be considered by the tourism practitioner.

The tourism industry in Africa, and most of the developing world, is tilted toward outdoor activities such as safaris, which are dependent on good weather conditions for a tourist to have a memorable experience. Any negative development, such as changes of temperatures below or above the norm, affects the performance of the industry, both in the short and long terms.

Performance of the Tourism Sector

The World Tourism Organization (UNWTO) estimates that international tourist arrivals will reach 1.56 billion by 2020 from the current level of 1,035 million reached in December 2012, according to the UNWTO World Tourism Barometer. Tourism is directly responsible for 5 percent of global GDP, accounts for 6 percent of the world's trade and employs one out of eleven people worldwide. The tourism industry is considered one of the fastest-growing international industries – having grown from 112 million people employed directly and indirectly in 1991 to 234.3 million in 2006. However, while this is a positive development, the world at large as well as planners should visualize the implications of this phenomenal growth. The growth of tourism should not be viewed in a negative way, as we are all aware that growth in the tourism industry brings with it a rising quality of life in many parts of the world. Simply, this means there is need for more planes, ships, trains, hotels, buses and coaches and all this means burning of fuel etc. Even at the destination, the activities, accommodation and land transport require energy. While tourism is one of the major economic sectors, the responsibility of reducing the effects of greenhouse gases (GHG) should be a shared responsibility across the sectors as the effects do not recognize borders nor are they segregated according to economic sector.

The Carbon Cycle, Climate Change and Impact on Tourism

Carbon is a key element of all living matter and it is released and re-absorbed in a continual process known as a carbon cycle. In general terms, the cycle involves an ongoing exchange of carbon dioxide (CO_2) between the atmosphere and the earth. Photosynthesis takes place when plants absorb carbon dioxide and then release it back into the air when the plant decays. In general, the cycle has been in balance; however, since the industrial revolution the heightened levels of carbon dioxide have led to a phenomenon referred to as climate change.

Tourism, by virtue of its rapid growth and being a key driver of economic development, is a significant contributor to climate change. Recent evidence suggests that the sector's contribution to global anthropogenic CO_2 emissions is in the order of 5% (in 2005), but may be higher (from 5% to 14%) if radiative forcing is considered, i.e., in this case the warming caused by CO_2 as well as other greenhouse gases. As this calculation only includes energy throughput, and does not consider the energy needed to construct hotels, airports, highways and runways, it needs to be seen as conservative. Furthermore, it deserves mention that only a minority of the world's population is responsible for these emissions, as, for instance, less than 2% of the world's population participates in international air travel on an annual basis. The UNWTO is therefore seeking ways for tourism's constituents, including tourists themselves as well as providers, both to adapt to climate change and to mitigate GHG emissions while reinforcing the contribution of the sector to socioeconomic development in general and the Millennium Development Goals in particular.

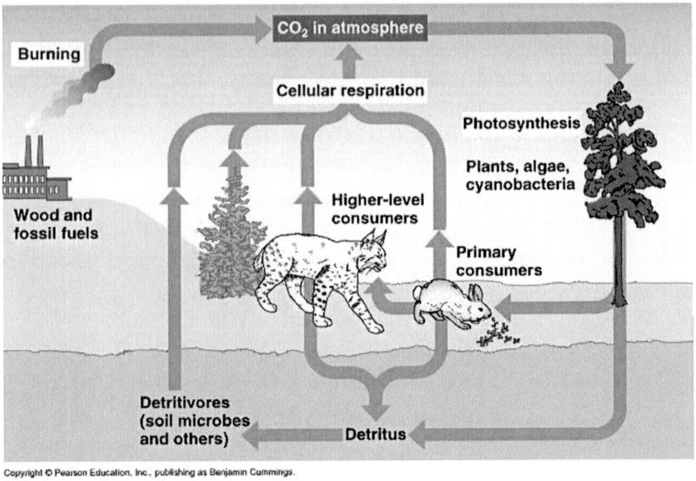

Figure 39-1. The carbon cycle (picture credit: Pearson Education/Benjamin Cummings).

Strategies for Mitigation of Climate Change

One assumption underlying this discussion is that there are already various efforts and strategies in place focused on reducing the impacts of climate change. These strategies should be tackled both at a macro and a micro level. In addition, an integrated approach from all players is required to ensure successful and effective implementation. It should be noted that emission reductions should be considered in the context of the development and poverty-reduction objectives of developing countries, considering that many of them depend on international tourism for their tourism revenues and economic development in general. The tourism sector has a responsibility to minimize harmful emissions by encouraging sustainable, carbon-neutral transport solutions and low-carbon market destination combinations, to improve the use of natural resources (water and energy) and to contribute to the conservation of natural areas. However, these strategies should be implemented in a manner that does not compromise the development of the industry by the developing countries that are the major beneficiaries of the growth in the sector.

In proposing the key greening tourism strategies for mitigating the effects of climate change, it is also imperative to address the issues concerning energy-efficient technologies and transportation. However, to do this it is important to first identify how climate change specifically impacts tourism in Africa and the developing world.

A Conceptual Framework

The overall objective of climate-change mitigation strategies, policies and activities in the tourism sector is to contribute to the achievement of "carbon neutrality" in the sector. For business and institutions, "carbon neutrality" can be defined by the entire set of policies that an institution or business uses when it estimates its known greenhouse gas (GHG) emissions, takes measures to reduce them and purchases carbon offsets to "neutralize" those emissions that remain. Carbon neutrality for a business or institution signifies an entity (organization) that has a zero net contribution of greenhouse gases to the atmosphere. This includes all activities directly controlled by an organization, including travel, purchasing of goods and services, and daily behavior of staff.

Carbon neutrality can be achieved by improving the way the organization operates (e.g., through modified procurement considerations), by improving efficiency of operations (e.g., communications and meetings) and equipment (e.g., vehicle fleets and building). Carbon neutrality also recognizes offsetting as a last resort among the options to achieving full neutrality. The tourism sector is composed of a wide range of businesses, from small, local operations that service a single local market to very large transport, hotel and tour operator companies that serve global markets across entire regions and sell or facilitate millions or tens of millions of tour packages to foreign destinations each year.

The industry provides tourists with products and services such as accommodation, transport, food and drink, attractions to visit and souvenirs to purchase. It is clear that the industry shapes demand through its marketing strategies, but consumers (tourists) ultimately make the final choices. Recognizing that tourists have an important role in creating business interest in sustainable tourism products, the sector should consider mitigation options and be proactive in addressing climate change. In response to the growing awareness of consumers as to the contribution of tourism to climate change, the sector should be looking more actively into exploring strategies for becoming carbon neutral since there is no universal blueprint for achieving such a goal in the sector.

At the macro level, governments can formulate regulatory procedures that provide for mainstreaming of the green issues in any policies on tourism development, such as regulations on registration and licensing of new facilities, greenhouse gases emissions, energy-efficiency standards for new products and compulsory environmental impact assessments for tourism projects, among other initiatives.

Governments should also incorporate climate-change issues into tourism legal frameworks and enabling policies, including the development of minimum standards in mitigation of GHG emissions and other pollutants, which should be met for infrastructural developments and equipment such as airplanes. It is recommended that incentives such as tax credits should also be put in place to encourage companies to adhere to these requirements and to even propose their own initiatives.

Environmental issues should be incorporated into the educational curriculum to create awareness of the climate-change challenges at an early age and with a broader scope. A deliberate effort should be made to engage the various forms of the media to spread the news about the challenges of climate change and the need to address the causes of this phenomenon.

In relation to the concept of carbon neutrality, a successful mitigation policy could consider four main steps that any tourism-related business or institution can implement as a practical response to climate change.

1. Eliminate the emission of greenhouse gases by refraining from certain activities that can be avoided without a considerable change to the tourism product or service quality.

2. Reduce the emission of greenhouse gases by focusing on energy-efficiency practices in specific activities.

3. Substitute practices that are responsible for a large amount of greenhouse gas emissions with practices that have a lower carbon footprint.

4. Institutions or business units can offset remaining emissions to achieve full carbon neutrality.

The industrial greening concept thus reduces the impact of the industry on the environment.

Tourism sector operators and professionals can adopt green tourism practices with emphasis on application in the context of profitability and efficiency and also through self-monitoring initiatives such as codes of conduct on environment management. Outlined below is a case study where creativity in terms of eco-friendliness was significant in the development of the Eco-tel in Mumbai, India.

Case Study: The Orchid, a Five-Star Eco-Tel Hotel, Mumbai, India

Tourism Destination and Situation: The Orchid, a five-star hotel, was opened in September 1997, focusing on economically viable pro-environmental measures. The hotel, belonging to Kamat Hotels Ltd., is interesting in that it takes a strictly profit-oriented approach to environmental measures, providing figures of financial savings through pro-environmental management on its website. Simultaneously, the approach taken is a whole-systems perspective, including all aspects of the environment, consistently focusing on best practice.
Climate Change Impact: The Orchid Hotel shows that it is economically feasible to implement a wide range of environmental practices, including virtually any aspect of the lifecycle of the hotel. This shows that five-star hotels can financially benefit from environmental management, make huge contributions to resource savings, create markets for innovative and green technology and involve their guests in pro-environmental management.

Mitigation Tools, Techniques, Policies or Measures

The hotel has engaged in a wide range of measures to reduce its impact on the environment, many of these unique.

Architecture and Construction

- Passive-energy conservation efforts in design.
- Depressions and protrusions in the facade reduce surface radiation.
- The building is designed with seventy-two rooms facing the atrium, reducing the heat load.
- A skylight detailed in a manner of doubly layered domes reduces heat load and noise levels while admitting maximum natural light in the atrium space.
- The swimming pool is located on the rooftop with four feet of water body, which acts as an insulator from the heat.

- Cement used for constructions is PPC (Portland Pozzalana Cement) containing 15–20% fly ash, as compared to OPC (Ordinary Portland Cement).
- Internal partitions are made from so-called “Quite Easily Done” wall panels, which are made from fertilizer waste, instead of using red bricks made from topsoil.
- Autoclaved Aerated Concrete is used for external walling and wet walling. AAC is eco-friendly as it is manufactured using approximately 60% fly ash. It has thermal insulation properties and a better sound-absorption coefficient than ordinary bricks.

Reduce, Reuse and Recycle

- All taps contain aerators, which increase the water’s force and reduce outflow, saving up to 50% of water. The restaurant has taps operating on timers.
- Geberit Concealed Cisterns use only 6 liters of water per toilet flush as against 15–20 liters used in conventional flushes, while the use of Geberit Urinal Flush Valve infrared detectors ensures a definite flush after every use preventing the unwanted flushing of timer-set systems.
- Wastewater is treated and then reused in areas like air-conditioning and gardening.
- Drip irrigation reduces water use for gardening.
- Aqua Zone is a system using ozone to destroy microorganisms including bacteria, virus, spores, fungi, etc. The water is safe for drinking and free from chlorine.

Interior Design

- Window frames, master control panels in the guestrooms and shutters are made from rubber wood. After producing rubber sap, the tree is cut down and cannot be used for other purposes.
- Medium Density Fiber Wood is used instead of real wood throughout the hotel. MDFW is manufactured from cotton stalks, which grow to a height of 5–6 feet, and are cut down after yield and are usually not used for other purposes.
- The restaurant uses recycled wood from old buildings.
- Triple-glazed windows with added reflective glass block heat from the sun, reducing energy for air-conditioning.

- All lamps are PL lamps or florescent tubes.
- All rooms have key card light switches so that all lights and air conditioning are switched off once the guest leaves the room.
- Mini bars sense the load inside the refrigerator and cool it accordingly. Mini bars are CFC-free.
- Air conditioning uses R22 instead of CFC refrigerants, a far more climate-friendly alternative.
- Attached to the air-conditioning system is a tank to store cold energy during off-peak hours. This stored energy is then used during the peak hours/periods reducing compressor overloading and cutting power consumption.
- The heat generated from the air conditioners provides hot water to the guestrooms, laundry, toilets and kitchen.
- Hangers in guestrooms are made from sawdust.
- The hotel only uses herbal products which have not been tested on animals.
- Paints used in the hotel are eco-friendly. The exterior of the hotel is painted with water-based paint with negligible VOC (Volatile Organic Compounds) content (only 0.0125%) and the paint used for interiors has 0% VOC.

Educating Customers

- The control panel in the guestrooms has a feature known as the green button. On pressing the green button the thermostat of the air-conditioning unit is stepped up by two degrees. The saving in electricity resulting from the two degrees increase in temperature is translated into rupees and displayed on the guest folio and profile. A certificate is then issued to the guest who has voluntarily participated in conserving energy.
- Guests who have left their emails are informed about ongoing environmental activities.

Garbage

- The Orchid seeks to become a "zero garbage" hotel and has taken various measures to reduce the waste being generated. This includes vermiculture on the hotel site to recycle kitchen waste.
- Reusable cloth laundry bags are used instead of standard paper or plastic bags.

- For shopping, cloth shopping bags are provided to guests.
- Garbage bags are made from recycled plastic.
- Cut flowers are used sparingly, and the hotel uses potted plants instead.

Use of Chemicals

- Anti-Cockroach Herbal Treatment has been recently introduced and is the first of its kind in India to eradicate cockroaches. The herbal paste poses no health hazards, as there is no use of any poisonous chemicals. No cleaning is required after the treatment is done hence there is no wastage of water, detergents, etc. It is applied in the form of a paste and placed on nooks and corners in the form of small globules.

Conclusion

The case study of the Orchid Eco-Tel Hotel has brought to light the fact that, at the micro level, a strategy of self-regulation of developing a code of conduct for tourism greening adopted by operators, if implemented in comprehensive manner, can play a positive role in the operations of a company. The Orchid became Asia's first five-star hotel to win the ECOTEL® certification shortly after opening in May of 1997, and as of January 2011 is the only Hotel in the world to win over seventy-four international/national awards in thirteen years from inception. Under the management of owner Vithal Kamat, the hotel has earned more environmental accolades than any other hotel in the world.

At the macro level, it should be noted that an appreciation of the whole picture of environment pollution or climate-change issues cannot be achieved through country-by-country analysis because these are transient geo-political boundaries drawn by human beings. Thus to ensure the mitigation of climate change at a global level requires a more holistic approach, such as international conventions that cascade down to continental, regional and eventually national level for effective implementation.

CHAPTER FORTY

WILDLIFE-BASED TOURISM AND CLIMATE: POTENTIAL OPPORTUNITIES AND CHALLENGES FOR BOTSWANA

NAOMI N. MOSWETE AND PAULINE O. DUBE[1]

Abstract

Tourism plays a major role in boosting the economies of most countries in Africa. Many African countries have begun to realize the limited successes of agricultural exports due to anthropogenic climate change and developmental pressures and have shifted their focus to tourism as a new source of growth. Nature-based tourism has become the most popular land-use tool and is a rural-development strategy in southern Africa, in particular Botswana where it has promoted the conservation of natural resources and improved the living standards of remote rural communities. This chapter aims to build/add knowledge and understanding of the effects of climate change on nature (e.g., wildlife) and highlight the relationship between wildlife resources in general and the climate-related risks to Botswana's lucrative nature-based tourism industry in particular. The paper identifies and discusses the likely impacts of climate on natural resources and notes how this will have negative effects on the wildlife/safari tourism. The paper concludes by discussing existing opportunities for diversifying the tourism-product base for Botswana away from sole dependence on wildlife resources to a form of tourism with nature and culture combined. Recommendations include the creation of well-rounded, all-year tourism packages that incorporate both wildlife and cultural tourism.

[1] Department of Environmental Sciences, University of Botswana.

Introduction

The economy of Botswana has prospered for more than forty years on mineral-based natural resources. Between 2001–2005, diamonds and copper-nickel accounted for an average of 86% of the country's total exports (Jefferies, 2009). While Botswana is deemed a middle-income country, it is sparsely populated with only 2 million people within a land surface area of 581,730km^2; it is semi-arid, with 80% of the surface area being the Kalahari Desert; and it is a land-locked country, dependent on a single non-renewable commodity (diamonds), in a region where more than half of the countries fall under the category of the world's poorest nations known as Least Developed Countries (LDCs) (SADC, 2008). This makes Botswana particularly vulnerable to global market pressures and the impacts of climate change. The growing pressure to diversify the economy is turning the focus to tourism, in particular nature-based tourism, which currently is one of Botswana's fastest-growing economic sectors (Government of Botswana, 2001; Knowles & Theron, 2010) and is considered to be a more sustainable economic option; however, a purely nature-based tourism industry may become unattainable in the future due to a combination of population and development pressures plus anthropogenic climate change. Land degradation and increased drying, leading to a decrease in mammal populations, are likely to reduce the attractiveness of the country's natural landscape and make wildlife tourism less viable (Dube & Kwerepe, 2000; Dube & Moswete, 2003).

This chapter aims to build the knowledge and understanding of the effects of climate change on nature (e.g., wildlife), and highlight the relationship between wildlife-based tourism resources in general and the climate-related risks to Botswana's lucrative nature-based tourism industry. The chapter also discusses the existing opportunities for diversifying the tourism product base for Botswana away from sole dependence on spectacular wildlife and scenery as attractions to other forms of tourism.

A review of the status of tourism in Botswana follows this introduction, after which the likely impacts of climate change and other stressors on Botswana's wildlife resources are discussed. The last section attempts to provide alternative avenues to reduce the negative impacts of climate change on the tourism industry in the future.

Tourism in Botswana

Tourism is the second-largest economic sector in Botswana (after mining) and is recognized as a major stimulus to the economy (BTDP, 2000; GOB, 2001, WTTC, 2007). The industry has been growing steadily throughout the years as evidenced by the number of visitors, which grew from 106,800 to 203,172 between 1993 and 1998 (Central Statistics Office CSO, 1998; 2003). An increase in visitors was observed in 2005 when the number approached that of the total population of Botswana (DOT, 2006; WTTC, 2007). The number of foreign investors in the tourism business has also increased from 331 in 2000 to approximately 550 in 2004 (Botswana Review, 2005).

About 90% of tourists who come to Botswana visit National Parks and Game Reserves (Magole & Gojamang, 2005; WTTC, 2007). Also, about 90% of Botswana tourists listed wildlife-related tourism activity, especially in Chobe and Moremi Game Reserves, as the greatest attraction (Magole & Gojamang, 2005). Approximately 40% of the country is designated as national parks and game reserves, and 22% of this has been set aside as Wildlife Management Areas (WMA) (GOB, 2003) (Figure 40-1). It is in these protected areas that wildlife-based tourism (photographic safaris, trophy hunting and filming) is mostly practiced (GOB, 2001; 2003), and many long-haul tourists visit Botswana to see charismatic species, including the Big Five (see Table 40-1); however, wildlife-based tourism and eco-tourism is not restricted to protected areas.

Wildlife-based tourism plays a significant economic role and contributes 70% of all Protected Areas (PAs) revenues to the total national economy (Gujadhur, 2001; WTTC, 2007). Primarily, commercial utilization of wildlife resources has been practiced via non-consumptive means, mainly in PAs, including Wildlife Management Areas (WMAs). In 2000, wildlife/safari-hunting activities generated about US$12.5 million (Arntzen, 2003). The industry has been growing in leaps and bounds in the past ten years as evidenced by growth in commercial lodging facilities (hotels, safari lodges and built campsites), and other infrastructural development in urban and rural towns/villages.

Table 40-1. Large mammals in the Okavango Delta, 2002.

Numbers calculated for an area of 20,000km^2 derived from ten aerial counts carried out in 1988–2002 by DWNP, Government of Botswana (Bonyongo, 2004). Numbers of impala have been corrected on ground counts (Ramberg et al., 2006: 329).

Species	Total number
Elephant, *Loxodonta Africana*	35,000
Zebra, *Equus burchelli*	14,000
Warthog, *Phacochoerus aethiopicus*	2,000
Hippopotamus, *Hippopotamus amphibious*	2,500
Giraffe, *Giraffa camelopardalis*	5,000
Wildebeest, *Connochaetes taurinus*	8,000
Tsessebe, *Damaliscus lunatus*	3,000
Impala, *Aepyceros melampus*	140,000
Buffalo, *Syncerus caffer*	60,000
Kudu, *Tragelaphus strepsiceros*	300
Sitatunga, *Tragelaphus spekei*	500
Red Lechwe, *Kobus leche*	60,000

Note: Department of Wildlife & National Parks.

Figure 40-1. NG 31 Controlled Hunting Area, Okavango area. Photo credit: Naomi Moswete (September 2011).

Wildlife-based tourism is also considered as the most appealing form of land use in the country (Barnes, 2001; GOB, 2002; 2007; MFDP, 2009), as the major activity of safari hunting occurs in remote parts of the country, which creates jobs for rural communities where poverty levels are high. Trophy-animal hunting (e.g., elephant or buffalo) accrue high revenue for safari hunters and provide income for the rural resident communities (Gujadhur, 2001; Mbaiwa, 2003; 2008; Thakadu et al., 2006). In all, wildlife-based tourism has been centered in and around the Okavango Delta and Chobe National Park, accounting for about 95% of visitations and 91% of revenue for all park-based eco-tourism activity in 2006 (WTTC, 2007).

There is growing concern about nature-based tourism as more evidence signals that changes in climate due to human activities may have a significant impact on natural resources globally (Alley et al., 2007; Becken & Hay, 2007). For example, the effect of the flooding in the Okavango Delta in 2000 resulted in the closure of the Moremi Game Reserve for a period of eight months because roads were impassable. During that period, there was a decline in the number of self-drive tourists in the Okavango Delta and tourism revenue went down (Mbaiwa & Mmopelwa, 2009). This might have had a direct impact on local livelihoods since the majority of the workers in tourism facilities are local people. The section below explores potential impacts of climate change on Botswana's wildlife resources, which may have implications for the sustainability of tourism in the country.

Climate Change and Wildlife Resources in Botswana

Expected Impacts of Climate Change

Indications are that impacts of climate change in southern Africa, where Botswana is located, will not be uniform. For instance, more drying is expected toward the west, with negative implications particularly on wildlife, which is currently the major tourist attraction in Botswana (Table 40-2). Work by Thuiller et al. (2006) identified the region centered on the Kalahari in arid Namibia, northern South Africa and Botswana as the second most likely (after central Africa) to experience significant wildlife species losses due to climate change. Similar predictions for Botswana were noted in Hulme (1996). This is because climate scenarios signal greater warming and drying throughout the western parts of southern Africa (see Table 40-2).

Table 40-2. Indicated future effects of climate change.

• The central-southern Africa landmass, which extends over Botswana, parts of north-western South Africa, Namibia and Zimbabwe, is likely to experience the greatest warming of 0.2°C to 0.5°C per decade. For Botswana, MWTC (2001) estimates are in the range of 1 to 3°C during the next 100 years. Because of the slow response of the oceans, further warming in the next two decades of about 0.1°C per decade is expected globally even if the concentrations of all GHGs and aerosols had been kept constant at year 2000 levels.
• Overall, the number of warm spells is likely to increase in the region, with a decrease in the extremely cold days (and these have already been observed in Botswana).
• An increase in rainfall variability and extreme events and a decrease in the growing season have been indicated and also observed from station data. Climate scenarios point to a potential for greater drying in western parts of southern Africa. Greater drying of the range of 30 to 40% throughout the winter season, June to August, has particularly been noted.
• More intense droughts are likely due to a number of factors, including the warming of the Indian Ocean and potential linkages between global warming and changes in patterns of El Niño/Southern Oscillation, which is known to influence climate variability in the region.
Sources: Becken & Hay, 2007; Desanker et al., 2001; Hulme, 1996; Hulme et al., 2001; Ministry of Works, Transport and Communications (MWTC), 2001; Scholes and Biggs, 2004; MFDP, 2009; Alley, et al., 2007; Boko et al., 2007; Christensen et al., 2007; Funk et al., 2008).

Climate Change and Wildlife Resources

As noted above, wildlife is the major source of tourism attraction in Botswana. Climate change will likely affect wildlife resources in Botswana in multiple ways, depending on the status and management of these resources. A recent aerial survey over the Okavango Delta concluded that wildlife species have shrunk in the past fifteen years, reaching as much as 95% for ostrich, 90% for wildebeest, 84% for antelope tsessebe and 81% for warthogs and kudus; although others, such as elephants and plains zebra, remained stable while hippos increased by 6% (Guardian-UK, 2011). The results are, however, based on a one-off aerial survey and need to be interpreted with caution, although others have signaled similar trends (see Perkins & Ringrose 1996; Rudee, 2011). A number of interactive factors, among which are drought, fires, habitat fragmentation and encroachment and poaching, are noted to contribute to population decline (Knight, 1995).

Wildlife is generally adapted to climate variability that is characteristic of Botswana's semi-arid environment where annual rainfall ranges from 200 to 600mm (Parida & Moalafhi, 2008). However, increasing evidence points to significant departure from the normal climate variability characteristic of southern Africa (Table 40-2). Droughts experienced since the 1970s, for instance 1983 to 1987 and 1991/92, have been linked to anthropogenic climate change, particularly changes in patterns of El Niño/Southern Oscillation phenomenon and the warming of the Indian Ocean (Desanker, 2001; Boko et al., 2007; Funk et al., 2008).

Prolonged climate-change-induced dry spells in Botswana may reduce moisture and nutrient content of forage. Most wildlife is physiologically adapted to arid conditions, for instance, in times of water scarcity springbok can meet their water needs from forage (Stapelberg et al., 2008), but there are limits to these coping mechanisms. Projections of increased dryness throughout the winter months (Christensen et al., 2007) will limit further availability of green forage in periods of great need (see Table 40-2). In South Africa, nearly all ungulate species in Kruger National Park (KNP) were observed to be extremely sensitive to a lack of rainfall during the dry season (Thuiller et al., 2006). Similarly, Raseroka (1975) linked the disappearance of buffalo in the southern parts of Botswana to the drying and loss of forest vegetation along the Molopo and Limpopo River systems. While mass mortality in Kalahari wildlife in 1985 was linked to the cumulative effects of droughts that persisted since 1977 (Knight 1995) and as noted in Table 40-2, these droughts were partly driven by climate change.

Several studies have pointed to a change in vegetation structure of Botswana and much of southern Africa in which few woody species, such as *acacia tortilis* (mosu) and *melefera* (mongana), grewia bicolor (moretlwa), *dichrostachys cinerea* (moselesele), dominate the rangelands. Previous studies have linked this to degradation of rangelands during the dry periods and change in fire regimes (Dube & Kwerepe, 2000; Reynolds & Stafford-Smith, 2002). But recently these changes have been linked to a gradual shift to a drier and warmer climate in addition to an increase in CO_2 in the atmosphere, which may favor woody plants (Bond, 2003; Dube, 2003).

Such changes have implications for wildlife habitat structure and dietary supply. When combined with projected/likely shortages of water and rising temperatures due to climate change, particularly in the western parts of the region that include Botswana where these conditions could be severely felt in future, this will most likely affect breeding patterns and wildlife species composition in a way that has not been witnessed before

(Hulme, 1996; Thuiller et al., 2006). Rudee (2011) noted a loss of compensatory breeding among springbok in Botswana that currently cannot be accounted for and is suspected to be a factor in the abnormally low rate of natural increase in the Kalahari. Springboks breed seasonally with a short six-month gestation period and can give birth twice a year and this is what is referred to as compensatory breeding, a form of adaptation enabling the springbok to recover quickly in cases of population declines (Skinner & Moss 2004).

Wildlife is widely known to cope with spatial and temporal variability of water and nutrient availability through migration and dispersal (Dube & Moswete, 2003). The seasonal migration of wildebeests in search of surface water and more nutritionally rich forage extended to 200,000km^2 throughout the Kalahari (Williamson et al., 1988; Child et al., 1971). Fragmentation of wildlife habitats and existence of barriers, such as veterinary fences and human settlements, have severely reduced this coping strategy, increasing vulnerability to human pressures and climate change.

Thuiller et al. (2006) noted a potential eastward species shift in line with the projected west–east temperature and precipitation gradients in the region that may result in the northeastern South Africa and adjacent southern Mozambique areas experiencing an increase in wildlife species richness of 50% and 80% for A2 2050 and A2 2080 time slices and climate change scenarios respectively. Western areas, including Botswana, will record losses and changes in species composition as wildlife shift to cooler and wetter areas. For example, Hulme (1996) noted that the Okavango Delta may be less favorable for elephants by 2050 but more attractive to species such as giraffe and warthogs, as the area gets drier. However, wildlife survival options under the projected changes will be obstructed by the inability to migrate first within a country and worse for the case of between countries, due largely to human-made barriers. West–east temperature and precipitation gradients also imply a potential shift in cultivation and general human population to more conducive climate, which will not favor flourishing wildlife, except in cases of direct-management interventions. Thuiller et al.'s (2006) study found that out of the 277 African mammalian species assessed for impacts of climate change, none were committed to extinction where there was unhindered migration, but a maximum of ten species faced extinction by 2080 under the A2 HadCM3 scenario in the case of constrained migration (Thuiller et al., 2006).

Veterinary fences are a common feature in Namibia, South Africa and Zimbabwe, and combined with international border fences have a negative effect on wildlife movements, although efforts are being made to establish trans-boundary conservation areas (for example, the Great Limpopo Transfrontier Park, which shares borders and management with Mozambique, South Africa and Zimbabwe (Peace Parks Foundation, 2003; Ramutsindela, 2007)). In Botswana, veterinary fences date back to before independence and have continued to expand despite the declining role of the livestock sector in the GDP, i.e., from more than 40% in 1968 to less than 2% in 2006 (Hillbom, 2008).

Wildlife in the south, central and western parts of Botswana are cut off from the more wet areas of the north that include the Okavango Delta, Makgadikgadi systems and Chobe-Kasane area by the Kuke Fence erected between 1954 and 1958 (Figure 40-2). The Kuke fence runs from the Namibian border in the west across to the northern boundary of the Central Kgalagadi Game Reserve (CKGR) where it is joined by the Phefodiafoka-Kuke Fence, which then extends to the Dibete/Lephepe/Kaka veterinary fence, which runs to the border with South Africa in the east (Mbaiwa & Mbaiwa, 2006). Wildebeest die-offs in 1964, the 1970s and the 1980s droughts at Lake Xau have been linked to constrained migration to perennial water sources caused by the Kuke Fence (Child, 1972).

Interventions have been made by the government to minimize the effects of drought on wildlife, by providing water pumped from boreholes, for instance; however, most of Botswana's underground water reserves are fossil resources that cannot be sustained for long-term water needs. Campbell and Child (1971) and van Vegten (1981) have raised concerns about the drop in water levels of aquifers in Botswana. Beekman and Xu (2003) have estimated that underground recharge in southern Africa is at 1% or less of the mean annual rainfall, far lower than the current rate of extraction.

The water situation will worsen with rising temperatures due to global warming elevating evapo-transpiration leading to rapid losses; evapo-transpiration rates increase by approximately 3 to 4% for every 1°C rise in temperature (Du Pisani & Partridge, 1990; Schulze et al., 1995). Using the core scenario, Hulme (1996) estimated that potential evapo-transpiration in the Okavango Delta might increase by 15% by 2050 resulting in a decrease in runoff of about 20%. These changes will intensify already experienced water shortages and demand competition with negative consequences on both humans and wildlife (Dube, 2003).

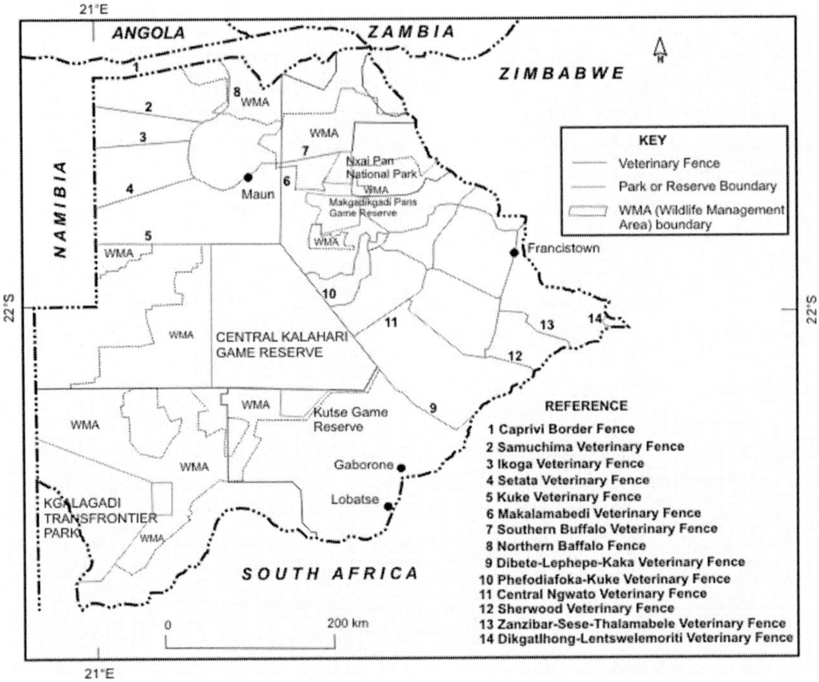

Figure 40-2. Map of Botswana. National Parks and Game Reserves (created by G. P. Koorutwe). Note: WMA – Wildlife Management Areas.

Periods of exceptionally high rainfall in between dry spells due to increased climate variability will provide conditions conducive for build up of biomass and the spread of invasive species. Accumulated biomass combined with hot temperatures, low humidity and windy conditions under climate change will increase fire risk (Dube, 2007). Wild-land fires are common in Botswana, affecting most wildlife areas during dry seasons following wet seasons. The total area affected by fire estimated from satellite data in 2006, 2008 and 2010 was approximately 5.7 million, 11.8 million and 13.6 million hectares respectively (DFRR, 2011). A decline in wildlife, especially a reduction in large herbivore biomass in the Kgalagadi system, has been noted to be a factor contributing to biomass accumulation leading to fires (Perkins et al., 2002).

Reduced wildlife migration and dispersion in dry periods may contribute to the degradation of resources in concentrated wildlife areas, such as the destruction of woodland by the large concentration of elephants in the Chobe district of northern Botswana (Hulme, 1996). Severely degraded areas tend to support fast-growing and more flammable vegetation species when rains return, which fuels wild-land fires, thus reinforcing the negative effects of climate-change-driven changes on wildlife habitats and dietary needs. Nearly 90% of fire ignitions are caused by humans, indicating that there is a chance to reduce burning where concerted efforts are made to engage communities and invest in fire management (Dube & Mafoko, 2009).

In addition to the above, another threat to wildlife linked to climate change is unprecedented disease outbreaks. The dynamics of wildlife diseases are generally not well understood in southern Africa. Disease outbreaks due to changes in temperature, rainfall and humidity will particularly be a threat under restricted migration. Hulme (1996) noted that ticks in the Okavango Delta will multiply by 2050 when temperatures increase by 1.5°C. The tick *Boophilus decoloratus* larvae was observed to peak twice in the same year in southern Kruger National Park (KNP) after an unusually warm winter in 1998, indicating the role of climate in the lifecycles of ticks or other parasites, which has negative implication for wildlife (Thuiller et al., 2006). In Uganda, between 2004–2005, one of the worst anthrax outbreaks occurred at the Queen Elizabeth National Park (QENP) resulting in the loss of 11% of the hippopotamus population and also affecting zebras, buffalos, warthogs and others (Wafula et al., 2007). An anthrax outbreak during an unusually dry season has been linked to declines in the population of buffalo after 1990 at Kruger National Park (Thuiller et al., 2006).

The major threat will occur where wildlife disease spreads to domestic animals or to humans. The need to control foot-and-mouth diseases, which result from contact between cattle and the buffalo, triggered most veterinary fences in Botswana where the beef industry is traditionally highly regarded. This is in contrast to Tanzania where tourism has more weight in the economy hence leading to less investment in cattle protection at the expense of wildlife. However, there are several cases of wildlife diseases that may spread to humans, e.g., anthrax; thus, an increase in such cases is likely to spark conflicts to the detriment of wildlife (where culling and/or barriers are used to reduce contact between the two, for example (Cleaveland et al., 2005)).

Diversifying Tourism in Botswana

The above discussion shows that a tourism industry that is strongly based on climate-sensitive resources (such as wildlife) may not be sustainable in the future due to climate change. There is need to diversify the product away from dependence on nature-based attractions to other, all-inclusive forms of tourism.

The history of tourism in Botswana is short; hence, many types and/or forms of tourism have not yet been explored to diversify the industry. Culture-based attractions offer great potential for diversification given the diverse cultures that co-exist peacefully in the country and the ongoing efforts to revive indigenous-knowledge practices (culture-based tourism relies on numerous attractions such as heritage resources, cultural sites, history, anthropology, cultural landscape, archaeological remains, poetry and music, arts and crafts, and other forms of cultures).

Figure 40-3. Giraffe in a natural setting in a Photographic Concession area (NG 31) in the Okavango Area. Photo credit: Naomi Moswete (September 2011).

The World Tourism Organization (1985, cited in Mckercher & Du Cros, 2002: 3) describes cultural-heritage tourism as a form of tourism that deals with the movement of tourists and/or persons toward essentially cultural motivations, such as performing arts, attending cultural festivals and events, visits to sites and monuments, travel to memorial sites, archaeological sites and battle fields, folklore, art or pilgrimages (see Mckercher & Du Cros, 2002; Timothy & Boyd, 2006). Cultural heritage tourism is one of the most notable and widespread types of tourism (Akama & Sterry, 2002; Timothy & Boyd, 2006). Culture-based tourism is the fastest-growing sub-sector of the tourism industry because many people are becoming more interested in experiencing and appreciating different cultures of nations globally (Akama & Sterry, 2002; Nicholas, Thapa & Ko, 2009; Timothy & Boyd, 2006).

Cultural-heritage tourism is widely seen as one of the mainstays of rural- or countryside-tourism worldwide (Frost, 2003; Robinson, 2001). Research has revealed that culture-based tourism can have positive impacts on the environment, social/cultural institutions and economy of host destinations (Gratton & Taylor, 1995; Nicholas, Thapa & Ko, 2009; Smith, 2003; Terry, 1994). It can promote indigenous resource conservation (Timothy & Boyd, 2006), and contribute to rural economic development through production and sale of local and/or indigenous goods (Frost, 2003; Moswete, Thapa & Lacey, 2009; Tohmo, 2005). Above all, culture-based tourism has the potential to revive the world's dying cultures, via music and dance, initiation schools, traditional healing rituals, rain-making practices, indigenous medicine and handicrafts (Moswete, Thapa & Lacey, 2009; Nicholas, Thapa & Ko, 2009; Smith, 2003; Tohmo, 2005).

The Botswana tourism industry has had little or no emphasis on culture-based tourism in the past. Although culture and heritage have been mentioned as attractions in some of the literature used to market the country, cultural heritage tourism has never been strongly marketed as one of the key tourist resources internationally. For that reason, many heritage sites, including unique monuments, historical sites, stonewall villages, traditions and customs that could be of interest to tourists (GOB, 2001; 2003), are neglected and/or poorly developed for tourism (BTDP, 1999; Mbaiwa, 2004). Yet the potential of this form of tourism has been demonstrated; for instance, a number of heritage sites have been exploited for culture-based tourism before. These include but are not limited to the renowned world-heritage sites: Tsodilo hills and Lekhubu Island (cultural landscape and history) in Ngamiland district; Manyana rock paintings and David Livingstone Memorial site in the Kweneng district; Lepokole hills near the copper and nickel mining townships of Selibe Phikwe and Fort Motloutse; and Solomon's Wall in the east. All these have visitor interpretation centers, trails, signage and site personnel (GOB, 2001).

However, there are several other heritage sites that are locally unknown, and inadequately documented or exploited for tourism by institutions responsible for heritage management and tourism development – the unique Qcwihaba "historic" caves in the Kalahari sand veldt and Old Palapye ruins in the central district, for example. It is important that cultural-heritage resources, both tangible and intangible, are exploited for tourism to diversify away from over-dependence on wildlife-based tourism products that are sensitive to human and climate-change pressures.

The way of life of the people, daily village activities, traditional architecture (huts, kraals), dress, language, local cuisine and preparation, marriage and death ritual performances, arts and crafts, music and dance in Botswana should be packaged and marketed for tourism. Events and festivals (including indigenous and modern dance and music and cuisine) could be organized at different locations to showcase various cultural aspects of Botswana, like the wine-tasting tourism in South Africa (e.g., Cape Town). Some of the special-event tourism, as with festivals in Edinburgh (Scotland), attracts millions of domestic and overseas tourists (Gratton & Taylor, 1995; Smith, 2003).

Currently, parts of the country with large wildlife diversity, such as the Chobe and Okavango Delta regions, receive the majority of overseas, regional and domestic tourists. Thus, alternative tourism packages that include cultural heritage activities could be initiated and commenced in districts that are not endowed with wildlife resources (e.g., villages and towns in southern Botswana). The advantage of culture-based tourism is the potential to create employment opportunities for rural areas (Moswete, Toteng & Mbaiwa, 2009).

Conclusion

Tourism offers great potential for diversifying Botswana's economy away from mining. However, so far, the emphasis is on wildlife-resource-based tourism. All efforts are required to conserve these resources; in most parts of the world similar resources went extinct. There are indications that wildlife resources are declining due to human activities in southern Africa. In Botswana too, emphasis on the cattle industry has led to fragmentation of wildlife habitats and barriers to their movement in search of water and forage in response to changes in supply. This fact, combined with the indicated increased climate variability due to climate change, puts a dark shade on the sustainability of a wildlife-based tourism product, calling for the need to consider adaptation measures in this sector, and to seek for other alternative tourism attractions and/or products.

Culture-based tourism has the potential to diversify tourism and address the acute problem of unemployment in rural areas of Botswana. However, to fully realize this potential, revenues from minerals need to be invested in developing a well-rounded (all-year-round) tourism packages for the domestic, regional and the international market, and these packages should include activities built around nature (i.e., wildlife), culture and heritage resources of Botswana.

There are indications that the country's culture is gradually being revived, but this needs to be harnessed and carefully promoted for tourism purposes. The case for cultural-heritage tourism and its value to Botswana should be made and shared among tourism planners and decision makers with clear links to its socio-economic, environmental and educational benefits.

However, tourism and climate are intertwined (see Gossling & Hall, 2006). Neither culture-based nor nature/wildlife-based tourism products are totally immune to climate change. Exceptionally hot temperatures will drive tourists away (to alternative destinations) as is the case where there is acute shortages of water or outbreak of diseases due to climate extremes. Protection and maintenance of the clean environment, both locally and regionally, is key to a successful tourism industry.

References

Akama, J., & P. Sterry (2002). Cultural tourism in Africa: Strategies for the new millennium. Proceedings of the Atlas Africa International Conference, held in Kenya, December 2000, Kenya. Association of Tourism and Leisure Education.

Alley, R., T. Berntsen, N. L. Bindoff, Z. Chen, A. Chidthaisong, P. Friedlingstein, J. Gregory, G. Hegerl, M. Heimann, B. Hewitson, B. Hoskins, F. Joos, J. Jouzel, V. Kattsov, U. Lohmann, M. Manning, T. Matsuno, M. Molina, N. Nicholls, J. Overpeck, D. Qin, G. Raga, V. Ramaswamy, J. Ren, M. Rusticucci, S. Solomon, R. Somerville, T. F. Stocker, P. Stott, R. J. Stouffer, P. Whetton, R. A. Wood & D. Wratt (2007). *Climate Change 2007. The Physical Science Basis, Summary for Policymakers*. Intergovernmental Panel on Climate Change. Geneva: UNEP, WMO.

Arntzen, J. W. (2003). An economic view on wildlife management areas in Botswana. Occasional paper No. 10. IUCN/SNV CBNRM support program, Gaborone.

Barnes, J. I. (2001). Economic returns and allocation of resources in wildlife sector of Botswana. *South African Journal of Wildlife Research* 31, nos. 2 and 4: 141–153.

Becken, S., & J. E. Hay (2007). *Tourism and Climate Change: Risks and Opportunities*. Clevedon: Channel-view.

Beekman, H. E., & Y. Xu (2003). Review of groundwater recharge estimation in arid and semi-arid Southern Africa. A report to START within Phase I of the Framework for Recharge Estimation in Southern Africa project, www.start.org/projectpages/africagroundwater/frameworkrecharge.html.

Boko, M., I. Niang, C. Vogel, A. Githeko, M. Medany, B. Osman-Elasha, R. Tabo & P. Yanda (2007). Africa. In *Climate Change 2007: Impacts,*

Adaptation and Vulnerability: Contribution of Working Group II to the Fourth Assessment Report of the Intergovernmental Panel on Climate Change, ed. M. L. Parry, O. F. Canziani, J. Palutikof, P. J. van der Linden & C. E. Hanson, pp. 433–467. Cambridge: Cambridge University Press.

Bond, W. J., G. F. Midgley & F. I. Woodward (2003). The importance of low atmospheric CO_2 and fire in promoting the spread of grasslands and savannas. *Global Change Biology* 9: 973–982.

Bonyongo, M. C. (2004). The Ecology of Large Herbivores in the Okavango Delta, Botswana. PhD diss., Bristol University, Bristol, UK.

Botswana Review (2005). Botswana Review: Commerce and Industry, 25th ed. Botswana Export Development and Investment Authority (BEDIA). Gaborone, Botswana.

BTDP [Botswana Tourism Development Program] (2000). Botswana tourism master plan: Final draft, Ministry of Commerce and Industry, Department of Tourism. Gaborone.

CSO [Central Statistics Office] (1998). Tourism Statistics. Gaborone, Botswana.

—. (2003). Tourism Statistics. Gaborone, Botswana.

Campbell, A. C., & G. Child (1971). The impact of man on the environment in Botswana. *Botswana Notes and Records* 3: 91–111.

Child, G., R. Parris & E. Le Riche (1971). Use of mineralized water by Kalahari wildlife and its effects on habitats. *East African Wildlife Journal* 9, no. 1: 125–142.

Child, G. (1972). Observations on wildebeest die-off in Botswana. *Arnoldia* 5, no. 31: 1–13.

Christensen, J. H., B. Hewitson, A. Busuioc, X. Chen, I. Gao, R. Held, R. K. Jones, W. T. Kolli, R. Kwon, V. Laprise, L. Magaña Rueda, C. G. Mearns, J. Menéndez, A. Räisänen, A. Rinke, A. Sarr & P. Whetton (2007). Regional Climate Projections. In *Climate Change 2007: The Physical Science Basis. Contribution of Working Group I to the Fourth Assessment Report of the Intergovernmental Panel on Climate Change*, ed. S. Solomon, D. Qin, M. Manning, Z. Chen, M. Marquis, K. B. Averyt, M. Tignor & H. L. Miller. Cambridge: Cambridge University Press.

Cleaveland, S., K. Laurenson & T. Mlengeya (2005). Impacts of wildlife infections on human and livestock health with special reference to Tanzania: Implications for Protected Area Management. In *Conservation and Development Interventions at the Wildlife/Livestock Interphase: Implications for Wildlife, Livestock and Human Health*, ed.

S. Osofsky. Occasional Papers of the IUCN Species Survival Commission No. 30, pp. 147–152.

DOT [Department of Tourism] (2006). Tourism Statistics 2001–2005. Gaborone, Botswana.

Desanker, P., C. Magadza, A. Allali, C. Basalirwa, M. Boko, D. Dieudonne, T. E. Downing, O. P. Dube, A. Githeko, M. Githendu, P. Gonzalez, D. Gwary, B. Jallow, J. Nwafor & R. Scholes (2001). Africa. In *Climate Change 2001: Impacts, Adaptation, and Vulnerability: IPCC 2nd Assessment Report*, ed. J. J. MacCarthy, O. F. Canziani, N. A. Leary, D. J. Dokken & K. S. White, pp. 489–531. Cambridge: Cambridge University Press.

DFRR (2011). National wildland fire management strategy. Draft. Ministry of Environment, Wildlife and Tourism Department of Forestry and Range Resources. Gaborone.

Du Pisani, A. L., & T. C. Partridge (1990). Effects of global warming on crop production in South Africa. *South Africa Journal of Science* 86: 306–311.

Dube, O. P. (2003). Impacts of climate change, vulnerability and adaptation options: Exploring the case for Botswana through Southern Africa; A review. *Botswana Notes and Records* 35: 147–168.

—. (2007). Fire, weather and land degradation. In *Climate and Land Degradation*, ed. M. V. K. Sivakumar & Ndiang'ui, pp. 223–251. Berlin: Springer.

Dube, O. P., & R. M. Kwerepe (2000). Human-induced change in the Kgalagadi sands: Beyond the year 2000. In *Towards Sustainable Management in the Kalahari Region: Some Essential Background and Critical Issues*, ed. S. Ringrose & R. Chanda, pp. 244–258. Directorate of Research and Development, University of Botswana, Gaborone.

Dube, O. P., & Mafoko, J. G. (2009). Botswana. In *Africa Environmental Outlook Case Studies: Impacts of Fire on the Environment*. UNEP, DEWA, Nairobi.

Dube, O., & N. Moswete (2003). Tourism: Searching for adaptation options to climate change in Southern Africa. *AIACC Notes* 2, no. 1: 6–7.

Guardian-UK (2011). Drought and poachers take Botswana's natural wonder to brink of catastrophe. *The Guardian* (UK), 18 June 2011, http://www.guardian.co.uk/environment/2011/jun/18/botswana-natural-wonder-brink-of-catastrophe.

Frost, W. (2003). The financial viability of heritage tourism attractions: Three cases from rural Australia. *Tourism Review International* 7, no. 1: 13–22.

Funk, C., M. D. Dettinger, J. C. Michaelsen, J. P. Verdin, M. E. Brown, M. Barlow & A. Hoell (2008). Warming of the Indian Ocean threatens eastern and southern African food security but could be mitigated by agricultural development. *PNAS* [The National Academy of Sciences of the USA] 105, no. 32: 11081–11086.

GOB [Government of Botswana] (2001). Botswana National Atlas. Department of Surveys and Mapping, Botswana, Government Printers.

—. (2003). National Development Plan 9 2003/04–2008/09. Ministry of Finance and Development Planning. Government Printers, Gaborone.

Gossling, S., & M. Hall (2006). Uncertainties in predicting tourist flows under scenarios of climate change. *Climatic Change* 79: 163–173.

Gratton, C., & P. D. Taylor (1995). Impact of festival events: A case-study of Edinburgh. In *Tourism and Spatial Transformations: Implications for Policy and Planning*, ed. G. J. Ashworth & A. G. J. Dietvorst, pp. 225–238. CAB International.

Gujadhur, T. (2001). Joint venture options for communities and safari operators in Botswana. CBNRM support programmed occasional paper No.6, IUCN/SVN. Botswana.

Hulme, M., ed. (1996). Climate change and Southern Africa: An exploration of some potential impact and implications for the SADC region; A Report Commissioned by WWF International and Co-ordinated by Climate Research Unit. UEA, Norwich, UK.

Hulme, M., R. Doherty, T. Ngara, M. New & D. Lister (2001). African climate change: 1900–2100. *Climate Research* 17: 145–168.

Hillbom, E. (2008). Diamonds or development? A structural assessment of Botswana's forty years of success. *Journal of Modern African Studies* 46, no. 2 (2008): 191–214.

Jefferies, K. (2009). The role of TNCs in the extractive industry of Botswana: United Nations Conference on Trade and Development Division on Investment and Enterprise. UN New York and Geneva. *Transnational Corporations* 18, no. 1: 61–92.

Knight, M. H. (1995). Drought-related mortality of wildlife in the southern Kalahari and the role of man. *African Journal of Ecology* 33, no. 4: 377–394.

Knowles, T., & L. Theron (2010). A literature review and overview report pertaining to climate change adaptation in Botswana. A report submitted to Conservation International Southern Africa Wilderness Program. The Cirrus Group, June 2010, Botswana.

Magole, L. I., & O. Gojamang (2005). The dynamics of tourist visitation to national parks and game reserves in Botswana. *Botswana Notes and Records* 37: 80–96.

Mbaiwa, J. E. (2003). The socio-economic benefits and challenges of community-based safari hunting tourism in the Okavango Delta, Botswana. *The Journal of Tourism Studies* 15, no. 2: 37–50.

—. (2008). Tourism Development, Rural Livelihoods, and Conservation in the Okavango Delta, Botswana. PhD diss., Texas A&M University, USA.

Mbaiwa, J. E., & O. I. Mbaiwa (2006). The effects of veterinary fences on wildlife populations in Okavango Delta, Botswana. *International Journal of Wilderness* 12, no. 3: 17–41.

Mbaiwa, J. E., & Mmopelwa, G. (2009). Assessing the impact of climate on tourism activities and their economic benefits in the Okavango Delta. Okavango River Basin Trans-Boundary Diagnostic Analysis Technical report, Botswana Component: Climate Change and Tourism Development: iwlearn.net/iw-projects/842/reports/environmental-flow-assessment-reports/.

Mckercher, B., & H. Du Cros (2002). *Cultural Tourism: The Partnership between Tourism and Cultural Heritage Management*. New York: Haworth Hospitality Press.

Ministry of Finance and Development Planning (MFDP) (2009). National Development Plan 10. April 2009–March 2016, Vol. 1. Gaborone, Government Printing and Publishing Services.

Ministry of Works, Transport and Communications (2001). Botswana Initial National Communication to the United Nations Framework Convention on Climate Change. Gaborone: Bay Publishing Pty Ltd.

Moswete, N., B. Thapa & G. Lacey (2009). Village-based tourism and community participation: A case study of the Matsheng Villages in southwest Botswana. In *Sustainable Tourism in Southern Africa: Local Communities and Natural Resources in Transition*, ed. J. Saarinen, F. Becker, H. Manwa & D. Wilson, pp. 189–209. Clevedon, UK: Channelview.

Moswete, N., E. N. Toteng & J. Mbaiwa (2009). Resident involvement and participation in urban tourism development: A comparative study in Maun and Gaborone, Botswana. *Urban Forum* 19, no. 4: 381–394.

Nicholas, L., B. Thapa & Y. J. Ko (2009). Residents' perspectives of a World Heritage Site: The Pitons management area, St Lucia. *Annals of Tourism Research* 35, no. 3: 390–412.

Parida, B. P., & D. B. Moalafhi (2008). Regional rainfall frequency analysis for Botswana using L-Moments and radial basis function network. *Physics and Chemistry of The Earth Parts A B C* 33, nos. 8–13: 614–620.

Perkins, J. S., & S. M. Ringrose (1996). Development Cooperation's Objectives and the Beef Protocol: The Case of Botswana, a Study of Livestock/Wildlife/Tourism/Degradation Linkages. Study carried out for Metroeconomica Ltd., contract B-7-504094. University of Botswana, Gaborone.

Perkins. J. S., G. Stuart-Hill & B. Kgabung (2002). The impact of cattle-keeping on the wildlife, vegetation, and veldt products. In *Sustainable Livelihoods in the Kalahari Environments: A Contribution to Global Debates*, ed. D. Sporton & D. S. G. Thomas. Oxford: Oxford University Press.

PPF [Peace Park Foundation] (2003). Profile of the Peace Parks Foundation, http://www.peaceparks.org.

Ramberg, L., P. Hancock, M. Lindholm, T. Meyer, S. Ringrose, J. Sliva, J. Van As & C. VanderPost (2006). Species diversity of the Okavango Delta, Botswana. *Aquatic Sciences* 68: 310–337.

Ramutsindela, M. (2007). *Transfrontier Conservation in Africa: At the Confluence of Capital, Politics, and Nature*. Cabi International, UK.

Raseroka, B. H. (1975). Past and present distribution of Buffalo in Botswana. *Botswana Notes and Records* 7: 131–140.

Reynolds, J. F., & D. M. Stafford-Smith (2002). *Global Desertification: Do Humans Cause Deserts?* Berlin: Dahlem University Press.

Robinson, J. (2001). Tourism Development Framework (TDF) Draft. World Tourism Organization, Technical Mission to Botswana. Gaborone, Botswana.

Rudee, A. (2011). Comparative ecology and population dynamics of ungulates in Botswana. Bachelor of Arts Degree in Environmental Analysis 2010/11, Pomona College, Claremont, California.

SADC (2008). Southern Africa Sub-Regional Environmental Action Plan (SREAP) of the Environment Action Plan of NEPAD. The New Partnership for Africa's Development (NEPAD). Final Report, SADC/ESD/1/2008/2 Doc. Gaborone.

Scholes, R. J., & R. Biggs, eds. (2004). Ecosystem Services in Southern Africa: A Regional Assessment. Council for Scientific and Industrial Research, Pretoria.

Schulze, R. E., G. A. Kiker & R. P. Kunz, (1995). Global climate change and agricultural productivity in Southern Africa: Thought for food and food for thought. In *Climate Change and World Food Security*, ed. T. E. Downing, pp. 421–447. Berlin: Springer.

Skinner, J. D., & D. G. Moss (2004). Kgalagadi springbok (*Antidorcas marsupialis*): Bucking the trend. *Transactions of the Royal Society of South Africa* 59, no. 2: 119–121.

Smith, M. A. (2003). *Issues in Cultural Tourism Studies*. London: Routledge Taylor and Francis.

Stapelberg, H., M. W. Van Rooyen, J. P. Bothma, M. J. Van der Linde & H. T. Groeneveld (2008). Springbok behavior as affected by environmental conditions in the Kalahari. *Koedoe* 50, no. 1: 145–153.

Terry, M. E. (1994). The women handicraft industry: Moving from the 20th to the 21st century. In *Botswana in the 21st Century*, ed. J. Hemos & K. Ntetas, pp. 571–583. Gaborone: Botswana Society.

Thakadu, O. T., K. T. Mangadi, F. E. Bernard & J. E. Mbaiwa (2006). The economic contribution of safari hunting to rural livelihood in the Okavango: The case of Sankuyo village. *Botswana Notes and Records* 37: 22–39.

Thapa, B. (2010). Funding strategies for World Heritage Sites in Least Developed Countries. In *Heritage Resource Management, Policy, and Issues in Global Perspective*, ed. P. Messenger & G. Smith. Gainesville, FL: University of Florida Press.

Timothy, D. J., & S. W. Boyd (2006). Heritage tourism in the 21st century: Valued traditions and new perspectives. *Journal of Heritage Tourism* 1, no. 2: 1–17.

Thuiller, W., O. Broennimann, G. Hughes, J. Robert, M. Alkemade, G. F. Midgley & F. Corsi (2006). Vulnerability of African mammals to anthropogenic climate change under conservative land transformation assumptions. *Global Change Biology* 12: 424–440.

Tohmo, T. (2005). Economic impacts of cultural events on local economies: An input-output analysis of the Kaustinen Folk Music Festival. *Tourism Economics* 11, no. 3: 431–451.

Wafula, M. M., A. Patrick & T. Charles (2007). Managing the 2004/05 anthrax outbreak in Queen Elizabeth and Lake Mburo National Parks, Uganda. *African Journal of Ecology* 46: 24–31.

Williamson, D., J. Williamson & K. T. Ngwamotsoko (1988). Wildebeest migration in the Kalahari. *African Journal of Ecology* 26, no. 4: 269–280.

WTO [World Tourism Organization] (2002). Tourism and Poverty Alleviation. Madrid: WTO.

WTTC [World Travel & Tourism Council] (2007). Botswana: The Impact of Travel and Tourism on Jobs and the Economy. London: WTTC.

UNWTO (2011). Background paper: From Davos to Copenhagen and beyond: Advancing tourism's response to climate change, www.unwto.org/pdf.

Van Vegten, J. A. (1981). Man-made vegetation changes: An example from Botswana's savanna. National Institute of Research. University of Botswana, Gaborone.

CHAPTER FORTY-ONE

PERCEIVED EFFECTS OF CLIMATE CHANGE ON THE TOURISM BUSINESS IN THE OKAVANGO DELTA, BOTSWANA

JOSEPH E. MBAIWA
AND GAGOITSEOPE MMOPELWA[1]

Abstract

Climate change is bound to have an impact on the tourism business in the Okavango Delta, Botswana. The objective of this survey was to determine the perceptions of lodge managers/tour operators on how their tourism operations and output (turnover) would be affected by changes in flooding patterns in the Okavango Delta, Botswana. The study was carried out in the Okavango Delta located in northwestern Botswana. A sample of forty-eight tour operators based in Maun or lodge managers based in parts of the delta were surveyed using a short face-to-face interview to determine managers' perceptions regarding the impacts on their own business in the face of different flooding scenarios in the delta. The study found that changes in flood patterns in the Okavango Delta were perceived to have positive and negative impacts on turnover in the various tourism businesses. Mainly managers whose businesses are located relatively far away from flowing channels reported positive impacts of high flooding.

Introduction

Tourism is considered the world largest industry, contributing to significant growth of many countries' economies (Roe et al., 2004; Viner and Agnew, 1999). In most developing countries, tourism is increasingly becoming a significant contributor to gross domestic product (Roe and Khanya, 2001). In Botswana, for instance, tourism proved to be a potential

[1] Okavango Research Institute, University of Botswana.

engine for economic growth (Government of Botswana, 2000). It is a source of revenue, creates jobs, stimulates investment and powers sustainable growth around the world (Roe and Khanya, 2001). At the community level, tourism has contributed to the improvement of welfare of those communities that are situated in the proximity of natural resources such as wildlife, and hence its great potential for reducing rural poverty (Ijeomah and Aiyeloja, 2009; Roe and Khanya, 2001).

Worldwide climate change is impacting negatively and positively on tourism (Hein et al., 2009; Hamilton et al., 2005; Wall, 1998; Bigano et al., 2007; UNWTO, 2008). The direct negative impacts include, for example, shortening of tourism-dependent seasons and rising sea levels that can cause flooding, while indirect impacts include changes in water availability, biodiversity loss and reduced landscape aesthetic (UNWTO, 2008; Wall, 1998; UNWTO, 2008; Agnew and Palutikof 2001; Braun et al., 2007). The positive impacts include "push" and "pull" factors (Amelung et al., 2007). Push factors are mainly negative factors prevailing in tourists' places of origin that may induce them to leave their origin, while pull factors are those factors that exist in tourist destinations (mainly attractions) that may induce tourists to visit those destinations (Amelung et al., 2007). Climate change is among the various factors affecting tourism, especially nature-based tourism. The climate-change-related factors affecting tourism include temperature, precipitation, clouds, fog, wind and humidity (Nyaupane and Chhetri, 2009).

In Botswana, the impacts of these factors, especially flooding, have not been investigated. An investigation of perceptions on the impacts of flooding on the tourism turnover would give some insights into the changes in the contribution of tourism to Botswana's national GDP. The goal of this study therefore was to assess perceptions of tourism operators regarding the effects of climate change on output (turnover) of their own business in the face of changes of flooding in the Okavango Delta, Botswana. Specific objectives of this survey were as follows:

(a) To determine perceptions of lodge managers/tour operators on how their tourism operations and outputs would be affected by changes in flooding in the Okavango Delta, Botswana;
(b) To determine the most favorable flooding scenario for the lodge managers/tour operators using overall deviation averages for increase and decrease of turnover for each scenario overall; and
(c) To determine how lodge managers/tour operators are likely to adapt to the changing flooding patterns as a result of climate change.

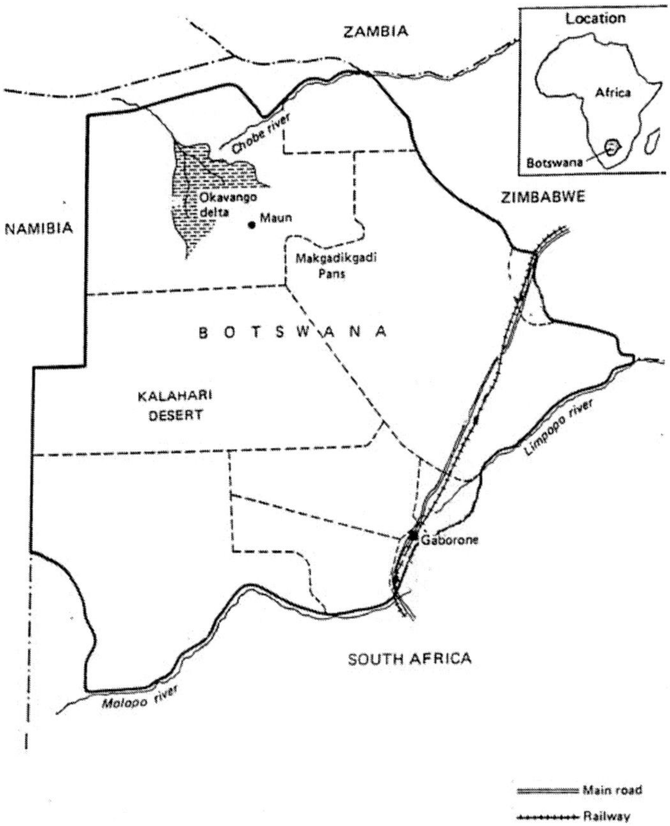

Figure 41-1. Map of Botswana showing the Okavango Delta.

Study Area

This study was carried out in the Okavango Delta, located in northwestern Botswana (Figure 41-1). The Okavango Delta is formed by the inflow of the Okavango River that originates from the Angolan Highlands. The river flows through Namibia and drains into northwestern Botswana to form the huge wetland known as the Okavango Delta. The Okavango Delta is characterized by large amounts of open water and grasslands that sustain human life, plant life, wildlife, birds, insects and various other organisms.

The Okavango Delta also sustains the livelihoods of local communities. Livelihood activities include collection of various plant products for different uses, fishing and flood-recession farming (Kgathi et al., 2004). Due to its rich wildlife diversity, permanent water resources, rich grasslands and forests and scenic landscapes, the Okavango Delta has become one of the key international tourism destinations in Botswana. Tourists come from the all over the world but mainly from South Africa and other African countries, North America, Australasia (Australia and New Zealand, the Caribbean and Europe (Mmopelwa et al., 2007)).

Methods

This study sampled forty-eight lodge managers/tour operators operating in the Okavango Delta. Lodge managers/tour operators were surveyed using a short face-to-face interview to determine their perceptions regarding the effects of the output (turnover) of their own business in the face of changes in flooding patterns in the Delta. We interviewed lodge managers and tour operators because they are the main information channels through which customer preferences can be determined and they know the levels and trends of supply and demand for their products (Budeanu, 2005).

Respondents were asked to assume (that all other things being equal) changes in flooding were as a result of climate change or variability. We asked them to assume this because there are several other factors involved, such as the current economic recession, political unrest or terrorism in other countries, prices of park fees, etc. Systematic sampling was used. An attempt was made to ensure that the sample covered lodges/camps that offer mostly water-based experiences; those that offer mostly land-based activities; and those that offer more or less of both. Each respondent was asked questions relating to possible flooding patterns and expected changes in business or economic output in their resort.

The respondents were asked to assume that the current annual turnover at that site is 100% and to suggest how turnover might change with each of the following long-term scenarios reflecting changes to the flooding patterns in their area, e.g., would it increase from 100% by 25% to 125%? Or decease by 40% to 60%? Their estimates of likely change in annual turnover output were made for each of the following scenarios shown in Figure 41-2.

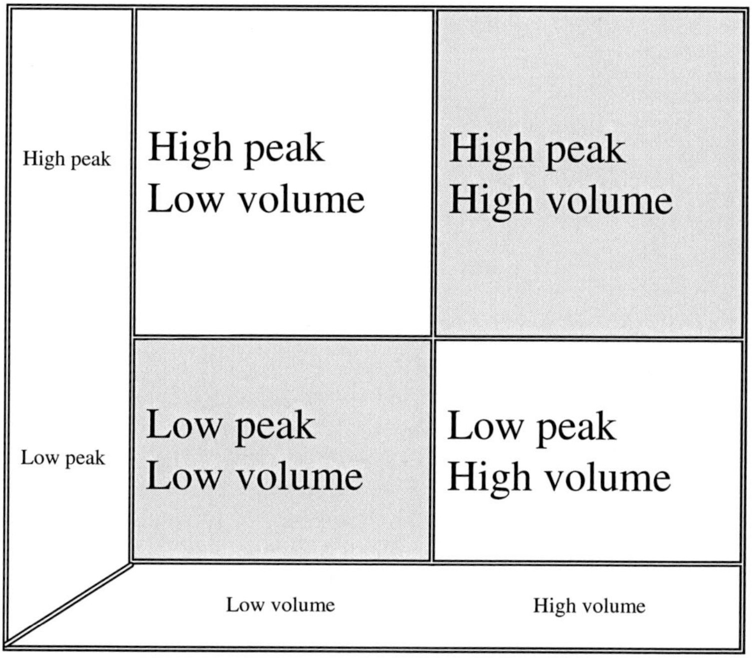

Figure 41-2. Characterization of flood categories according to flood height and flood volume.

Results

High-Level/High-Volume Flooding

Respondents were asked to state how their business turnover would be affected by flooding in the event that the frequently flooded parts of their area where their lodge happen to be located became permanently flooded; the occasionally flooded parts became frequently flooded; and the rarely flooded parts became occasionally flooded. Results on how turnover would be affected by high floods are shown in Table 41-1 below.

Table 41-1. Effect of high-level/high-volume flooding on location of business.

Location of business	Description of turnover			Total
	Increase	*Decrease*	*No change*	
Panhandle	2 (4)	3 (6)	0	5 (10)
Middle of Delta	2 (4)	14 (29)	8 (17)	24 (50)
Lower area of Delta	4 (8)	10 (21)	5 (10)	19 (40)
Total	8 (17)	27 (56)	13 (27)	48 (100)

Table 41-1 shows that 56% of the respondents noted that their turnover would decrease in the event of flooding levels going high. The majority (29%) of these had business located in the middle of the Delta. Asked about the extent at which turnover would decrease, 29.2% of the respondents indicated it would decrease by 25%, 21% said it would decrease by 50%, while 4% said it would decrease by 100%. Respondents mentioned that as a result of high flooding, roads would be damaged making access to tourism facilities more difficult. As a result of damaged roads, the number of tourist visitors would be significantly reduced, leading to low turnover. Secondly, respondents mentioned that high flooding would make game viewing and bird watching difficult because of a lack of suitable points (due to damaged roads) for viewing wildlife. Thirdly, respondents mentioned that as all areas would be flooded, there would be no water concentration points for animals or birds (as is the case when there is no flooding). Tourists usually target these water concentration points for viewing wildlife. Lastly, respondents mentioned that during high floods, tourists often cancel their bookings because of the fear of increased incidence of diseases such as malaria and the risk associated with being killed in the floods.

On a similar note, tourism facilities like hotels in Maun and those situated far away from channels seemed to have a lower probability of being affected by high-flooding levels in the Okavango Delta. In fact, 45.8% of the respondents noted that turnover in their establishments would not be affected.

The effect of flooding on specialization of business is presented in Table 41-2. According to Table 41-2, the majority (23%) of respondents who indicated that there would be a reduction in tourism turnover were those with both water and land activities, followed by those who were with land-based tourism activities (21%). The main reason these respondents gave for the reduction in turnover was that when their areas are flooded, tourists cancel their bookings.

Table 41-2. Effect of high-level/high-volume flooding on specialization of business.

Specialization of business	Description of turnover			Total
	Increase	*Decrease*	*No change*	
Water-based tourism activities	2 (4)	3 (6)	2 (4)	7 (15)
Land-based tourism activities	2 (4)	10 (21)	6 (13)	13 (27)
Both water- and land- tourism activities	4 (8)	11(23)	4 (8)	19 (40)
Accommodation to everyone	0	3 (6)	1 (2)	4 (8)
Total	8 (17)	27 (56)	13 (27)	48 (100)

Low-Level/Low-Volume Flooding (main channels and lagoons remain permanently flooded)

Respondents were also asked to project their turnover changes, if any, in the event that flooding levels were low (low-level/low-volume flooding). They were specifically asked to indicate the extent to which their business turnover would change in the event that the permanently flooded parts of their area became only frequently flooded; the frequently flooded parts of their area became only occasionally flooded; and the occasionally flooded parts of their area became only rarely flooded and main channels and lagoons remained permanent.

Results on the extent at which turnover would be affected by low-flooding patterns in the Okavango Delta are shown in Table 41-3. Results in Table 41-4 show that the majority (42%) of respondents noted that the turnover would increase. Understandably, when the main channels and lagoons (which are located in the middle of the Delta) remain permanently flooded, the likely impact would be that the number of tourists visiting businesses in these areas would increase while there would be fewer tourists visiting drier sites. Of those who noted that their turnover would increase, 31% said it would increase by 25%; 4% said it would increase by 50%; while 6.2% said it would increase 100%.

Nineteen percent of the respondents said the turnover would decrease. The majority of these were from the lower area of the Delta (with no main channels and lagoons). About 40% of respondents indicated that there would be no change in turnover. Most of these respondents had businesses located in the middle of the Delta and the reasons given for the increase in turnover might also apply here.

Table 41-3. Effect of low-level/low-volume flooding (main channels and lagoons remain permanently flooded) on location of business.

Location of business	Description of turnover			Total
	Increase	*Decrease*	*No change*	
Panhandle	1 (2)	2 (4)	2 (4)	5 (10)
Middle of Delta	13 (27)	1 (2)	10 (21)	24 (50)
Lower area of Delta	6 (13)	6 (13)	7 (15)	19 (40)
Total	20 (42)	9 (19)	19 (40)	48 (100)

The effect of low-level/low-volume flooding on specialization of business is shown in Table 41-4. The positive impact of low-level/low-volume flooding (increased turnover) would be felt by those with both water and land activities because main channels would provide increased boating experience as well as a concentration water points for better wildlife viewing.

About 19% with both water- and land-based tourism activities did not think that that their business would be affected. This was because the majority of these facilities were located in seasonal flooding areas. As such, there is likelihood of less flooding in these areas when compared to permanent flooded areas.

Table 41-4. Effect of low-level/low-volume flooding (main channels and lagoons remain permanently flooded) of specialization of business.

Specialization of business	Description of turnover			Total
	Increase	*Decrease*	*No change*	
Water-based tourism activities	2 (4)	3 (6)	2 (4)	7 (15)
Land-based tourism activities	7 (15)	3 (6)	9 (19)	18 (38)
Both water- and land- tourism activities	10 (21)	2 (4)	6 (13)	19 (40)
Accommodation to everyone	1 (2)	1 (2)	2 (4)	4 (8)
Total	20 (42)	9 (19)	19 (40)	48 (100)

Low-Level/Low-Volume Flooding (only deeper channels and lagoons remain flooded)

Under this scenario, respondents were asked to assume that flooding patterns in the Delta became low such that only deeper main channels and lagoons remain permanently flooded while the rest of the areas became seasonally flooded. Specifically, respondents were asked to indicate the extent of the change in business turnover under the low level of flooding where the permanently flooded parts of their area became only occasionally flooded; the frequently flooded parts of their area became only rarely flooded; and the rarely flooded parts of their area became never flooded. In this scenario, only the deeper main channels and lagoons remained permanent while the rest became seasonal only.

As shown in Table 41-5, an equal number of respondents (29.2%) said that the turnover would increase and decrease, respectively. For respondents who noted that their turnover increased, 25% noted that their turnover would go up by 25%, while only 4.2% said it would increase by 100%. A majority (23%) of those who indicated an increased turnover had businesses located in the middle of the Delta, while the majority (17%) of those who indicated a decrease had businesses in the lower area of the Delta. Again, as the water level decreased, it would likely become easier to view wildlife in concentration points, although 18% and 19% of respondents with businesses in the middle Delta and lower areas of the Delta, respectively, indicated that there would be no change in tourism business.

Table 41-5. Effect of low-level/low-volume flooding (only deeper channels and lagoons remain flooded) on location of business.

Location of business	Description of turnover			Total
	Increase	*Decrease*	*No change*	
Panhandle	1 (2)	1 (2)	3 (6)	5 (10)
Middle of Delta	11 (23)	5 (10)	8 (18)	24 (50)
Lower area of Delta	2 (4)	8 (17)	9 (19)	19 (40)
Total	14 (29)	14 (29)	20 (42)	48 (100)

Low-level/low-volume flooding would also have positive and negative impacts in terms of specialization of tourism businesses. The majority of those who indicated an increase in turnover had both water- and land-based tourism activities, while those who indicated a decrease had water-based tourism activities only (6%), and land-based tourism activities only 6% (Table 41-6). A higher percentage of respondents indicated that there would be no change in turnover.

According to respondents, the reduction of floods would lead to a reduction in water-based tourism activities because without water some channels would not be accessible; it would be hard to find game or wildlife for viewing since it would migrate to areas distant from their lodge. Some camps might become seasonal in operation.

Table 41-6. Effect of low-level/low-volume flooding (only deeper channels and lagoons remain flooded) on specialization of business.

Specialization of business	Description of turnover			Total
	Increase	*Decrease*	*No change*	
Water-based tourism activities	2 (4)	3 (6)	2 (4)	7 (15)
Land-based tourism activities	7 (15)	3 (6)	9 (19)	18 (38)
Both water- and land- tourism activities	10 (21)	2 (4)	6 (13)	19 (40)
Accommodation to everyone	1 (2)	1 (2)	2 (4)	4 (8)
Total	20 (42)	9 (19)	19 (40)	48 (100)

Low-Level/Low-Volume Flooding (main channels and lagoons are seasonal and dry)

Respondents were asked to state how their turnover would change if flooding levels become low such that the main channels and lagoons become seasonal and dry. Respondents were specifically asked to assume that flooding levels were low to the extent that the permanently flooded parts of their area became only rarely flooded; frequently, occasionally or rarely flooded parts of their area became never flooded; and the main channels and lagoons became seasonal and dry for much of the year. Responses based on this scenario are shown in Table 41-7.

Results in Table 41-7 show that 33.3% of the respondents noted that turnover would increase. The majority of those who said that turnover would increase had businesses located in the middle of the Delta. Thirty-eight percent of the respondents indicated that there would be an increase in tourism turnover, and the majority of these had business located in the lower areas of the Delta. For those who said turnover would decrease, 10.4% said it would decrease by 25%, 25.0% said it would decrease by 50%, while 4.2% said it would decrease by 100%. Twenty-nine percent of respondents said there would be no change in the turnover. Respondents noted that turnover would decrease because game would migrate to water-fed points, water and land activities would become inaccessible, the environment would become dusty and clients might resort to other options. Camps would be run seasonally or closed down.

Table 41-7. Effect of low-level/low-volume flooding (main channels and lagoons are seasonal and dry) on location of business.

Location of business	Description of turnover			Total
	Increase	*Decrease*	*No change*	
Panhandle	2 (4)	2 (4)	1 (2)	5 (10)
Middle of Delta	11 (23)	4 (8)	9 (19)	24 (50)
Lower area of Delta	3 (6)	12 (25)	4 (8)	19 (40)
Total	16 (33)	18 (38)	14 (29)	48 (100)

The impact of low-level/low-volume flooding on specialization of business is shown in Table 41-8. According to the table, a majority of those who indicated an increase in turnover had both water- and land-based tourism activities, while a majority of those who indicated a turnover decrease had land-based tourism activities only. Different levels of flooding in particular areas of the Okavango Delta caused these differences.

Table 41-8. Effect of low-level/low-volume flooding (main channels and lagoons are seasonal and dry) on specialization of business.

Specialization of business	Description of turnover			Total
	Increase	*Decrease*	*No change*	
Water-based tourism activities	2 (4)	4 (8)	1 (2)	7 (15)
Land-based tourism activities	5 (10)	6 (13)	7 (15)	18 (38)
Both water- and land- tourism activities	8 (17)	5 (10)	6 (13)	19 (40)
Accommodation to everyone	1 (2)	3 (6)	0	4 (8)
Total	16 (33)	18 (38)	14 (29)	48 (100)

Low-Level/Low-Volume Flooding (no flooding at all, boreholes drilled)

Finally, respondents were asked to state how their turnover would change in the event that there was no flooding such that boreholes would have to be drilled for the supply of water. That is, low-volume flooding to the extent that no flooding took place in their area, and groundwater had to be tapped to provide water points for game and to supply the camp. Results on how turnover was perceived to change under low-volume flooding are shown in Table 41-9.

Table 41-9. Effect of low-level/low-volume flooding (no flooding at all, boreholes drilled) on location of business.

Location of business	Description of turnover			Total
	Increase	*Decrease*	*No change*	
Panhandle	1 (2)	4 (8)	0	5 (10)
Middle of Delta	13 (17)	9 (19)	2 (4)	24 (50)
Lower area of Delta	2 (4)	15 (31)	2 (4)	19 (40)
Total	16 (33)	28 (58)	4 (8)	48 (100)

Results in Table 41-10 show that the majority (58%) of respondents noted that turnover would decrease. The majority of those indicating a decrease in turnover had their business located in the lower area of the Delta. For those who said the turnover would decrease, 2.1% said it would decrease by 25%, about 25% said it would decrease by 50% and 10.4% said it would decrease by 100%. Respondents noted that turnover would decrease because a lack of water would affect water-based activities; game would migrate to other areas; and most water- and land-based activities could not be carried out if there was no water.

The effect of low-level/low-volume flooding on specialization of business is show in Table 41-10. According to the table, the majority (17%) of those with both land- and water-based tourism activities thought that tourism turnover would increase, while those with land-based tourism activities only thought that tourism activities would decrease.

Table 41-10. Effect of low-level/low-volume flooding (no flooding at all, boreholes drilled) on specialization of business.

Specialization of business	Description of turnover			Total
	Increase	*Decrease*	*No change*	
Water-based tourism activities	2 (4)	5 (10)	0	7 (15)
Land-based tourism activities	5 (10)	11 (23)	2 (4)	18 (38)
Both water- and land- tourism activities	8 (17)	9 (19)	2 (4)	19 (40)
Accommodation to everyone	1 (2)	3 (6)	0	4 (8)
Total	16 (33)	28 (58)	4 (8)	48 (100)

Table 41-11. Overall deviation averages for the for the "increase" and "decrease" scenarios.

Scenario	Deviation Averages (%): Increase	Deviation Averages (%): Decrease	Overall Deviation (%)
1. High-level/high-volume flooding	125	59.6	84.38
2. Low-level/low-volume flooding (main channels and lagoons remain permanently flooded)	113	55.5	99.47
3. Low-level/low-volume flooding (only the deeper main channels and lagoons remain permanent while the rest become seasonal only)	121.43	69.23	95.83
4. Low-level/low-volume flooding (the main channels and lagoons become seasonal and dry for much of the year)	115	51.32	84.42
5. Low-level/low-volume flooding (no flooding takes place in their area)	119.23	58.93	75

Table 41-11 shows the overall deviation averages for the "increase" and "decrease" scenarios. The smaller the overall deviation to the benchmark (100%) the more favorable the scenario is. As Table 41-11 shows, the scenarios can be ranked in the following order, starting with the most favorable: 2. Low-level/low-volume flooding (main channels and lagoons remain permanently flooded); 3. Low-level/low-volume flooding (only the deeper main channels and lagoons remain permanent while the rest become seasonal only); 4. Low-level/low-volume flooding (the main channels and lagoons become seasonal and dry for much of the year). 1. High-level/high-volume flooding and 5. Low-level/low-volume flooding (no flooding takes place in their area).

Discussion and Conclusion

This survey indicates that the majority of the accommodation facilities that would be affected by change in flooding patterns are located in the middle parts of the Okavango Delta. Most of these facilities offer both land- and water-based tourism activities. Results indicate that high-level/high-volume flooding, where the frequently flooded parts of their area become permanently flooded, the occasionally flooded parts in their area become frequently flooded, and the rarely flooded parts in their area become occasionally flooded, would negatively affect the tourism turnover in the Okavango Delta.

Factors that operators noted contributing to a downturn in turnover include:

- When floods are high, roads become damaged and access to tourism facilities becomes difficult;
- Game viewing and bird watching becomes poor due to road conditions;
- Game also becomes scarce due to the lack of concentration of these species in particular areas as is the case when flooding is not a problem; and
- Tourists cancel their bookings for fear of diseases and/or being injured or killed by the floods.

Survey results indicate that most of the facilities located in the middle and upper parts of the Okavango Delta are mostly those that would have a low turnover due high floods. Based on the perceptions of tourism operators, high-level/high-volume flooding is not a desirable phenomenon for the tourism industry. The issue of turnover going down can be illustrated by flooding patterns that took place in the Okavango Delta in 2000. Flooding in the Okavango Delta in 2000 resulted in the closure of Moremi Game Reserve (located within the inner parts of the Delta) for eight months because roads were impassable (Mbaiwa, 2005).

During this period, there was a decline in self-drive tourists in the Okavango Delta and tourism revenue went down. The results of flooding in the Okavango Delta at this time resulted in retrenchments of workers in the Okavango Delta since turnover was low in most accommodation establishments. A similar scenario is likely to happen in the event of high-flooding patterns in the Okavango Delta. Retrenchments suggest that income and livelihoods for most people working in tourism businesses in the Okavango Delta would suffer.

Results in this survey also indicate that turnover would also go down drastically in the low-level/low-volume flooding scenario where it would be so dry that boreholes would be needed to supply water. This scenario indicates a very dry period where groundwater has to be tapped to provide water points for game and to supply lodges and camps. In fact, operators noted that turnover would be reduced much more greatly than in the high-level/high-volume flood scenario.

Operators indicated that turnover would go down because:

- The lack of water would affect water-based activities;
- Game would migrate to other areas; and
- Most land-based activities cannot be carried out when there is no water.

Again this was not a favorable scenario for tourism operators. As with the high-level/high-volume flooding scenario, low tourism turnover would lead to layoff of workers and negative impacts on livelihoods.

Studies in desiccation of the western parts of the Okavango Delta (e.g., by Kgathi et al., 2004) have shown that dryness affects many economic activities of which tourism is one. Tourists visiting the Okavango Delta come mainly to enjoy the wilderness nature and scenic beauty of the wetland (see Mmopelwa et al., 2007), and a very dry environment is likely to be unattractive to tourists. Under such situations the tourism industry could conceivably collapse.

The favored scenario appears to be the low-level/low-volume flooding. This is a scenario where the permanently flooded parts of the Delta become only frequently flooded, the frequently flooded parts of their area become only occasionally flooded, the occasionally flooded parts of the area become only rarely flooded, and main channels and lagoons remain permanent.

It is in this scenario that most operators noted a likely increase in tourism turnover. In particular, under this scenario:

- Birds return to their areas of concentration;
- Fishing becomes more prolific;
- Hippos and crocodiles become visible all year round; and
- These all contribute to a more desirable tourism experience.

This study therefore shows that tourism in the Okavango Delta would do well where there is permanent water in channels and lagoons to allow for both water-based and land-based activities. It is in this scenario where income generation and employment opportunities are likely to be high. In this regard, local communities engaged in the tourism industry are likely to benefit more in terms of revenue generation and employment opportunities.

This study suggests that adaptation strategies in the event of high-level/high-volume flooding or low-level/low-volume flooding patterns, which seem to negatively affect turnover in the tourism industry, should be the subject of future tourism research.

References

Agnew, M. A., & J. P. Palutikof (2001). Climate impacts on the demand for tourism. In *Proceedings of the First International Workshop on Climate, Tourism and Recreation*, ed. A. Matzarakis & C. R de Freitas. Report of a workshop held in Halkidiki, Greece, 5–10 October 2003. Commission on Climate Tourism and Recreation.

Amelung, B., S. Nicholls & D. Viner (2007). Implications of global climate change for tourism flows and seasonality. *Journal of Travel Research* 45: 285–296.

Ashton, P. J., L. Nordin & L. E. Alonso (2003). Introduction to the Okavango Delta and the AquaRap Expedition. In *A Rapid Biological Assessment of the Aquatic Ecosystems of the Okavango Delta, Botswana: High Water Survey*, ed. L. E. Alonso & L. Nordin, pp. 29–37. RAP Bulletin of Biological Assessment No. 27. Washington, DC: Conservation International.

Bigano, A., J. M. Hamilton & R. S. J. Toll (2007). The impact of climate change on domestic and international tourism: A simulation study. *The Integrated Assessment Journal* 7, no. 1: 25–29.

Braun, O. L., M. Lohmann, O. Maksimovic, M. Meyer, A. Merkovic, E. Riedel Messerschmidt & M. Turner (1999). Potential impact of climate change effects on preference for tourism destinations: A psychological pilot study. *Climate Research* 11: 247–254.

Budeanu, A. (2005). Impacts and responsibilities for sustainable tourism: A tour operators' perspective. *Journal of Cleaner Production* 13: 89–97.

CSO [Central Statistics Office] (2002). National Population and Housing Census. Ministry of Finance and Development Planning, Gaborone.

Dilys Roe, D., C. Ashley, S. Page & D. Meyer (2004). Tourism and the Poor: Analyzing and interpreting tourism statistics from a poverty perspective. PPT Working Paper No. 16, http://www.dfid.gov.uk/r4d/PDF/Outputs/Mis_SPC/R8120-PPT16.pdf.

Hein, L., M. J. Metzger & A. Moreno (2009). Potential impacts of climate change on tourism: A case study for Spain. *Current Opinion in Environmental Sustainability* 1: 170–178.

Ijeomah, H. M., & A. A. Aiyeloja (2009). The impact of climate change on sustainable tourism management in Plateau State, Nigeria. *Journal of Sustainable Development in Africa* 11, no. 1: 149–175.

Kgathi, D. L., H. Bendsen, P. Blaikie, J. Mbaiwa, B. N. Ngwenya & J. Wilk (2004). Rural Livelihoods, Indigenous Knowledge Systems and Political Economy of Access to Natural Resources in the Okavango Delta. Botswana, HOORC, Maun.

Mbaiwa, J. E. (2005). Wildlife resource utilization at Moremi game reserve and Khwai community area in the Okavango Delta, Botswana. *Journal of Environmental Management* 77, no. 2: 144–156.

Mmopelwa, G., D. L. Kgathi & L. Molefhe (2007). Tourists' willingness to pay and their perception on park fees: A case study of self-drive and clients of mobile tour operators in Moremi game reserve. *Botswana Tourism Management* 28: 1044–1056.

Okavango Delta Management Plan (2006). Department of Environmental Affairs, Ministry of Environment, Wildlife and Tourism.

Roe, D., & P. U. Khanya (2001). Pro-poor Tourism: Harnessing the World's Largest Industry for the World's Poor. London: International Institute for Environment and Development.

UNWTO [United Nations World Tourism Organization] (2008). Climate Change and Tourism: Responding to Global Challenges. Nairobi: UNEP.

—. (2003). Climate Change and Tourism: Proceedings of the 1st International Conference on Climate Change and Tourism, Djerba, Tunisia, 9–11 April 2003.

Viner, D., & M. Agnew (1999). Climate Change and its Impacts on Tourism. Report Prepared for WWF-UK. Climatic Research Unit. University of East Anglia, http://www.wwf.org.uk/filelibrary/pdf/tourism_and_cc_full.pdf.

Wall, G. (1998). Implications of global climate change for tourism and recreation in wetland areas. *Climate Change* 40: 371–389.

Wolski, P. & M. Murry-Hudson (2007). An investigation of permanent and transient changes in flood distribution and outflows in the Okavango Delta, Botswana. *Physics and Chemistry of the Earth* 33: 157–164.

CHAPTER FORTY-TWO

WATER RESOURCES IN JAPAN FROM THE PERSPECTIVE OF "WATER FOR THE OCEAN"

SATOQUO SEINO[1]

Introduction

Appropriate allocation of water resources and compatibility between development and water environment conservation are problems that are both old and new. In Japan, many approaches have been tried to solve these problems, including research and creation of consensus-building mechanisms. However, in order to clearly highlight the issues involved, an unconventional, large-scale perspective is now needed.

"Water for the ocean" is one such perspective.

The worsening problems of the coastal environment are sending a wake-up call about the state of water resources on land. Problems in Japan's coastal areas started during its high-economic growth period, including water pollution, land reclamation and so on. These were problems that could be easily located since the environment was affected dramatically over a short period. Recently, problems have expanded into longer-term, more chronic phenomena, such as ecosystem change, coastal erosion and depletion of fishing resources.

Recently, river management is attracting attention as the cause of these problems. Conventional coastal environmental studies on river management used to focus mainly on reducing the inflow burden and preventing water pollution. But recently the issue has expanded to encompass the "artificial transformation of water circulation" caused by construction of dams and other river structures together with excessive water intake. Researchers had to start studying the already-changing system without enough baseline data on the dynamics of sediment and nutrient salts being transported in water systems.

[1] Kyushu University, Japan.

For these researchers, deductive approaches, with hypothetical thinking and control experiments, could have been possible if these problems had been predicted. As it turned out, inductive approaches had to be taken by integrating individuals' knowledge about river basins and coasts in order to understand past phenomena. For this reason, cases of long-term fixed-point observations, including ones conducted by local government fisheries experiment stations, turned out to be very important.

Meanwhile, the largest stakeholder group concerned with coastal environmental problems is people in fisheries. Although river fisheries in Japan have become too minor today to support anyone's livelihood, sea fisheries constitute an industry employing about 200,000 fishers throughout Japan. For example, in the Ariake Sea coastal area, changes in the ecosystem and natural conditions have been affecting the lives and economy of residents in the coastal area, and this has become a serious social issue with political implications.

Many cases of similar fishing-industry declines caused by decreases in natural resources have been occurring inside and outside Japan, but most cases have not taken on the form of a social issue. This is largely because this process takes many decades, during which fishers continue to gradually stop fishing or start doing other work. Since fisheries form an industry, compensation issues are involved, and it is realistically difficult for them to disclose their data completely without social implications.

"River management from the perspective of ecosystems and environment" is a concept that can change the stalled situation in the Japanese social system. As discussed later, "water for environmental use" will be more acceptable to society than "water for fishing use," because environmental water is not water for industry, but is "water for the public."

Under these circumstances, the research environment in Japan has long required researchers to deliberately verify that "rivers affect the coastal environment," despite the fact that this is an empirically well-known natural phenomenon. Strangely, in Japanese coastal fisheries, the impacts of water pollution caused by artificially created muddy water and toxic water discharges have been clarified, but the impacts of an artificially transformed water cycle caused by dams, weirs and estuary barrages have been put aside as issues too difficult to handle.

In this paper, I will discuss how the "ocean" is conceptualized and why this concept has not been incorporated into the social system, as well as how consensus-building and countermeasures have been achieved in individual river basins. I would then like to discuss the future of rebuilding the system for dealing with water issues so that it includes the ocean, which has remained a blank domain.

"Water Resources" Means Freshwater Only

Only supplies of freshwater, which is useful for humans, have been considered as water resources. Freshwater, which can be used for agriculture, industry, drinking and power generation, loses its value if salt is mixed in. Through history, it has been taken for granted in society that freshwater resources should be fully distributed for maintenance of land-based industries and cities, while only "residual water" would be discharged into river channels when there was no further use for it, where it will flow into the sea and "become useless once salt is mixed in."

So much water has been taken from rivers even while they are flowing across the land's surface that there is little water left in some rivers. Even though society has witnessed this situation, it has been slow to see the need for improving it. The necessity for freshwater inflow into brackish water and marine areas, which is more invisible, is still not recognized at all inside and outside Japan.

Although it has been known that brackish water environments support many creatures as habitat and breeding grounds, and that they are critical for migratory organisms, such awareness has not led to the further step of directly reviewing human water use on land.

Land-based water resources present clear economic benefits as this water is used to generate hydroelectric power, supply the needs of cities, and produce agricultural and industrial products. Therefore, the necessity for users to pay for it is obvious. The principle of paying costs has created consensus-building opportunities and sophisticated systems, where awareness about water has been raised through regular industrial activities. Correspondence between the size of industries and the amount of necessary water has also been well elaborated.

However, among coastal and estuary fishing industries, the idea that "water resources are necessary for the ocean too," has failed to evolve into a concept or system. In particular, the lack of quantification of the amount of water resources needed for fishing constitutes a vital difference from the case of land-based industries. They have not tried to secure water resources as aggressively as land-based people have.

In river water-resource development projects involving dams or dikes, water users bear some of the costs in addition to public outlays by national or local governments. The fishing industry may have received compensation for lost fishing grounds, etc., but it has never considered itself or been considered by others as a sector that can claim an interest in water use. As river development cut down the inflow of water, causing biological disturbance and other changes to natural rhythms, the fishing industry has tended to become fixed into a position of the victimized party, and has never been a party with a positive say in water-resource allocation.

Ways to Promote "Water for the Ocean"

The River Act was amended in 1997 to include environment conservation as one of its purposes, and the Fisheries Basic Act was enacted in 2001. Enactment of the basic law on fisheries was realized forty years after equivalent laws were enacted for agriculture and forestry. Possible measures for institutionally realizing "water for the ocean" are discussed below.

(1) "Water for Environmental Use"

In order to maintain the habitats of useful fish and to conserve river ecosystems, ecological approaches have been tried, and concepts of "water for environmental use" have been formed and implemented on an international basis. When reaching agreement in terms of industrial use by the fishing industry proves difficult, a consideration of the conservation of aquatic wildlife habitat becomes more important.

In Japan, too, an awareness of the need to secure minimum flow volumes arose, because dry sections appeared in rivers as early as the beginning of the twentieth century and flow depletion dramatically accelerated during the high-economic growth period. The situation was serious enough to bring about an amendment to the River Act in 1964 explicitly specifying the maintenance of flowing water. In the context of dam development projects, amounts set aside as "unspecified volumes" can be effectively allocated to environmental sustainability, which is after all a public use. This can be regarded as Japanese-style "water for environmental use"; it was a hard-won countermeasure under the straitened conditions of an under-developed environmental legal system.

However, environmental water set aside in dam construction will possibly be a counter-productive measure in terms of environmental health, as it may cause major problems, such as disruption of biological and material movements and spatial disconnection. It should not be prematurely included in plans before thorough investigations are made about the propriety of water-resource allocation and the volumes to be increased.

(2) Water Rights Claims Based on Indigenous Status of Fishing People

As stated earlier, water for fishing use needs to be conceptualized pursuant to a review of the conventional water institutions. Indigenous status is a factor of consideration in claims of rights to natural resources. It has been learned from ancient shell mounds, and corroborated by references to coastal fishing activities in the "Gishiwajinden" (a third-century Chinese historical record), that people in the Japanese archipelago have harvested water creatures since ancient times.

With the advent of technologies for securing exclusive use of water through the construction of facilities, in contrast to a situation of naturally flowing water, institutional systems for water use became necessary. The fishing industry has no such system even now because fishing can be carried out whenever there is water and so there has been almost no need to claim an exclusive right to water. Measures that can appeal to society aimed at promoting a review of fishing people's potential rights can be based on detailed research of their historical background.

(3) Environment-conservation Type Fisheries and the Multifaceted Functions and Public Nature of the Fishing Industry

It is also necessary to explain through easily understandable examples that fishing people are not only recipients of ecosystem services, but that they also contribute to marine environment conservation through environmental monitoring and preventing over-fishing through the system of fishing rights. The multifaceted function and public nature of the fishing industry are important factors in securing water for the ocean as water for environmental use, and for claiming indigenous status.

In addition, since the postwar reconstruction era, national policy on fisheries has mainly focused on deep-sea and offshore fisheries, while coastal fisheries have continued to decline as they compete with industrialization and urbanization.

In the twenty-first century, a realignment of industrial structures is likely to take place in relation to environment and food supply aspects.

Water rights are renewed once every few decades. If an agreement is reached that the fishing industry should officially take part in the system the next time they are renewed, a change may possibly take place.

(4) Problem Solving and Systematization in an Individual River Basin

There are some cases in which a response has been obtained to the maximum extent possible within the limits of the conventional system, even though the relevant social institutions do not really suffice. When individual cases bring problems to light, they can be incorporated into systems in the form of policies and laws.

Responding in 2001 to poor *nori* (laver seaweed) harvests in Ariake Bay, and also to poor *nori* harvests in the Seto Inland Sea, discharges from dams and barrages on the respective Chikugo River in Kyushu and Yoshii River in Okayama Prefecture were carried out as special operations during the *nori* cultivation season. Similarly, a discharge from an estuary dam on the Yodo River in Osaka Prefecture is being considered in order to improve the habitat environment of Shijimi clams

These are the cases where fishing administration or river-management authorities have accepted the fishing industry's claims, and conduct "flexible operations" in coordination with water for urban, power-generation, industrial and agricultural uses.

These marine areas are characteristically governed by individual sea-based laws, such as the Law Concerning Special Measures for Conservation of the Environment of the Seto Inland Sea (1973) and the Act on Special Measures concerning Rejuvenation of Ariake Sea and Yatsushiro Sea (2002). In these cases, information has been collected, and so consensus among stakeholders and administrative authorities can be relatively promptly formed.

In addition, as *nori* farmers and Shijimi fishers target non-migratory creatures, their fishing grounds are limited to certain areas. Therefore it is relatively easy to determine the numbers of target creatures and quantitatively monitor their growth. It is also a major merit that data for determining a cause-and-effect relationship with river-water volume can to a certain extent be obtained.

(5) Improvement of Legal Systems and Precautionary Measures

The Basic Act on Ocean Policy, enacted in 2007, organized formerly un-integrated ocean policies across disciplines and governing authorities for the first time from a holistic-ocean point of view. Language about taking a precautionary stance toward conservation of the marine environment is included in Article 18, while language dealing with impacts from the land in coastal-zone management is included in Article 25.

Clear numbers and cause-and-effect relationships are required in water-source allocation. However, although quantification and engineering approaches to habitat environments and landscapes incorporating aquatic wildlife are evolving, more investigative studies are still needed to attain the same level of accuracy as in similar data sets for land areas. But we cannot just sit and wait for progress in this field to catch up to reality. What we can do now is to take precautionary measures, which is the approach currently favored in environmental problems or even in medical care, as the goal of our discussion of "water for the ocean."

(6) The Public Nature of Seacoasts and the Ocean

If land-based people have to be forced to put up with some inconvenience for the sake of the marine environment, great impetus will be needed to gain consensus.

The sea around Japan is legally a publicly owned water body and the seacoasts are public space under national or local government control as natural public property. The environment itself is also a public commodity, and water is common property. Now that we have learned a great deal more about the coastal environment, we should open a new discussion on the public nature of the ocean.

A widespread understanding needs to be attained that "water for the ocean" will not only benefit the people in fishing businesses, but also will also contribute to the welfare of all people and benefit them through their enjoyment of marine products and waterfront environments. The term "ecosystem service" is widely known and a viewpoint of appreciating the value of creatures other than human beings has been growing. At the same time, fishing people who depend on the coastal environment, its wildlife and natural resources, are also recipients of marine-area environmental services where abundant river water is supplied.

(7) How to Think About Natural Resources in the Era of Global Environment

Regarding global environmental problems, the focus is now on responses to major disasters or adaptation to humanly uncontrollable circumstances. At the same time, the relationship between climate change on land and marine dynamics is becoming clarified. As present trends progress, a social system assuming a limited range of fluctuation, as in the case of land-based legal systems, will no longer be sufficient.

In this regard, we need to review the very methodology of managing things by simply excluding hard-to-handle factors, like the ocean factor in terms of the water-resource regime. Learning to incorporate a factor as dynamic as the ocean into the water-resource system will be a good exercise or test in the era of global environment.

In some water-resource issues that include conflicts over dam construction or environment deterioration, directions that cannot be recommended from a scientific and technological point of view seem to be the ones selected as a result of priority being given to human sentiment and circumstances. "The right answer" can fluctuate depending on social context. The scientific literacy and ability of decision-makers to make decent, viable judgments are put to the test, and the consequences are very serious and important.

For Japanese water-resource issues, "water for the ocean" is a challenge to the existing system, which has formed as a complex of scientific technologies and social circumstances. What are required for science are not only observation methods, data accumulation and technology development, but also improvement of the judgment and consensus-formation skills of the whole society with respect to how the results will affect nature and the society.

Recent Events

On March 11, 2011, a magnitude eight earthquake occurred on the floor of the Pacific Ocean, east of Japan's Tohoku ("northeast") district. A terrifying tsunami attacked its coast; more than 10,000 people lost their lives and nearly 10,000 more are still missing. The effects are illustrated in the images at the end of this paper (Figures 42-1 to 42-4).

Japan sunk deeply in sorrow due to the losses sustained in this disaster, and all of us are battling the fear that similar-scale earthquakes and tsunamis will happen again during our lifetimes.

This situation has clarified the need to substantially overhaul the way our country is being run, which has been very focused on land-based activities.

Many people are now considering the philosophical side of the question of how humankind is to conduct its relationship with the sea – in both its bountiful and fearsome aspects.

When Japan left its feudal society behind and entered the modern era, its economic system took shape centered squarely on rice cultivation. Wetlands were filled in to expand the area under wet-rice agriculture. As modernization progressed, even more coastal areas were filled in to provide land for industrial, urban and residential development.

We are now being forced to take a good, hard look at how our insatiable greed led us to advance too far into territory strongly subject to the external forces of the sea.

For the first time in 400 years, we are being called on to re-design our land- and water-management structures that have been focused on land. Studies will be renewed concerning the border zone between land and sea, which was defenseless and destroyed in the disaster. Of particular importance will be securing "buffer zones."

This paper looks at how the basic approaches of Japan's water-management systems have been biased towards on-land activities.

Should the future bring climate change, sea-level rise, etc., events similar to the earthquake/tsunami disaster Japan is now experiencing might arise in other parts of the world.

Prevention measures are of course vital, but in the course of preparing for and setting up such measures, knowledge of Japan's present difficulties could be a great help. I would particularly like to appeal for the re-examination of social systems in view of their relationship with the sea.

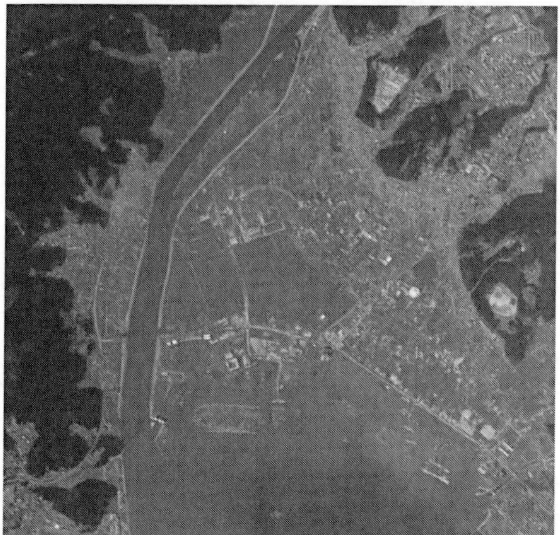

Figure 42-1. Aerial photograph of Rikuzen Takada, Iwate prefecture, Japan, after the earthquake and tsunami disasters (photo credit: Japan Geographic Survey).

Figure 42-2. Broken coastal dikes and facilities (June 6, 2011).

Figure 42-3. The residential and rural areas along the lower reaches of the river were washed out by the tsunami (June 6, 2011).

Figure 42-4. The residential areas were washed out by tsunami and still subsided (June 6, 2011).

PART VIII

DECLARATIONS

LUSAKA DECLARATION ON SUSTAINABLE TOURISM DEVELOPMENT, CLIMATE CHANGE AND PEACE

We, the more than 440 participants from 36 countries of Africa, Asia, Europe, North America and the Caribbean, assembling for the **5th IIPT African Conference: "Meeting the Challenges of Climate Change to Tourism in Africa and the Developing World**," at the Hotel InterContinental, Lusaka, Zambia, 15–20 May, 2011, in support of the U.N. Millennium Development Goals –

Considering:

World population has increased from 1.6 billion in 1900 to 6.0 in 2000 and will reach 9.0 billion by 2050, with 90% of this growth in developing countries.

Growth in population, global consumption and the global economy has stressed the Earth's ecological systems and depleted much of the world's natural resources.

It now takes the Earth one year and six months to regenerate what our world population consumes in one year, resulting in collapsing fisheries, depleting forests, species extinction, water shortages and crop failures – all with disproportionate impacts on the poor regions of the world.

U.N. Secretary-General Ban Ki-moon has stated, "Climate change is the pre-eminent geopolitical and economic issue of the 21st century. It rewrites the global equation for development, peace and prosperity."

Acknowledging:

The year 2010 ranked as the warmest on record, together with 1998 and 2005; the ten warmest years on record have all occurred since 1998; Arctic sea-ice cover in December 2010 was the lowest on record; and a growing number of extreme weather events have occurred in regions throughout the world in the past two years.

African and other developing countries are the least responsible for climate change, yet are particularly vulnerable to the effects, including reduced agricultural production, threats to food security, both drought and flooding, sea level rise, coastal erosion, coral bleaching, deforestation, loss of wildlife, reduced fresh water availability, spread of malaria and other diseases and increased risk of conflict over scarce land and water resources.

Climate change and the combined stresses on ecological systems can hinder progress made on reducing poverty and the U.N. Millennium Development Goals.

Noting that within this context:

World military expenditures are in excess of U.S. $1.6 trillion, more than the GDP of any nation except eight – approximately U.S. $225 for every man, woman and child in the world.

When 1 of every 6 survives on $1 a day or less; another 2 of every 6 on $2 a day or less; 1 of every 6 go to bed hungry; 1 of every 4 never get a clean glass of water to drink; 2 of every 5 lack proper sanitation; and 35,000 children die each day from preventable diseases – many of them from water-born diseases.

Recognizing:

Research conducted by the Institute for Economics and Peace and Economists for Peace and Security estimated the monetary value of peace on the global economy within a four-year period, 2006 to 2009, would have been U.S. $18.5 trillion, not including an additional U.S. $9.8 trillion in expenditures that would have been diverted from violence-related activities to areas benefiting society – a total of U.S. $28.3 trillion within a four-year period.

Participants in the 5th IIPT African Conference conclude that:

The global challenges facing humanity call for global solutions and cooperation at all levels – and by all sectors, public and private, and civil society as a whole – on an unprecedented scale.

Peace will be an essential prerequisite to achieve this level of cooperation and will also generate the additional economic wealth to fund these solutions.

Travel and tourism – the world's largest industry, affecting every sector of society, in virtually every nation – an industry that is founded on peace and that contributes to peace and understanding – has a key leadership role to play in achieving these solutions.

The travel and tourism industry has a central role to play in the creation of jobs with dignity, foreign exchange earnings, the development of disadvantaged areas, poverty reduction and the promotion of understanding, peace, love, unity and progress among all peoples, communities and nations on the African continent and throughout the world.

And therefore call on all sectors of the travel and tourism industry to:

1. Support the leadership role of UNWTO, UNEP and WMO in the specific actions called for in the Davos Declaration: Climate Change and Tourism – Responding to Global Challenges.
2. Give emphasis to energy conservation and aggressively make the transition to renewable and clean energy sources as a foundation for carbon/GHG neutral development strategies and a "Green Growth" transformation.
3. Continue to implement socially and environmentally responsible tourism practices that benefit local destinations and enhance both cultural and natural heritage, giving particular emphasis to fair trade in tourism practices, tourism that contributes to poverty reduction and tourism that contributes to the restoration of ecological systems.
4. Provide for quality tourism experiences that include meaningful interactions between visitors (guests) and the host community and a greater understanding and appreciation of local cultures, customs and traditions.
5. Implement environmentally responsible development and operational practices, particularly in vulnerable areas, including coastlines, low-lying areas and mountain regions.
6. Develop and implement waste management plans that give emphasis to reducing, recycling and re-using.
7. Incorporate quadruple bottom line reporting with climate change alongside environmental stewardship, social responsibility and economic value.
8. Actively engage local communities in planning and decision-making, empower women and youth and embrace the wisdom, knowledge and values of indigenous peoples.
9. For the transport sector, and particularly the airline industry, continue efforts to improve engine efficiencies and ultimately develop engines that operate with clean and renewable energy sources.
10. Recognize the important role of media, educational and religious institutions as agents of change.

Governments and International Development Agencies to:

1. Provide incentives to tourism projects that utilize green designs/technologies and contribute to creating sustainable livelihoods, poverty reduction and the U.N. Millennium Development Goals.
2. Only approve projects that adhere to globally recognized sustainable tourism certification standards and that incorporate construction designed to withstand severe storms and other effects of climate change.
3. Not allow heavy water-consuming projects where water is in limited supply.
4. Develop Disaster Management Plans that include phases for risk reduction, readiness, response and recovery.
5. Recognize that certain forms of tourism ARE an adaptation strategy.
6. Encourage more research and monitoring of emerging trends related to the inter-connectedness of tourism, environment and climate change – and provide web-based tools for the sharing and integration of research findings.
7. Design and implement broad public awareness campaigns focusing on the linkages of tourism, environment, climate change, sustainable development and broader society. Provide corresponding educational programs beginning at the primary level – through high school – and into universities.

And travelers to:

1. Appreciate that we are all custodians of our one common home – planet Earth – and we all share a common responsibility in preserving the abundant beauty and resources of God's creation for future generations.
2. Be selective in modes of travel and minimize/offset the carbon footprint of each trip.
3. Select accommodations, transport, tour operators and activities with eco-labels or that have been recognized for their green practices and strive to contribute to the social, cultural, economic and environmental enhancement of their destination.
4. Recognize that every traveler is potentially an "Ambassador for Peace" and a messenger of transformative change toward a "Green Growth" global village.

5. With Gandhi and Tolstoy, come to discover that "Life finds meaning in service to others," and
6. "By my spirit, words and actions, encourage others to travel the world in peace."

IIPT Credo of the Peaceful Traveler

We support the UNWTO, UNEP and WMO Davos Declaration in calling for U.N., international, financial and bilateral agencies to support the governments of developing countries, and in particular the least developed countries, in their efforts to address and adapt to the adverse effects of climate change.

We commend the International Institute for Peace through Tourism for its organization of this Conference and for giving scope to the vision of peace through tourism.

And express appreciation to our Host, the Government of the Republic of Zambia, Ministry of Tourism, Environment and Natural Resources; **Conference Patron**, His Excellency Rupiah B. Banda, President, Republic of Zambia; **Conference Partners and Sponsors**, including the World Tourism Organization (UNWTO), U.N. Development Program (UNDP), Pacific Asia Travel Association (PATA), Regional Tourism Organization of Southern Africa (RETOSA), World Travel Market, Zambia Tourism Board, EcoBank, South African Airways, Hotel InterContinental Lusaka, Livingstone International University of Tourism Excellence and Business Management (LIUTEBM), Zambia Ethno Tourism and Cultural Tourism Alliance (ZECUTA), Children and Youth Welfare Foundation, Sustainable Travel International (STI), The Blue Yonder, Center for Responsible Tourism (CREST), Garden Group Hotels, Blue Nile Inn, Church Road Lodge and Golden Bridge Hotel; **Media Partners**, including eTurboNews, World Tourism Directory, TravelVideo.TV, Travel Mole, India Tourism eCatalog, World Tourism Directory, Lights of Africa, Society of American Travel Writers (SATW) and Africa Diaspora Tourism; the many **Conference Supporters;** and the **People of Zambia** for their smiles, welcoming spirit and hospitality.

Adopted in Lusaka, Zambia, 20 May, 2011

By the more than 440 participants of the 5th IIPT African Conference on Peace through Tourism, from 36 countries in Africa, Asia, Europe, North America and the Caribbean, including Ministers of Tourism, senior representatives of U.N. agencies, senior government officials, senior representatives from diverse sectors of the industry and NGOs, educators, students and leading experts in tourism and climate change.

ASIA PACIFIC YOUTH DECLARATION ON CLIMATE CHANGE AND SUSTAINABLE DEVELOPMENT

The future belongs to us and we, the youth from 15 countries of the Asia-Pacific region are aware of the urgency and threats posed by different environmental problems including climate change. We believe that the current global economic models are affecting the overall environment and natural resources, as a result the global ecological footprint has exceeded by 30% contributed by and seen largely in high carbon emissions, climate change and global warming.

We realize that the institutional frameworks and mechanisms prepared at the Rio Summit, most importantly Agenda 21 and UNFCCC, have failed to meet the high expectations from them, mainly due to weak political commitment, weak institutional mechanisms, less efficient governance, and lack of human and financial capital.

We strongly advocate for the justice, equity (intergenerational equity and north-south equity) and global sustainability.

We are here to claim our safe future and we demand from the global leaders and global citizens to:

1. Be aware of the impact of climate change and vulnerability of the poor people in Asia-Pacific region ranging from top of the world to low land coastal areas. We are facing the increasing frequency and severity of disasters, melting of glaciers and Glacier Lake Outburst Flood (GLOF), declining water resources, loss of forest coverage and biodiversity, declining agricultural productivity and increasing food insecurity, health impact, sea level rise and coastal flooding, environmental refugees and overall socio economic problems.

2. Act urgent and immediately to mitigate the climate change by reducing emission of GHGs at sources, the global temperature rise should be limited to 1.5 degree C from the pre-industrial level and CO_2 concentration should return to 350 PPM.

3. Immediate commitment for the mitigation of emission of Black Carbon and develop strategies to minimize the Atmospheric Brown Cloud (Trans-boundary Air Pollution) within the region.

4. Strong commitment from the relatively higher GHGs emitting nations within the region (like India and China) and globally to revise their development path and make transition to a low-carbon development path.

We also urge to our respective governments to plan and follow the low-carbon development path.

5. Provide adequate and long-term financial support as a grant to the poor, most vulnerable and marginalized community for the climate change adaptation (NAPA implementation) and mitigation at the national and local level. Any funds related to climate change to be channeled via globally agreed framework within the UN process. We are strongly against any financial assistance as a loan (climate loan) and the finance should be free of climate corruption and should follow the principles of good governance i.e., authority, responsibility, accountability and transparency.

6. Respect our right to move towards prosperous future and support the low-carbon development path through the development and transfer of clean technologies, research, conservation, promotion and sharing of traditional indigenous knowledge, belief and values.

7. Significantly increase investment in promoting green technologies and economic sectors like alternative energy resources, waste management, clean energy transport system, development of green parks, organic agriculture, urban farming and recreational sites in urban areas.

8. Develop flexible, transparent and accountable mechanisms and common strategy and platforms at regional level with a multi-stakeholder approach which should address all the dimensions of sustainable development i.e., social, environment and economic.

9. [Agree to develop standards for all environment goods and services and setup certification mechanisms when they go the market.]

10. Agree to establish simple but efficient environmental governance from central to local levels in all member countries and in the development arena to ensure environmental issues are better coordinated, and more reflective and responsive to public interest. Agree to facilitate establishment of local, decentralized economies where ownership of the natural resources and business profits are equitably shared amongst local stakeholders.

11. Develop compensatory mechanisms for resource-sharing between upstream and downstream resource managers/beneficiaries and as well as people living in the rural areas for their contribution in environmental conservation and have benefit-sharing mechanisms in place.

12. Promote research, scientific study and information sharing within the region and globally to understand the problem more clearly and develop sustainable solutions.

13. Enhance the capacity and ensure the role of youth to contribute to sustainable development by including them in the decision-making process locally, nationally and globally.

14. Finally, we urge to the people around the globe, especially youth, to develop environmental voluntarism within themselves, adopt the sustainable lifestyle, hold their government accountable to safeguard future.

DAVOS DECLARATION

CLIMATE CHANGE AND TOURISM: RESPONDING TO GLOBAL CHALLENGES

Davos, Switzerland, 3 October 2007

The international community is taking concerted action against climate change around a commonly agreed framework led by the United Nations. This UN framework will seek to establish a long term post-Kyoto roadmap with rapid deployment and targeted milestones. The tourism sector has an important place in that framework, given its global economic and social value, its role in sustainable development and its strong relationships with climate.

To support this action the UN World Tourism Organization (UNWTO), jointly with the United Nations Environment Programme (UNEP) and the World Meteorological Organization (WMO), with the support of the World Economic Forum (WEF) and the Swiss Government, convened the Second International Conference on Climate Change and Tourism, in Davos, Switzerland, from 1 to 3 October 2007. This event, building on the results of the First International Conference organised on this topic in Djerba, Tunisia in 2003, gathered 450 participants from over 80 countries and 22 international organizations, private sector organizations and companies, research institutions, NGOs and the media, with the aim of responding in a timely and balanced way to climate change imperatives in the tourism sector. In preparation of this Conference the organizers commissioned a report to provide an extensive review of current impacts and analyse options for possible actions.

The Conference agreed that:

- climate is a key resource for tourism and the sector is highly sensitive to the impacts of climate change and global warming, many elements of which are already being felt. It is estimated to contribute some 5% of global CO_2 emissions.
- tourism - business and leisure - will continue to be a vital component of the global economy, an important contributor to the

Millennium Development Goals and an integral, positive element in our society.

- given tourism's importance in the global challenges of climate change and poverty reduction, there is a need to urgently adopt a range of policies which encourages truly sustainable tourism that reflects a "quadruple bottom line" of environmental, social, economic and climate responsiveness.
- the tourism sector must rapidly respond to climate change, within the evolving UN framework and progressively reduce its Greenhouse Gas (GHG) contribution if it is to grow in a sustainable manner; This will require action to:
    - o mitigate its GHG emissions, derived especially from transport and accommodation activities;
    - o adapt tourism businesses and destinations to changing climate conditions;
    - o apply existing and new technology to improve energy efficiency;
    - o secure financial resources to help poor regions and countries.

The Conference calls for the following actions

1) Governments and International Organizations:

- Incorporate tourism in the implementation of existing commitments under the United Nations Framework Convention on Climate Change (UNFCCC) and its Kyoto Protocol, and respond to the call by the United Nations Secretary-General for launching, at the 13th session of the UNFCCC Conference of the Parties in Bali, December 2007, an effective and comprehensive climate change framework for the post-2012 period.
- Implement concrete, simultaneous actions for mitigation, adaptation, technology and financing, consistent with the Millennium Development Goals.
- Provide financial, technical and training support to tourism destinations and operators in developing countries (especially in the least developed countries and Small Island Developing States) to ensure that they can participate in the global climate response framework, through established initiatives, such as the Clean Development Mechanism.

- Promote, at all levels, interdisciplinary partnerships, networks and information exchange systems essential to sustainable development of the sector.
- Collaborate in international strategies, policies and action plans to reduce GHG emissions in the transport (in cooperation with ICAO and other aviation organizations), accommodation and related tourism activities.
- Introduce education and awareness programs for all tourism stakeholders – public and private sector – as well as consumers.
- Develop regional and local climate information services tailored to the tourism sector and promote their use among tourism stakeholders. Build capacities for interpretation and application of this information, strengthening collaboration with WMO's National Meteorological Services.
- Implement policy, regulatory, financial, managerial, educational, behavioural, diversification, research and monitoring measures, for effective adaptation and mitigation.

2) Tourism Industry and Destinations

- Take leadership in implementing concrete measures (such as incentives) in order to mitigate climate change throughout the tourism value chain and to reduce risk to travellers, operators and infrastructure due to dynamic climate variability and shift. Establish targets and indicators to monitor progress.
- Promote and undertake investments in energy-efficiency tourism programmes and use of renewable energy resources, with the aim of reducing the carbon footprint of the entire tourism sector.
- Integrate tourism in the formulation and implementation of regional, national and local level adaptation and mitigation strategies and implementation plans. The Nairobi Work Programme on Impacts, Vulnerability and Adaptation to Climate Change, coordinated by UNFCCC, represents an important opportunity for the tourism sector to enhance knowledge, increase capacities and stimulate action.
- Strive to conserve biodiversity, natural ecosystems and landscapes in ways which strengthen resilience to climate change and ensure a long-term sustainable use of the environmental resource base of tourism - in particular those that serve as "earth lungs" (carbon sinks), sequestering GHGs through forest management and other biological programmes, or that protect coastlines (e.g. mangroves and coral reefs).

- Seek to achieve increasingly carbon free environments by diminishing pollution through design, operations and market responsive mechanisms.
- Implement climate-focused product diversification, to reposition destinations and support systems, as well as to foster all-season supply and demand.
- Raise awareness among customers and staff on climate change impacts and engage them in response processes.

3) Consumers:

- In their choices for travel and destination, tourists should be encouraged to consider the climate, economic, societal and environmental impacts of their options before making a decision and, where possible to reduce their carbon footprint, or offset emissions that cannot be reduced directly.
- In their choices of activities at the destination, tourists should also be encouraged to opt for environmentally-friendly activities that reduce their carbon footprint as well as contribute to the preservation of the natural environment and cultural heritage.

4) Research and Communications Networks:

- Encourage targeted, multi-disciplinary research on impacts of climate change in order to address regional gaps in current knowledge, develop tools for risk assessment and cost-benefit analyses with which to gauge the feasibility of various responses.
- Include environmental and climate specific subjects in the study curricula of tourism training programmes and extend these to broader educational systems.
- Promote responsible travel that supports "quadruple bottom line" sustainable tourism, incorporating climate, environmental, social and economic considerations.
- Raise awareness on tourism's economic role as a tool for development, and present information on causes and effects of climate change based on sound science, in a fair, balanced and user-friendly manner.

The Conference

- sets out a range of specific actions to be taken by all stakeholders in the sector to immediately begin to establish and implement a long range carbon-neutral roadmap;
- invites governments and international organizations; the tourism industry; consumers; research and communications networks to implement these recommendations, with concrete commitments and action plans, and to use the UNWTO on-line Climate Change and Tourism Information Exchange Service as a platform, for committed stakeholders to register their pledges and activities toward adaptation and mitigation on an on-going basis;
- stresses the need that UNWTO, in collaboration with UNEP and WMO, continue to lead this process, and to consider convening a Third Conference on Climate Change and Tourism, at an appropriate time in the future, to review progress, to maintain response levels and to identify further needs and actions;
- urges action by the entire tourism sector to face climate change as one of the greatest challenges to sustainable development, and to the Millennium Development Goals in the 21st Century.

***

The Davos Declaration and results of this conference will provide the basis for the UNWTO Minister's Summit on Tourism and Climate Change, scheduled at the World Travel Market, London, UK, 13 November 2007. It will be submitted for adoption at the UNWTO General Assembly in Cartagena de Indias, Colombia, 23–29 November 2007, and also will be presented at the United Nations Climate Change Conference in Bali, Indonesia, in December 2007.